LONDON MATHEMATICAL SOCIETY LECTURE NOTE SERIES

Managing Editor: Professor J.W.S. Cassels, Department of Pure Mathematics and Mathematical Statistics, University of Cambridge, 16 Mill Lane, Cambridge CB2 1SB, England

The books in the series listed below are available from booksellers, or, in case of difficulty, from Cambridge University Press.

34 Representation theory of Lie groups, M.F. ATIYAH *et al*
36 Homological group theory, C.T.C. WALL (ed)
39 Affine sets and affine groups, D.G. NORTHCOTT
40 Introduction to H_p spaces, P.J. KOOSIS
43 Graphs, codes and designs, P.J. CAMERON & J.H. VAN LINT
45 Recursion theory: its generalisations and applications, F.R. DRAKE & S.S. WAINER (eds)
46 p-adic analysis: a short course on recent work, N. KOBLITZ
49 Finite geometries and designs, P. CAMERON, J.W.P. HIRSCHFELD & D.R. HUGHES (eds)
50 Commutator calculus and groups of homotopy classes, H.J. BAUES
51 Synthetic differential geometry, A. KOCK
57 Techniques of geometric topology, R.A. FENN
59 Applicable differential geometry, M. CRAMPIN & F.A.E. PIRANI
62 Economics for mathematicians, J.W.S. CASSELS
66 Several complex variables and complex manifolds II, M.J. FIELD
69 Representation theory, I.M. GELFAND *et al*
74 Symmetric designs: an algebraic approach, E.S. LANDER
76 Spectral theory of linear differential operators and comparison algebras, H.O. CORDES
77 Isolated singular points on complete intersections, E.J.N. LOOIJENGA
78 A primer on Riemann surfaces, A.F. BEARDON
79 Probability, statistics and analysis, J.F.C. KINGMAN & G.E.H. REUTER (eds)
80 Introduction to the representation theory of compact and locally compact groups, A. ROBERT
81 Skew fields, P.K. DRAXL
82 Surveys in combinatorics, E.K. LLOYD (ed)
83 Homogeneous structures on Riemannian manifolds, F. TRICERRI & L. VANHECKE
86 Topological topics, I.M. JAMES (ed)
87 Surveys in set theory, A.R.D. MATHIAS (ed)
88 FPF ring theory, C. FAITH & S. PAGE
89 An F-space sampler, N.J. KALTON, N.T. PECK & J.W. ROBERTS
90 Polytopes and symmetry, S.A. ROBERTSON
91 Classgroups of group rings, M.J. TAYLOR
92 Representation of rings over skew fields, A.H. SCHOFIELD
93 Aspects of topology, I.M. JAMES & E.H. KRONHEIMER (eds)
94 Representations of general linear groups, G.D. JAMES
95 Low-dimensional topology 1982, R.A. FENN (ed)
96 Diophantine equations over function fields, R.C. MASON
97 Varieties of constructive mathematics, D.S. BRIDGES & F. RICHMAN
98 Localization in Noetherian rings, A.V. JATEGAONKAR
99 Methods of differential geometry in algebraic topology, M. KAROUBI & C. LERUSTE
100 Stopping time techniques for analysts and probabilists, L. EGGHE

101 Groups and geometry, ROGER C. LYNDON
103 Surveys in combinatorics 1985, I. ANDERSON (ed)
104 Elliptic structures on 3-manifolds, C.B. THOMAS
105 A local spectral theory for closed operators, I. ERDELYI & WANG SHENGWANG
106 Syzygies, E.G. EVANS & P. GRIFFITH
107 Compactification of Siegel moduli schemes, C-L. CHAI
108 Some topics in graph theory, H.P. YAP
109 Diophantine Analysis, J. LOXTON & A. VAN DER POORTEN (eds)
110 An introduction to surreal numbers, H. GONSHOR
111 Analytical and geometric aspects of hyperbolic space, D.B.A.EPSTEIN (ed)
112 Low-dimensional topology and Kleinian groups, D.B.A. EPSTEIN (ed)
113 Lectures on the asymptotic theory of ideals, D. REES
114 Lectures on Bochner-Riesz means, K.M. DAVIS & Y-C. CHANG
115 An introduction to independence for analysts, H.G. DALES & W.H. WOODIN
116 Representations of algebras, P.J. WEBB (ed)
117 Homotopy theory, E. REES & J.D.S. JONES (eds)
118 Skew linear groups, M. SHIRVANI & B. WEHRFRITZ
119 Triangulated categories in the representation theory of finite-dimensional algebras, D. HAPPEL
121 Proceedings of *Groups - St Andrews 1985*, E. ROBERTSON & C. CAMPBELL (eds)
122 Non-classical continuum mechanics, R.J. KNOPS & A.A. LACEY (eds)
124 Lie groupoids and Lie algebroids in differential geometry, K. MACKENZIE
125 Commutator theory for congruence modular varieties, R. FREESE & R. MCKENZIE
126 Van der Corput's method for exponential sums, S.W. GRAHAM & G. KOLESNIK
127 New directions in dynamical systems, T.J. BEDFORD & J.W. SWIFT (eds)
128 Descriptive set theory and the structure of sets of uniqueness, A.S. KECHRIS & A. LOUVEAU
129 The subgroup structure of the finite classical groups, P.B. KLEIDMAN & M.W.LIEBECK
130 Model theory and modules, M. PREST
131 Algebraic, extremal & metric combinatorics, M-M. DEZA, P. FRANKL & I.G. ROSENBERG (eds)
132 Whitehead groups of finite groups, ROBERT OLIVER
133 Linear algebraic monoids, MOHAN S. PUTCHA
134 Number theory and dynamical systems, M. DODSON & J. VICKERS (eds)
135 Operator algebras and applications, 1, D. EVANS & M. TAKESAKI (eds)
136 Operator algebras and applications, 2, D. EVANS & M. TAKESAKI (eds)
137 Analysis at Urbana, I, E. BERKSON, T. PECK, & J. UHL (eds)
138 Analysis at Urbana, II, E. BERKSON, T. PECK, & J. UHL (eds)
139 Advances in homotopy theory, S. SALAMON, B. STEER & W. SUTHERLAND (eds)
140 Geometric aspects of Banach spaces, E.M. PEINADOR and A. RODES (eds)
141 Surveys in combinatorics 1989, J. SIEMONS (ed)
142 The geometry of jet bundles, D.J. SAUNDERS
143 The ergodic theory of discrete groups, PETER J. NICHOLLS
144 Introduction to uniform spaces, I.M. JAMES
145 Homological questions in local algebra, JAN R. STROOKER
146 Cohen-Macaulay modules over Cohen-Macaulay rings, Y. YOSHINO
147 Continuous and discrete modules, S.H. MOHAMED & B.J. MÜLLER
148 Helices and vector bundles, A.N. RUDAKOV et al
149 Solitons, nonlinear evolution equations and inverse scattering, M.A. ABLOWITZ & P.A. CLARKSON
150 Geometry of low-dimensional manifolds 1, S. DONALDSON & C.B. THOMAS (eds)
151 Geometry of low-dimensional manifolds 2, S. DONALDSON & C.B. THOMAS (eds)
152 Oligomorphic permutation groups, P. CAMERON
153 L-functions in Arithmetic, J. COATES & M.J. TAYLOR
154 Number theory and cryptography, J. LOXTON (ed)
155 Classification theories of polarized varieties, TAKAO FUJITA
156 Twistors in mathematics and physics, T.N. BAILEY & R.J. BASTON (eds)

London Mathematical Society Lecture Note Series. 151

Geometry of Low-dimensional Manifolds

2: Symplectic Manifolds and Jones-Witten Theory

Proceedings of the Durham Symposium, July 1989

Edited by
S. K. Donaldson
Mathematical Institute, University of Oxford
C.B. Thomas
Department of Pure Mathetmatics and mathematical Statistics, University of Cambridge

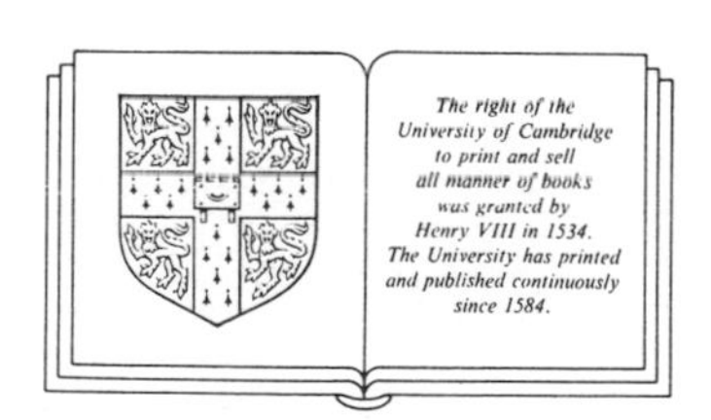

CAMBRIDGE UNIVERSITY PRESS
Cambridge
New York Port Chester Melbourne Sydney

Published by the Press Syndicate of the University of Cambridge
The Pitt Building, Trumpington Street, Cambridge CB2 1RP
40 West 20th Street, New York, NY 10011, USA
10, Stamford Road, Oakleigh, Melbourne 3166, Australia

First published 1990

Printed in Great Britain at the University Press, Cambridge

Library of Congress cataloguing in publication data available

British Library cataloguing in publication data available

ISBN 0 521 40001 5

CONTENTS

CONTENTS OF VOLUME 1

CONTRIBUTORS

I. R. Aitchison, Department of Mathematics, University of Melbourne, Melbourne, Australia

M. F. Atiyah, Mathematical Institute, 24-29 St. Giles, Oxford OX1 3LB, UK

F. E. Burstall, School of Mathematical Sciences, University of Bath, Claverton Down, Bath, UK

Ralph E. Cohen, Department of Mathematics, Stanford University, Stanford CA 94305, USA

S. K. Donaldson, Mathematical Institute, 24-29 St. Giles, Oxford OX1 3LB, UK

Yakov Eliashberg, Department of Mathematics, Stanford University, Stanford CA 94305, USA

Ronald Fintushel, Department of Mathematics, Michigan State University, East Lansing, MI 48824, USA

A. Floer, Department of Mathematics, University of California, Berkeley CA 94720, USA

Mikio Furuta, Department of Mathematics, University of Tokyo, Hongo, Tokyo 113, Japan, *and*, Mathematical Institute, 24-29 St. Giles, Oxford OX1 3LB, UK

A. B. Givental, Lenin Institute for Physics and Chemistry, Moscow, USSR

Robert E. Gompf, Department of Mathematics, University of Texas, Austin TX, USA

D. H. Hartley, Department of Physics, University of Lancaster, Lancaster, UK

N. J. Hitchin, Mathematical Institute, 24-29 St. Giles, Oxford OX1 3LB, UK

H. Hofer, FB Mathematik, Ruhr Universität Bochum, Universitätstr. 150, D-463 Bochum, FRG

Lisa Jeffrey, Mathematical Institute, 24-29 St. Giles, Oxford OX1 3LB, UK

F. A. E. Johnson, Department of Mathematics, University College, London WC1E 6BT, UK

J. D. S. Jones, Mathematics Institute, University of Warwick, Coventry CV4 7AL, UK

Robion Kirby, Department of Mathematics, University of California, Berkeley CA 94720, USA

Dieter Kotschick, Queen's College, Cambridge CB3 9ET, UK, *and,* The Institute for Advanced Study, Princeton NJ 08540, USA

Matthias Kreck, Max-Planck-Institut für Mathematik, 23 Gottfried Claren Str., Bonn, Germany

N. S. Manton, Department of Applied Mathematics and Mathematical Physics, University of Cambridge, Silver St, Cambridge CB3 9EW, UK

Dusa McDuff, Department of Mathematics, SUNY, Stony Brook NY, USA

Paul Melvin, Department of Mathematics, Bryn Mawr College, Bryn Mawr PA 19010, USA

Christian Okonek, Math Institut der Universität Bonn, Wegelerstr. 10, D-5300 Bonn 1, FRG

J. H. Rubinstein, Department of Mathematics, University of Melbourne, Melbourne, Australia, *and,* The Institute for Advanced Study, Princeton NJ 08540, USA

Ronald J. Stern, Department of Mathematics, University of California, Irvine CA 92717, USA

L. R. Taylor, Department of Mathematics, Notre Dame University , Notre Dame IN 46556, USA

C. B. Thomas, Department of Pure Mathematics and Mathematical Statistics, University of Cambridge, 16, Mill Lane, Cambridge CB3 9EW, UK

K. P. Tod, Mathematical Institute, 24-29 St. Giles, Oxford OX1 3LB, UK

R. W. Tucker, Department of Physics, University of Lancaster, Lancaster, UK

Edward Witten, Institute for Advanced Study, Princeton NJ 08540, USA

John C. Wood, Department of Pure Mathematics, University of Leeds, Leeds, UK

Names of Participants

N. A'Campo (Basel)
M. Atiyah (Oxford)
H. Azcan (Sussex)
M. Batchelor (Cambridge)
S. Bauer (Bonn)
I.M. Benn (Newcastle, NSW)
D. Bennequin (Strasbourg)
W. Browder (Princeton/Bonn)
R. Brussee (Leiden)
P. Bryant (Cambridge)
F. Burstall (Bath)
E. Corrigan (Durham)
S. de Michelis (San Diego)
S. Donaldson (Oxford)
S. Dostoglu (Warwick)
J. Eells (Warwick/Trieste)
Y. Eliashberg (Stanford)
D. Fairlie (Durham)
R. Fintushel (MSU, East Lansing)
A. Floer (Berkeley)
M. Furuta (Tokyo/Oxford)
G. Gibbons (Cambridge)
A. Givental (Moscow)
R. Gompf (Austin, TX)
C. Gordon (Austin, TX)
J-C. Hausmann (Geneva)
N. Hitchin (Warwick)
H. Hofer (Bochum)
J. Hurtebise (Montreal)
D. Husemoller (Haverford/Bonn)
P. Iglesias (Marseille)
L. Jeffreys (Oxford)
F. Johnson (London)
J. Jones (Warwick)
R. Kirby (Berkeley)
D. Kotschick (Oxford)
M. Kreck (Mainz)
R. Lickorish (Cambridge)
J. Mackenzie (Melbourne)
N. Manton (Cambridge)
G. Massbaum (Nantes)
G. Matić (MIT)
D. McDuff (SUNY, Stony Brook)
M. Micallef (Warwick)
C. Okonek (Bonn)
P. Pansu (Paris)
H. Rubinstein (Melbourne)
D. Salamon (Warwick)
G. Segal (Oxford)
R. Stern (Irvine, CA)
C. Thomas (Cambridge)
K. Tod (Oxford)
K. Tsuboi (Tokyo)
R. Tucker (Lancaster)
C.T.C. Wall (Liverpool)
S. Wang (Oxford)
R. Ward (Durham)
P.M.H. Wilson (Cambridge)
E. Witten (IAS, Princeton)
J. Wood (Leeds)

INTRODUCTION

In the past decade there have been a number of exciting new developments in an area lying roughly between *manifold theory* and *geometry*. More specifically, the principal developments concern:

(1) geometric structures on manifolds,
(2) symplectic topology and geometry,
(3) applications of Yang-Mills theory to three- and four-dimensional manifolds,
(4) new invariants of 3-manifolds and knots.

Although they have diverse origins and roots spreading out across a wide range of mathematics and physics, these different developments display many common features—some detailed and precise and some more general. Taken together, these developments have brought about a shift in the emphasis of current research on manifolds, bringing the subject much closer to geometry, in its various guises, and physics.

One unifying feature of these geometrical developments, which contrasts with some geometrical trends in earlier decades, is that in large part they treat phenomena in specific, low, dimensions. This mirrors the distinction, long recognised in topology, between the flavours of "low-dimensional" and "high-dimensional" manifold theory (although a detailed understanding of the connection between the special roles of the dimension in different contexts seems to lie some way off). This feature explains the title of the meeting held in Durham in 1989 and in turn of these volumes of Proceedings, and we hope that it captures some of the spirit of these different developments.

It may be interesting in a general introduction to recall the the emergence of some of these ideas, and some of the papers which seem to us to have been landmarks. (We postpone mathematical technicalities to the specialised introductions to the six separate sections of these volumes.) The developments can be said to have begun with the lectures [T] given in Princeton in 1978-79 by W.Thurston, in which he developed his "geometrisation" programme for 3-manifolds. Apart from the impetus given to old classification problems, Thurston's work was important for the way in which it encouraged mathematicians to look at a manifold in terms of various concomitant geometrical structures. For example, among the ideas exploited in [T] the following were to have perhaps half-suspected fall-out: representations of link groups as discrete subgroups of $PSL_2(\mathbf{C})$, surgery compatible with geometric structure, rigidity, Gromov's norm with values in the real singular homology, and most important of all, use of the theory of Riemann surfaces and Fuchsian groups to develop a feel for what might be true for special classes of manifolds in higher dimensions.

Meanwhile, another important signpost for future developments was Y. Eliashberg's proof in 1981 of "symplectic rigidity"– the fact that the group of symplectic diffeomorphisms of a symplectic manifold is C^0-closed in the full diffeomorphism group.

This is perhaps a rather technical result, but it had been isolated by Gromov in 1970 as the crux of a comprehensive "hard versus soft" alternative in "symplectic topology": Gromov showed that if this rigidity result was not true then any problem in symplectic topology (for example the classification of symplectic structures) would admit a purely algebro-topological solution (in terms of cohomology, characteristic classes, bundle theory etc.) Conversely, the rigidity result shows the need to study deeper and more specifically geometrical phenomena, beyond those of algebraic topology.

Eliashberg's original proof of symplectic rigidity was never fully published but there are now a number of proofs available, each using new phenomena in symplectic geometry as these have been uncovered. The best known of these is the "Arnol'd Conjecture" [A] on fixed points of symplectic diffeomeorphisms. The original form of the conjecture, for a torus, was proved by Conley and Zehnder in 1982 [CZ] and this established rigidity, since it showed that the symplectic hypothesis forced more fixed points than required by ordinary topological considerations. Another demonstration of this rigidity, this time for contact manifolds, was provided in 1982 by Bennequin with his construction [B] of "exotic" contact structures on $\mathbf{R}^3$.

Staying with symplectic geometry, but moving on to 1984, Gromov [G] introduced "pseudo-holomorphic curves" as a new tool, thus bringing into play techniques from algebraic and differential geometry and analysis. He used these techniques to prove many rigidity results, including some extensions of the Arnol'd conjecture and the existence of exotic symplectic structures on Euclidean space. (Our "low-dimensional" theme may appear not to cover these developments in symplectic geometry, which in large part apply to symplectic manifolds of all dimensions: what one should have in mind are the crucial properties of the *two-dimensional* surfaces, or pseudo-holomorphic curves, used in Gromov's theory. Moreover his results seem to be particularly sharp in low dimensions.)

We turn now to 4-manifolds and step back two years. At the Bonner Arbeitstagung in June 1982 Michael Atiyah lectured on Donaldson's work on smooth 4-manifolds with definite intersection form, proving that the intersection form of such a manifold must be "standard". This was the first application of the "instanton" solutions of the Yang-Mills equations as a tool in 4-manifold theory, using the moduli space of solutions to provide a cobordism between such a 4-manifold and a specific union of $\mathbf{CP}^2$'s [D]. This approach again brought a substantial amount of analysis and differential geometry to bear in a new way, using analytical techniques which were developed shortly before. Seminal ideas go back to the 1980 paper [SU] of Sacks and Uhlenbeck. They showed what could be done with non-linear elliptic problems for which, because of conformal invariance, the relevant estimates lie on the borderline of the Sobolev inequalities. These analytical techniques are relevant both in the Yang-Mills theory and also to pseudo-holomorphic curves. Other important and influential analytical techniques, motivated in part by Physics, were developed by C.Taubes [Ta].

Combined with the topological h-cobordism theorem of M. Freedman, proved shortly before, the result on smooth 4-manifolds with definite forms was quickly used to deduce, among other things, that $\mathbf{R}^4$ admits exotic smooth structures. Many different applications of these instantons, leading to strong differential-topological conclusions, were made in the following years by a number of mathematicians; the other main strand in the work being the definition of new invariants for smooth 4-manifolds, and their use to detect distinct differentiable structures on complex algebraic surfaces (thus refuting the smooth h-cobordism theorem in four dimensions).

From an apparently totally different direction the *Jones polynomial* emerged in a series of seminars held at the University of Geneva in the summer of 1984. This was a new invariant of knots and links which, in its original form [J], is defined by the traces of a series of representations of the Braid Groups which had been encountered in the theory of von Neumann algebras, and were previously known in statistical mechanics. For some time, in spite of its obvious power as an invariant of knots and links in ordinary space, the geometric meaning of the Jones invariant remained rather mysterious, although a multitude of connections were discovered with (among other things) combinatorics, exactly soluble models in statistical physics and conformal field theories.

In the spring of the next year, 1985, A. Casson gave a series of lectures in Berkeley on a new integer invariant for homology 3-spheres which he had discovered. This Casson invariant "counts" the number of representations of the fundamental group in $SU(2)$ and has a number of very interesting properties. On the one hand it gives an integer lifting of the well-established Rohlin $\mathbf{Z}/2$ μ-invariant. On the other hand Casson's definition was very geometric, employing the moduli spaces of unitary representations of the fundamental groups of surfaces in an essential way. (These moduli spaces had been extensively studied by algebraic geometers, and from the point of view of Yang-Mills theory in the influential 1982 paper of Atiyah and Bott [AB].) Since such representations correspond to flat connections it was clear that Casson's theory would very likely make contact with the more analytical work on Yang-Mills fields. On the other hand Casson showed, in his study of the behaviour of the invariant under surgery, that there was a rich connection with knot theory and more familiar techniques in geometric topology. For a very readable account of Cassons work see the survey by A. Marin [M].

Around 1986 A. Floer introduced important new ideas which applied both to symplectic geometry and to Yang-Mills theory, providing a prime example of the interaction between these two fields. Floer's theory brought together a number of powerful ingredients; one of the most distinctive was his novel use of ideas from Morse theory. An important motivation for Floer's approach was the 1982 paper by E. Witten [W1] which, among other things, gave a new analytical proof of the Morse inequalities and explained their connection with instantons, as used in Quantum Theory.

In symplectic geometry one of Floer's main acheivements was the proof of a generalised form of the Arnol'd conjecture [F1]. On the Yang-Mills side, Floer defined new invariants of homology 3-spheres, the *instanton homology groups* [F2]. By work of Taubes the Casson invariant equals one half of the Euler characteristic of these homology groups. Their definition uses moduli spaces of instantons over a 4-dimensional tube, asymptotic to flat connections at the ends, and these are interpreted in the Morse theory picture as the gradient flow lines connecting critical points of the Chern-Simons functional.
Even more recently (1988), Witten has provided a quantum field theoretic interpretation of the various Yang-Mills invariants of 4-manifolds and, in the other direction, has used ideas from quantum field theory to give a purely 3-dimensional definition of the Jones link invariants [W2]. Witten's idea is to use a functional integral involving the Chern-Simons invariant and holonomy around loops, over the space of all connections over a 3-manifold. The beauty of this approach is illustrated by the fact that the choices (quantisations) involved in the construction of the representations used by Jones reflect the need to make this integral actually defined. In addition Witten was able to find new invariants for 3-manifolds.
It should be clear, even from this bald historical summary, how fruitful the cros-fertilisation between the various theories has been. When the idea of a Durham conference on this area was first mooted, in the summer of 1984, the organisers certainly intended that it should cover Yang-Mills theory, symplectic geometry and related developments in theoretical physics. However the proposal was left vague enough to allow for unpredictable progress, sudden shifts of interest, new insights, and the travel plans of those invited. We believe that the richness of the contributions in both volumes has justified our approach, but as always the final judgement rests with the reader.

References

[A] Arnold, V.I. *Mathematical Methods of Classical Mechanics* Springer, Graduate Texts in Mathematics, New York (1978)

[AB] Atiyah, M.F. and Bott, R. *The Yang-Mills equations over Riemann surfaces* Phil. Trans. Roy. Soc. London, Ser. A **308** (1982) 523-615

[B] Bennequin, D. *Entrelacements et équations de Pfaff* Astérisque **107-108** 1983) 87-91

[CZ] Conley, C. and Zehnder, E. *The Birkhoff-Lewis fixed-point theorem and a conjecture of V.I. Arnold* Inventiones Math.**73** (1983) 33-49

[D] Donaldson, S.K. *An application of gauge theory to four dimensional topology* Jour. Differential Geometry **18** (1983) 269-316

[F1] Floer, A. *Morse Theory for Lagrangian intersections* Jour. Differential Geometry **28** (1988) 513-547

[F2] Floer, A. *An instanton invariant for 3-manifolds* Commun. Math. Phys. **118** (1988) 215-240

[G] Gromov, M. *Pseudo-holomorphic curves in symplectic manifolds* Inventiones Math. **82** (1985) 307-347

[J] Jones, V.R.F. *A polynomial invariant for links via Von Neumann algebras* Bull. AMS **12** (1985) 103-111

[M] Marin, A. (after A. Casson) *Un nouvel invariant pour les spheres d'homologie de dimension trois* Sem. Bourbaki, no. 693, fevrier 1988 (Astérisque **161-162** (1988) 151-164)

[SU] Sacks, J. and Uhlenbeck, K.K. *The existence of minimal immersions of 2-spheres* Annals of Math. **113** (1981) 1-24

[T] Thurston, W.P. *The Topology and Geometry of 3-manifolds* Princeton University Lecture Notes, 1978

[Ta] Taubes, C.H. *Self-dual connections on non-self-dual four manifolds* Jour. Differential Geometry **17** (1982) 139-170

[W1] Witten, E. *Supersymmetry and Morse Theory* Jour. Differential Geometry **17** (1982) 661-692

[W2] Witten, E. *Some geometrical applications of Quantum Field Theory* Proc. IXth. International Congress on Mathematical Physics, Adam Hilger (Bristol) 1989, pp. 77-110.

Acknowledgements

We should like to take this opportunity to thank the London Mathematical Society and the Science and Engineering Research Council for their generous support of the Symposium in Durham. We thank the members of the Durham Mathematics Department, particularly Professor Philip Higgins, Dr. John Bolton and Dr. Richard Ward, for their work and hospitality in putting on the meeting, and Mrs. S. Nesbitt and Mrs. J. Gibson who provided most efficient organisation. We also thank all those at Grey College who arranged the accommodation for the participants. Finally we should like to thank Dieter Kotschick and Lisa Jeffrey for writing up notes on some of the lectures, which have made an important addition to these volumes.

PART 1

SYMPLECTIC GEOMETRY

In this section we gather together papers on symplectic and contact geometry. Recall that a symplectic manifold (M, ω) is a smooth manifold M of even dimension $2n$ with a closed, nondegenerate, 2-form ω i.e $d\omega = 0$ and ω^n is nowhere zero. A contact structure is an odd-dimensional analogue; a contact manifold (V, H) is a pair consisting of a manifold V of odd dimension $2n+1$ with a field H of $2n$-dimensional subspaces of the tangent bundle TV which is maximally non-integrable, in the sense that if α is a 1-form defining H, then $d\alpha^n \wedge \alpha$ is non-zero (i.e. $d\alpha$ is non-degenerate on H).

In their different ways, all the articles in this section are motivated by the work of M. Gromov, and in particular by his paper [G2] on pseudo-holomorphic curves. Here the idea is to replace a complex manifold by an almost-complex manifold with a compatible symplectic structure, and to study the generalisations of the complex curves—defined by this almost-complex structure. The paper of McDuff below gives a direct application of this method by showing that a minimal 4-dimensional symplectic manifold containing an embedded, symplectic, copy of $S^2 = \mathbf{CP}^1$ is either $\mathbf{CP}^2$ or an S^2 bundle over a Riemann surface, with the symplectic form being non-degenerate on fibres. The uniqueness of the structure in the minimal case can be thought of as an example of rigidity.

Another important symplectic notion investigated by Gromov is that of "squeezing". He proves for example that the polycylinder $D^2(1) \times \cdots \times D^2(1)$ (n factors) cannot be symplectically embedded in $D^2(R) \times \mathbf{R}^{2n-2}$ if the radius R of the disc is less than 1. H. Hofer (see below) approaches this and other questions from the point of view of a symplectic capacity: we can summarise his definition as follows.

A capacity *c is a function defined on all subsets of symplectic manifolds of a given dimension $2n$ taking values in the positive real numbers augmented by ∞ and satisfying the axioms;*

(C0) *If f is a symplectic DEM defined in a neighbourhood of a subset $S \subset (M, \omega)$ then $c(S, \omega) = c(f(S), f_*(\omega))$.*

(C1)(Conformality) *If $\lambda > 0$ then $c(S, \lambda\omega) = \lambda c(S, \omega)$.*

(C2)(Monotonicity) *If $(S, \omega) \subset (T, \omega)$ then $c(S, \omega) \leq c(T, \omega)$.*

(C3)(Normalisation) *$c(D^{2n}(1) = c(D^2(1) \times D^{2n-2}(r)) = \pi$ for all $r \geq 1$, with respect to the standard symplectic form on $\mathbf{C}^n = \mathbf{C} \times \mathbf{C}^{n-1}$.*

An example of a capacity is provided by the "displacement energy"—heuristically, given two disjoint bounded subsets S, S' of $\mathbf{R}^{2n}$ how much energy does one need to deform S into S'? As Hofer shows, this capacity can be used to prove the squeezing theorem above, and furthermore Axioms C0-C2 suffice to recover the rigidity theorem (mentioned in the general introduction) namely that the symplectic DEM group is C^0 closed in $4\mathit{Diff}(M)$.

The existence of periodic solutions for Hamiltonian systems leads to a whole family of independent capacities, which provide a framework in which to discuss the existence of a closed integral curves for the vector field associated with an arbitrary contact structure on S^{2n-1}. More generally still there is the possibility of using Floer's symplectic instanton homology groups to define sequence-valued capacities. These homology groups were the tool used by Floer to prove Arnol'd's conjecture on the fixed points of symplectic diffeomorphisms for a wide class of manifolds. We refer to the notes by Kotschick in the first volume of these proceedings for the definition of the symplectic instanton homology groups, and their relation to Floer's homology groups for 3-manifolds. Gromov's paper provides many parallels between the theory of Yang-Mills instantons and pseudo-holomorphic curves, and between the results derived from the two. For example, the existence of a symplectic form on $\mathbf{R}^{2n}$ which is not the restriction of the standard form under some embedding of $\mathbf{R}^{2n}$ in itself is reminiscent of the exotic smooth structures on $\mathbf{R}^4$.

The Arnol'd conjecture can also be discussed in the framework of the "Nonlinear Maslov Index" developed by A. Givental and described in his article below. The definition can be summarised as follows: the linear Maslov index of a loop γ in the manifold Λ_n of linear Legendrian subspaces of $\mathbf{RP}^{2n-1}$ is its class in the $\pi_1(\Lambda_n) \equiv \mathbf{Z}$. The nonlinear invariant is obtained by replacing Λ_n by the infinite-dimensional homogeneous space of all Legendrian embeddings of $\mathbf{RP}^{n-1}$ in $\mathbf{RP}^{2n-1}$.

The various fixed-point theorems now in the literature illustrate the "hard" aspect of symplectic geometry (for this terminology see [G3]). This rests on the Eliashberg-Gromov rigidity theorem to which we have referred above, and in the general introduction. In another striking parallel with C^∞ theory Gromov has shown that 4-dimensional symplectic theory has some of the flavour of smooth surfaces. For example, if $M = S^2 \times S^2$ has the standard symplectic structure $\omega \oplus \omega$ coming from the Kähler forms on the factors, then $Diff_{\omega\oplus\omega}(M)$ contracts onto the isometry group of M, which is a $\mathbf{Z}/2\mathbf{Z}$ extension of $SO(3) \times SO(3)$. *Question*: does rigidity give rise to similar results for contact manifolds in dimension 5, for example S^5 or $S^2 \times S^3$?

Perhaps the most important test between the hard and soft approaches to symplectic geometry lies in the problem of the existence and classification of symplectic forms on a closed manifold M^{2n}. At present, no counterexample is known to the obvious soft conjecture that a global symplectic form exists whenever TM^{2n} has structural group reducible to $U(n)$ and one prescribes a class $x \in H^2(M;\mathbf{R})$ such that x^n is compatible with the complex orientation. Given the early work of Gromov [G1] on geometric structures on open manifolds, the problem is to find obstructions to extending a symplectic structure defined near the boundary ∂D^{2n} over the whole ball. This shows the importance of what Eliashberg has called "fillable" structures; some of the difficulties which arise are illustrated as follows. Let Σ be a 2-dimensional surface embedded in the three-dimensional boundary of an almost-complex manifold M^4, and consider the natural foliations of the 3-dimensional cylinder $Z = D^2 \times [0,1]$ (or S^3 minus two poles) by holomorphic discs. If Σ is diffeomorphic to ∂Z or ∂D^3, and if the embedding of Σ in ∂M can be extended to an embedding of Z or D^3 in

M^4 which is holomorphic on the leaves of the foliation, then we say that Σ is fillable by holomorphic discs. Under certain conditions, explained below in the article of Eliashberg, these extensions exist and this can be used, for example, to provide a necessary condition (not "overtwisted") for a contact structure on ∂M to bound a symplectic structure on M.
We conclude this introduction with a few remarks about contact manifolds. Here is one reason for believing that contact manifolds may be "softer" than symplectic: if V_1 and V_2 are contact, then their connected sum admits a contact structure agreeing with the original forms on V_1, V_2 outside small discs. This cannot occur for symplectic manifolds—the basic difference being that the odd-dimensional sphere S^{2n+1} admits a contact form α such that $d\alpha$ is the pull-back of the standard symplectic form on $\mathbf{CP}^n$. It follows that 0-surgeries can be performed on contact manifolds, and $2n$ surgeries seem to fit into the same framework. The situation for $2n+1$ surgeries is more delicate—at least under certain conditions a 1-surgery can be carried out, as Thurston and Winkelnkemper showed in dimension 3. And by using a rag-bag of special tricks it is possible to prove the soft realisation conjecture, in the contact case, for a large class of $n-1$-connected $2n+1$-manifolds. Thus at the time of writing the situation is tantalisingly similar to that for codimension-1 foliations before the major contribution of Thurston; see [T] for a summary of what was known a few years back.
As Eliashberg's paper shows, much work has been done in dimension 3, stimulated not only by Gromov but also by the work of Bennequin. It would be very interesting to see if contact geometry can be applied to classification problems in 3-dimensional topology. For example, A. Weinstein conjectures that if $H_1(V^3; \mathbf{Z})$ is finite (and in particular if V^3 has universal cover diffeomorphic to S^3), then the characteristic foliation of an s-fillable contact form contains at least one closed orbit. Under what conditions is it possible to use the existence of such orbits to construct a Seifert fibering of V? Such a manifold would then be elliptic, completing part of the geometrisation programme.

[G1] Gromov, M. *Partial differential relations* Springer Berlin-Heidelberg (1986)

[G2] Gromov, M. *Pseudoholomorphic curves in symplectic manifolds* Inventiones Math. **82** (1985) 307-347

[G3] Gromov, M. *Soft and hard Symplectic Geometry* Proc. Int. Congress Mathematicians, Berkeley 1986, Vol. I, 81-98

[T] Thomas, C.B. *Contact structures on $(n-1)$-connected $(2n+1)$-manifolds* Banach Centre Publications **18** (1986) 254-270

Rational and Ruled Symplectic 4–Manifolds

DUSA MCDUFF(*)

State University of New York at Stony Brook

1. INTRODUCTION

This note describes the structure of compact symplectic 4-manifolds (V, ω) which contain a symplectically embedded copy C of S^2 with non-negative self-intersection number. (Such curves C are called "rational curves" by Gromov: see [G].) It turns out that there is a concept of minimality for symplectic 4-manifolds which mimics that for complex surfaces. Further, a minimal manifold (V, ω) which contains a rational curve C is either symplectomorphic to $\mathbb{C}P^2$ with its usual Kähler structure τ, or is the total space of a "symplectic ruled surface" i.e. an S^2–bundle over a Riemann surface M, with a symplectic form which is non-degenerate on the fibers. It follows that if a (possibly non-minimal) (V,ω) contains a rational curve C with $C \cdot C > 0$, then (V, ω) may be blown down either to $S^2 \times S^2$ with a product form or to $(\mathbb{C}P^2, \tau)$, and hence is birationally equivalent to $\mathbb{C}P^2$ in Guillemin and Sternberg's sense: see [GS]. (In analogy with the complex case, we will call such manifolds rational.) Moreover, if V contains a rational curve C with $C \cdot C = 0$, then V may be blown down to a symplectic ruled surface. Thus, symplectic 4-manifolds which contain rational curves of non-negative self-intersection behave very much like rational or ruled complex surfaces.

It is natural to ask about the uniqueness of the symplectic structure on the manifolds under consideration: more precisely, if ω_0 and ω_1 are cohomologous symplectic forms on V which both admit rational curves of non-negative self-intersection, are they symplectomorphic? We will see below that the answer is "yes" if the manifolds in question are minimal. In the general case, the most that is known at present is that any two such forms may be joined by a family ω_t, $0 \le t \le 1$, of (possibly non-cohomologous) symplectic forms on V. Since the cohomology class varies here, this does not imply that the forms ω_0 and ω_1 are symplectomorphic: cf [McD 1]. Similarly, all the symplectic

---

(*) partially supported by NSF grant no: DMS 8803056

1980 Mathematics Subject Classification (revised 1985): 53 C 15, 57 R 99

key words: symplectic manifold, 4-manifolds, pseudo-holomorphic curves, almost complex manifold, blowing up.

forms under consideration are Kähler for some integrable complex structure J on V, provided that V is minimal. In the general case, we know only that ω may be joined to a Kähler form by a family as above. Note also that there might be some completely different symplectic forms on these manifolds which do not admit rational curves.

The present work was inspired by Gromov's result in [G] that if (V,ω) is a compact symplectic 4-manifold whose second homology group is generated by a symplectically embedded 2-sphere of self-intersection +1, then V is $\mathbb{C}P^2$ with its usual Kähler structure. Our proofs rely heavily on his theory of pseudo-holomorphic curves. The main innovation is a homological version of the adjunction formula which is valid for almost complex 4-manifolds. (See Proposition 2.9 below.) This gives a homological criterion for a pseudo-holomorphic curve in an almost-complex 4-manifold to be embedded, and is a powerful mechanism for relating the homological properties of a symplectic manifold V to the geometry of its pseudo-holomorphic curves. We also use some new cutting and pasting techniques to reduce the ruled case to the rational case.

Proofs of the results stated here appear in [McD 3,4]. I wish to thank Ya. Eliashberg for many stimulating discussions about the questions studied here. I am also grateful to MRSI for its hospitality and support during the initial stages of this work.

2. STATEMENT OF RESULTS

We will begin by discussing blowing up and blowing down. All manifolds considered will be smooth, compact and, unless specific mention is made to the contrary, without boundary.

By analogy with the theory of complex surfaces, we will say that (V,ω) is minimal if it contains no exceptional curves, that is, symplectically embedded 2-spheres Σ with self-intersection number $\Sigma\cdot\Sigma = -1$. We showed in [McD 2] Lemma 2.1 that every exceptional curve Σ has a neighbourhood N_ε whose boundary $(\partial N_\varepsilon, \omega)$ may be identified with the boundary $(\partial B^4(\lambda+\varepsilon), \omega_0)$ of the ball of radius $\lambda+\varepsilon$ in $\mathbb{C}P^2$, where $\pi\lambda^2 = \omega(\Sigma)$ and $\varepsilon > 0$ is sufficiently small. Hence Σ can be blown down by cutting out N_ε and gluing in the ball $B^4(\lambda+\varepsilon)$, with its standard form ω_0. It is easy to check that the resulting manifold is independent of the choice of ε, so that there is a well-defined blowing down operation, which is inverse to symplectic blowing up.

The following result is not hard to prove: its main point is that one blowing down operation suffices.

2.1 Theorem
Every symplectic 4-manifold (V,ω) *covers a minimal symplectic manifold* (V', ω') *which may be obtained from* V *by blowing down a finite collection of disjoint exceptional curves. Moreover, the induced symplectic form* ω' *on* V' *is unique up to isotopy.*

There is also a version of Theorem 1 for manifold pairs (V, C) where C is a symplectically embedded compact 2-manifold in V. We will call such a pair <u>minimal</u> if $V - C$ contains no exceptional curves.

2.2 Theorem
Every symplectic pair (V, C, ω) *covers a minimal symplectic pair* (V', C, ω') *which may be obtained by blowing down a finite collection of disjoint exceptional curves in* $V-C$. *Moreover, the induced symplectic form* ω' *on* V' *is unique up to isotopy* (rel C).

2.3 Note
If C is a closed subset of V, two symplectic forms ω_0 and ω_1 are said to be isotopic (rel C) if they can be joined by a family of cohomologous symplectic forms whose restrictions to C are all equal. If C is symplectic, Moser's theorem then implies that there is an isotopy g_t of V which is the identity on C and is such that $g_1^*(\omega_1) = \omega_0$.

It is well-known that the diffeomorphism type of V' is not uniquely determined by that of V. For example, because $(S^2 \times S^2) \# \overline{\mathbb{C}P}^2$ is diffeomorphic to $\mathbb{C}P^2 \# \overline{\mathbb{C}P}^2 \# \overline{\mathbb{C}P}^2$, the manifold $V = (S^2 \times S^2) \# \overline{\mathbb{C}P}^2$ may be reduced to $\mathbb{C}P^2$ as well as to $S^2 \times S^2$. However,this is essentially the only ambiguity, and V' is determined up to diffeomorphism if we fix the homology classes of the curves which are blown down.

Conversely, one can ask to what extent the minimal manifold (V', ω') determines its blowing up (V, ω). Since each exceptional curve Σ in (V, ω) corresponds to an embedded ball in V' of radius λ, where $\omega(\Sigma) = \pi\lambda^2$, this question is related to properties of the space of symplectic embeddings of $\coprod B(\lambda_i)$ into (V', ω'), where $\coprod B(\lambda_i)$ is the disjoint union of the symplectic 4-balls $B(\lambda_i)$ of radius λ_i. We discussed the corresponding question for manifold pairs in [McD 2]. We showed there that, if $C \cdot C = 1$, and if V is diffeomorphic to $\mathbb{C}P^2$ with k points blown up, there is a unique symplectic structure on (V, C) in the cohomology class a if and only if the space of symplectic embeddings of $\coprod B(\lambda_i)$ into $\mathbb{C}P^2 - \mathbb{C}P^1$ is connected, where $\pi\lambda_1^2, \ldots, \pi\lambda_k^2$ are the values of a on the exceptional curves in V. Unfortunately nothing is known about this space of embeddings per se. In fact, the information we have goes the other way: we proved in [McD 2] that the structure on (V, C) is unique when $k = 1$, which implies that the corresponding space of embeddings is connected. (Because $\mathbb{C}P^2 \# \overline{\mathbb{C}P}^2$ is ruled, this uniqueness statement is closely related to the results in Theorem 2.4 below.) It is not clear what happens when $k \geq 2$. However, because any two embeddings of $\amalg B(\lambda_i)$ into (V', ω') are isotopic when restricted to the union $\amalg B(\varepsilon_i)$ of suitably small subballs, any two forms on V which blow-down to ω' may be joined by a family of non-cohomologous forms.

Thus, the problem is essentially reduced to understanding the minimal case. The next ingredient is a result on the structure of symplectic S^2-bundles (symplectic ruled surfaces).

2.4 Theorem *Let* V *be an oriented* S^2*–bundle* $\pi: V \to M$ *over a compact oriented surface* M *with fiber* F.
(i) *The cohomology class* a *of any symplectic form on* V *which is non-degenerate on each fiber of* π *satisfies the conditions:*

(a) $a(F)$ *and* $a^2(V)$ *are positive, and*

(b) $a^2(V) > (a(F))^2$ *if the bundle is non-trivial.*

(ii) *Any cohomology class* $a \in H^2(V; \mathbb{Z})$ *which satisfies the above conditions may be represented by a symplectic form* ω *which is non-degenerate on each fiber of* π. *Moreover, this form is unique up to isotopy.*

The existence statement in (ii) above is well-known. It is obvious if the bundle is trivial. If it is non-trivial, one can think of V as the suspension of a circle bundle of Euler class 1 with the corresponding S^1–action, and can then provide V with an invariant symplectic form in any class a which satisfies (i) (a,b) since these conditions correspond to requiring that a be positive on each of the two fixed point sets of the S^1-action. (See [Au].) The other statements are more delicate. Consider first the case when the base manifold M is S^2. Gromov showed in [G] that any symplectic form on $S^2 \times S^2$ which admits symplectically embedded spheres in the classes $[S^2 \times pt]$ and $[pt \times S^2]$ is isotopic to a product (or split) form. (In fact, Gromov assumed that the form has equal integrals over the two spheres, but it is not hard to remove this condition.) A corresponding uniqueness result when $V = \mathbb{C}P^2 \# \overline{\mathbb{C}P}^2$ (which is the non-trivial S^2–bundle over S^2) was proved in [McD 2]. Here one requires the existence of just one symplectically embedded sphere, but it must be in the class of a section of self-intersection 1, not of the fiber. Gromov showed that this hypothesis also implies that there must be a symplectically embedded sphere in the class of the blown-up point, which is equivalent to condition (i)(b). In the present situation, we have less information since we start with only one symplectically embedded sphere, the fiber. Following an idea of Eliashberg's, we can construct a symplectic section of π (i.e. a section on which the symplectic form ω does not vanish) and so reduce to the previously considered case. In the process, we have to change the form ω by adding $\pi^*(\sigma)$ where σ is a 2-form on M such that $\sigma(M) > 0$. The argument is then completed by the following lemma, the proof of which uses the theory of holomorphic curves.

2.5 Lemma
Let V *be an* S^2*–bundle over a Riemann surface* M *and suppose that* $\omega_t, 0 \le t \le 1$, *is a family of (non-cohomologous) symplectic forms on* V *which are non-degenerate on one fiber of* V. *Then, if* ω_1 *admits a symplectic section, so does* ω_0.

In the general case, one cuts the fibration open over the 1–skeleton of M in order to reduce to the case $M = S^2$.

We can now state the classification theorem for minimal symplectic pairs.

2.6 Theorem
Let (V, C, ω) *be a minimal symplectic 4–dimensional pair where* C *is a 2–sphere with self-intersection* $C \cdot C = p \ge 0$. *Then* (V, ω) *is symplectomorphic either to* $\mathbb{C}P^2$ *or to a symplectic* S^2*–bundle over a compact surface* M. *Further, this symplectomorphism may be chosen so that it takes* C *either to a complex line or quadric in* $\mathbb{C}P^2$, *or to a fiber of the* S^2*–bundle, or (if* M *is* S^2*) to a section of this bundle.*

Thus, if p is odd and ≥ 3, (V, ω) is symplectomorphic to the Kähler manifold $\mathbb{C}P^2 \# \overline{\mathbb{C}P}^2$; if $p = 1$, (V, ω) is $\mathbb{C}P^2$ with its standard Kähler form; if p is even and ≥ 2, (V, ω) is the product $S^2 \times S^2$ with a product symplectic form (or, if $p = 4$, it could be $\mathbb{C}P^2$); and, if $p = 0$, (V, ω) is a symplectic S^2-bundle. From this, it is easy to prove:

2.7 Corollary
(i) *If* (V, C, ω) *is as above, the diffeomorphism type of the pair* (V, C) *is determined by* p *provided that* $p \ne 0, 4$.

(ii) *When* $p = 4$, *there are two possibilities for* (V, C): *it can be either* $(\mathbb{C}P^2, Q)$ *or* $(S^2 \times S^2, \Gamma_2)$, *where* Q *is a quadric and where* Γ_2 *is the graph of a holomorphic self-map of* S^2 *of degree* 2. *When* $p = 0$, C *is a fiber of a symplectic* S^2*–bundle.*

(iii) (V, C, ω) *is determined up to symplectomorphism by the cohomology class of* ω.

2.8 Corollary
A minimal symplectic 4-manifold (V,ω) *which contains a rational curve* C *with* $C \cdot C > 0$ *is symplectomorphic either to* $\mathbb{C}P^2$ *or to* $S^2 \times S^2$ *with the standard form.*

The main tool in the proof of Theorem 2.6 is the following version of the adjunction formula. We will suppose that J is an almost complex structure on V with first Chern class c_1, and that $f: S^2 \to V$ is a J-holomorphic map (ie $df \circ J_0 = J \circ df$, where J_0 is the usual almost complex structure on S^2) which represents the homology class $A \in$

$H_2(V; \mathbb{Z})$. The assumption that f is somewhere injective rules out the multiply-covered case, and implies that f is an embedding except for a finite number of multiple points and a finite number of "critical points", i.e. points where df_z vanishes.

2.9 Proposition
If f *is somewhere injective, then*

$$A \cdot A \geq c_1(A) - 2$$

with equality if and only if f *is an embedding.*

This is well-known if J is integrable: the quantity $\frac{1}{2}(A \cdot A - c_1(A) + 2)$ is known as the "virtual genus" of the curve $C = \text{Im } f$. It is also easy to prove if f is an immersion. For in this case $c_1(A) = 2 + c_1(\nu_C)$ where ν_C is the normal bundle to C, and $c_1(\nu_C) \leq A \cdot A$, with equality if and only if f is an embedding. In the general case, one has to show that each singularity of C contributes positively to $A \cdot A$. This is not hard to show for the simplest kind of singularities, and, using the techniques of [NW], one can reduce to these by a rather delicate perturbation argument.

With this in hand, we prove Theorem 2.6 by showing that V must contain an embedded J–simple curve of self-intersection +1 or 0. (J-simple curves do not decompose, so that their moduli space is compact.) It then follows by arguments of Gromov that V is $\mathbb{C}P^2$ in the former case and a symplectic S^2-bundle in the latter.

2.10 Note
Given an arbitrary symplectic 4-manifold one can always blow up some points to create a manifold (W, ω) which contains a symplectically embedded 2-sphere with an arbitrary negative self-intersection number. Hence, the existence of such a 2-sphere gives no information on the structure of (W, ω).

Corollary 2.7 may be understood as a statement about the uniqueness of symplectic fillings of certain contact manifolds. Indeed, consider an oriented (2n-1)-dimensional manifold Δ with closed 2-form σ. We will say that (Δ,σ) has contact type if there is a positively oriented contact form α on Δ such that $d\alpha = \sigma$. It is easy to check that the contact structure thus defined is independent of the choice of α. Following Eliashberg [E], we say that the symplectic manifold (Z,ω) fills (Δ, σ) if there is a diffeomorphism $f: \partial Z \to \Delta$ such that $f^*(\sigma) = \omega|\partial Z$. Further the filling (Z, ω) is said to be minimal if Z contains no exceptional curves in its interior.

As Eliashberg points out, information on symplectic fillings provides a way to distinguish between contact structures: if one constructs a filling of (Δ,σ_2) which does not have a certain property which one knows must be possessed by all fillings of (Δ, σ_1), then the contact structures on Δ defined by σ_1 and σ_2 must be different. In particular, it is interesting to look for manifolds of contact type which have unique minimal fillings. Obvious candidates are the lens spaces L_p, $p > 1$, which are obtained as the quotients of $S^3 \subset \mathbb{C}^2$ by the standard diagonal action of the cyclic subgroup $\Gamma_p \subset S^1$ of order p on $\mathbb{C}^2$, and whose 2-form σ is induced by ω_0.

It is not hard to see that if (Z,ω) fills (L_p, σ) we may quotient out $\partial Z = L_p$ by the Hopf map to obtain a rational curve C_p with self-intersection p in a symplectic manifold (V,ω) without boundary. Hence Corollary 2.7 implies:

2.11 Theorem
The lens spaces L_p, $p \geq 1$, *all have minimal symplectic fillings. If* $p \neq 4$, *minimal fillings* (Z,ω) *of* (L_p,σ) *are unique up to diffeomorphism, and up to symplectomorphism if one fixes the cohomology class* $[\omega]$. *However,* (L_4, σ) *has exactly two non-diffeomorphic minimal fillings.*

In higher dimensions, one cannot hope for such precise results. However, in dimension 6 there are certain contact-type manifolds (such as the standard contact sphere S^5) which impose conditions on any filling (Z,ω), even though they may not dictate the diffeomorphism type of minimal fillings. In dimensions > 6, one must restrict to "semi-positive" fillings to get analogous results. See [McD 5].

References

[Au] Audin, M. : Hamiltoniens périodiques sur les variétés symplectiques compactes de dimension 4, Preprint IRMA Strasbourg, 1988.

[E] Eliashberg, Ya.: On symplectic manifolds which are bounded by standard contact spheres, and exotic contact structures of dimension > 3, preprint, MSRI, Oct. 1988.

[G] Gromov, M. : Pseudo-holomorphic curves in symplectic manifolds, Invent. Math. 82 , 307-347 (1985)

[GS] Guillemin, V. and Sternberg, S.: Birational Equivalence in the symplectic category, preprint 1988

[McD 1] McDuff, D. : Examples of symplectic structures, Invent. Math 89, 13-36 (1987).

[McD 2] McDuff, D. : Blowing up and symplectic embeddings in dimension 4, to appear in Topology (1989/90).

[McD 3] McDuff, D. : The structure of rational and ruled symplectic 4-manifolds, preprint, Stony Brook, 1989.

[McD 4] McDuff, D. : The local behaviour of holomorphic curves in almost complex 4-manifolds, preprint, Stony Brook, 1989.

[McD 5] McDuff, D. : Symplectic manifolds with contact-type boundaries, in preparation, 1989.

[NW] Nijenhuis, A. and Woolf, W.: Some integration problems in almost–complex and complex manifolds, Ann. of Math. 77 (1963), 424 – 489.

Symplectic Capacities

H. HOFER
FB Mathematik
Ruhr Universität Bochum
Federal Republic of Germany

August 3, 1990

1 Introduction

In 1985 M. Gromov proved in his seminal paper *Pseudoholomorphic Curves in Symplectic Geometry* , [1], a striking rigidity result, the so–called squeezing theorem.

Consider the symplectic vectorspace $\mathbf{R}^{2n} := (\mathbf{R}^2)^n$ with the coordinates $z = (z_1, \ldots, z_n), z_i = (x_i, y_i)$, and the symplectic form σ, defined by

$$\sigma(z, z') = \sum_{i=1}^{n} x_i y_i' - x_i' y_i.$$

Denote by $\mathbf{B}^{2n}(r)$ the Euclidean r–ball in $\mathbf{R}^{2n}$ and by $Z^{2n}(r)$ the symplectic cylinder defined by

$$Z^{2n}(r) = \{z \in \mathbf{R}^{2n} \mid\mid z_1 \mid < r\}.$$

Gromov proved, that given a symplectic embedding

$$\Psi : \mathbf{B}^{2n}(r) \hookrightarrow Z^{2n}(r'),$$

we necessarily have the inequality $r \leq r'$. His proof of this fact relied on his existence theory for pseudoholomorphic curves, [1].

Coming from the variational theory of Hamiltonian Dynamics, I. Ekeland and the author observed in [2,3], that using Hamiltonian dynamics the squeezing theorem not only can be proved as well, but in fact many interesting symplectic invariants, so–called *Symplectic Capacities* can be constructed.

In this way a fruitful ***hook up*** between Symplectic Geometry and Hamiltonian Dynamics - with their different methods - has been achieved. This merger for example allows to see the nonobvious but very deep relationship between certain aspects of Hamiltonian Dynamics and Symplectic rigidity. The aim of this paper is to survey this relationship. We start with an axiomatic approach, then we survey several constructions of a symplectic capacity.

Acknowledgement

I would like to thank I. Ekeland for many stimulating discussions.

2 Axioms and Consequences

Consider the category $Symp^{2n}$, consisting of all $(2n)$–dimensional symplectic manifolds, together with the morphisms being the symplectic embeddings. We denote by S a subcategory of $Symp^{2n}$. We do not (!) assume that S is a full subcategory. The following *scaling axiom* is imposed:

(S) If $(M, \omega) \in S$ then also $(M, \alpha\omega) \in S$ for $\alpha \in \mathbf{R} \setminus \{0\}$

Denote by $\Gamma = (0, +\infty) \cup \{+\infty\}$ the extended positive half line. The obvious ordering on Γ is denoted by $\leq$. In the usual way we consider $(\Gamma, \leq)$ as a category.

Definition 2.1 *A symplectic capacity for S is a covariant functor $c : S \to \Gamma$ satisfying the axioms*

(N) $c\,(Z^{2n}(1), \sigma) < +\infty$

(C) $c\,(M, \alpha\omega) = \mid \alpha \mid c\,(M, \omega)$.

Here (N) stands for nontriviality and (C) for conformality.

Several remarks are in order.

Remark 2.2 **1.** *The map $(M, \omega) \to (\int_M \omega^n)^{\frac{1}{n}}$ satisfies (C) but not (N) if $n \geq 2$. So if $n \geq 2$ the nontriviality axiom excludes volume–related invariants. However, if $n = 1$, then $(M, \omega) \to vol(M, \omega)$ is a symplectic capacity.*

2. *Besides $S = Symp^{2n}$ there are several other interesting subcategories. For example let (V, ω) be a symplectic manifold. Denote by $Op\,(V, \omega)$ the collection of all symplectic manifolds of the form $(U, \alpha\omega \mid U)$ where U is a nonempty open subset of V and $\alpha \in \mathbf{R} \setminus \{0\}$. As morphisms we take the symplectic embeddings. Clearly $S = Op\,(V, \omega)$ is an admissible category. Next assume G is a compact Lie group.*

Denote by S_G the collection of all smooth $(2n)$–dimensional symplectic G–spaces, where G acts symplectically. As morphisms we take the symplectic G–embeddings.

3. *One might look for functors $c : S \to \Gamma$ which transform differently. For example $c(M, \alpha\omega) = |\alpha|^k c(M, \omega)$ for some $k \neq 1$. To exclude certain pathologies one could impose the following axiom system. Let $k \in \{1, \ldots, n\}$. A symplectic k-capacity is a covariant functor $c : S \to \Gamma$ satisfying*

$$(C)_k \quad c(M, \alpha\omega) = |\alpha|^k c(M, \omega), \quad \alpha \neq 0$$

$$(N)_k \begin{cases} c\left(B^{2(k-1)}(1) \times \mathbf{R}^{2(n-k+1)}, \sigma\right) = +\infty \\ c\left(B^{2k}(1) \times \mathbf{R}^{2(n-k)}, \sigma\right) \quad < +\infty \end{cases}$$

We note that for $k = 1$ we precisely recover our definition of a symplectic capacity. In fact the first part of $(N)_1$ follows from $(C)_1$ and the fact that $c(\mathbf{B}^{2n}(1), \sigma) > 0$. For $k = n$ the volume map is an example of a n–capacity. So far no examples are known for intermediate capacities $(1 < k < n)$. It is quite possible that they do not exist. Some evidence for this possibility are given by the fact that there is an enormous amount of flexibility for symplectic embeddings $M \hookrightarrow N$ with $\dim M \leq \dim N + 2$, see Gromov's marvellous book [4]. For example, the existence of a symplectic embedding

$$B^2(\varepsilon) \times \mathbf{R}^{2n-2} \hookrightarrow B^{2n-2}(1) \times \mathbf{R}^2$$

for some $\varepsilon > 0$, would immediately contradict the axioms for a k–capacity $1 < k < n$. As a consequence of results in [3], concerning symplectic capacitics, the inequality $\varepsilon^{-2} \geq n - 1$ is a necessary condition for the existence of such a symplectic embedding.

Next we derive some easy consequences of the axioms. Set $S = Symp^{2n}$ and assume $c : S \to \Gamma$ is a symplectic capacity.

Lemma 2.3 *Assume there exists a symplectic embedding $\Psi : Z^{2n}(r_1) \hookrightarrow Z^{2n}(r_2)$. Then $r_1 \leq r_2$.*

PROOF : First we note that the map

$$(Z^{2n}(r), \sigma) \ni z \to \frac{1}{r} z \in (Z^{2n}(1), r^2\sigma)$$

is a symplectic diffeomorphism. Hence

$$\begin{aligned} c(Z^{2n}(r), \sigma) &= c(Z^{2n}(1), r^2\sigma) \\ &= r^2 c(Z^{2n}(1), \sigma) \\ &=: \tau r^2. \end{aligned}$$

Here $\tau \in (0, +\infty)$ as a consequence of the axioms. If now $\Psi : Z^{2n}(r_1) \hookrightarrow Z^{2n}(r_2)$ we obtain, using the axioms, $\tau r_1^2 \leq \tau r_2^2$.

An even stronger result is

Lemma 2.4 *There exists a constant $R_0 \in (0, +\infty)$ such that there is no symplectic embedding*

$$\mathbf{B}^{2n}(R) \hookrightarrow Z^{2n}(1)$$

for $R \geq R_0$.

PROOF : If $\mathbf{B}^{2n}(R)$ embeds symplectically into $Z^{2n}(1)$ we obtain

$$R^2 c\left(\mathbf{B}^{2n}(1)\right) \leq c\left(Z^{2n}(1)\right)$$

Hence

$$R \leq \left(c\left(Z^{2n}(1)\right)/c\left(\mathbf{B}^{2n}(1)\right)\right)^{\frac{1}{2}}$$

■

Remark 2.5 *Gromov's Squeezing Theorem implies in fact, that there is a capacity c with*

$$c\left(Z^{2n}(1)\right) = c\left(\mathbf{B}^{2n}(1)\right).$$

Hence $R_0 = 1$ is the best possible choice.

So far we have only studied trivial consequences. The following corollaries are substantially deeper. We call a subset $E \subset \mathbf{R}^{2n}$ an ellipsoid if there exists a positive definite quadratic form q such that $E = \{z \in \mathbf{R}^{2n} \mid q(z) < 1\}$. Assume c is a capacity defined on $S = Op(\mathbf{R}^{2n}, \sigma)$. We extend c to a map $2^{\mathbf{R}^{2n}} \times (\mathbf{R} \setminus \{0\}) \to [0, +\infty]$ by defining

$$(1) \qquad c\,(U, \alpha) = \inf\left\{c\,(V, \alpha\sigma) \mid V \supset U,\ (V, \alpha\sigma) \in S\right\}$$

Let us also write $c\,(U) := c\,(U, 1)$ for $U \subset \mathbf{R}^{2n}$. It has been proved in [2,3]:

Proposition 2.6 *Let $T : \mathbf{R}^{2n} \to \mathbf{R}^{2n}$ be a linear map. Assume*

$$c\,(E) = c\,(T(E))$$

for every ellipsoid. Then T is symplectic or antisymplectic, i.e.

$$T^*\omega_0 = \omega_0 \qquad or \qquad T^*\omega_0 = -\omega_0.$$

In [2,3] the normalisation $c\,(\mathbf{B}^{2n}(1),\sigma) = c\,(Z^{2n}(1),\sigma) = \pi$ had been assumed throughout the paper. However, it had not been used for the above proposition, which only depends on (N) and (C). By the definition (1) we have meanwhile obtained a map still denoted by c:

$$c : 2^{(\mathbf{R}^{2n})} \to [0,+\infty]$$

such that for every $\Psi \in \mathit{Diff}\,(\mathbf{R}^{2n},\sigma), t \in \mathbf{R}$:

$$(2) \qquad \begin{cases} c\,(\Psi(S)) = c\,(S) \\ c\,(tS) = t^2 c\,(S) \\ c\,(S) \leq c\,(T) \quad \text{if } S \subset T \\ 0 < c\,(\mathbf{B}^{2n}(1)) \leq c\,(Z^{2n}(1)) < +\infty \end{cases}$$

We call this map again a symplectic capacity. Next we obtain the following local C^0–rigidity result, see [3] or [5].

Theorem 2.7 (Local Rigidity) *Assume $\Phi_k : \mathbf{B}^{2n}(1) \to \mathbf{R}^{2n}$ is a sequence of symplectic embeddings, converging uniformly to a continuous map $\Phi : \mathbf{B}^{2n}(1) \to \mathbf{R}^{2n}$, which is differentiable at $0 \in \mathbf{B}^{2n}(1)$ with derivative $\Phi'(0)$. Then $\Phi'(0)$ is symplectic.*

Using this theorem we recover the famous Eliashberg – Gromov rigidity result, [4,6].

Corollary 2.8 *Let (M,ω) be a symplectic manifold. Then $\mathit{Diff}_{\omega}(M)$ is C^0–closed in $\mathit{Diff}(M)$.*

The local rigidity result can be proved only using (2). One shows that $\Phi'(0) : \mathbf{R}^{2n} \to \mathbf{R}^{2n}$ and $\Phi'(0) \times Id_{\mathbf{R}^2} : \mathbf{R}^{2n+2} \to \mathbf{R}^{2n+2}$ preserve capacities $c_{R^{2n}}$ and $c_{R^{2n+2}}$ for ellipsoids. The linear algebra proposition then implies the desired result.

So far we surveyed results which can be obtained from the existence of a capacity. The following results are different. Their proofs depend on capacities with additional features.

3 Some Special Symplectic Capacities

Displacement Energy

The first capacity we consider here is very closely related to Hamiltonian Dynamics. Let us introduce some notation. We denote by $\mathcal{C}$ the vector space of all smooth Hamiltonians $H : [0,1] \times \mathbf{R}^{2n} \to \mathbf{R}$ having compact support. Denote by

X_{H_t} the time–dependent Hamiltonian vector field associated to H by the formula $dH_t(x)(*) = \sigma(X_{H_t}(x), *)$. Denote by Ψ_H the time–1–map associated to the Hamiltonian System

$$\dot{x} = X_{H_t}(x)$$

We define a norm $\| * \|$ on $\mathcal{C}$ by

$$\|H\| = \left(\sup_{[0,1]\times\mathbf{R}^{2n}} H \right) - \left(\inf_{[0,1]\times\mathbf{R}^{2n}} H \right).$$

By $\mathcal{D}$ we denote the collection of all Ψ_H, $H \in \mathcal{C}$. One easily verifies that $\mathcal{D}$ is a group.

Let S be a bounded subset of $\mathbf{R}^{2n}$. We study the importance of energy by analyzing the following problem: How much energy do I need to deform S into a set S' which is disjoint from S? To make this precise we define the displacement energy $d(S)$ by

$$d(S) = \inf \{\|H\| \mid \Psi_H(\bar{S}) \cap \bar{S} = \phi\}.$$

For an unbounded set $T \subset \mathbf{R}^{2n}$ let

$$d(T) = sup \; \{d(S) \mid S \subset T, \; S \text{ bounded}\}.$$

It is clear that $d(\Phi(T)) = d(T)$ for all $\Phi \in \mathcal{D}$ and $T \subset \mathbf{R}^{2n}$. What is not clear is that $d(T) \neq 0$. In order to stress this point consider $S = \mathbf{B}^2(1) \times \mathbf{B}^{2n-2}(R), R \geq 1$. To disjoin S from itself it is enough to disjoin $\mathbf{B}^2(1) \subset \mathbf{R}^2$ from itself. Now $\mathbf{B}^2(1)$ is via an element in $\mathcal{D}$ symplectomorphic to an almost square with round corners and sidelength approximately $\sqrt{\pi}$. By taking a translation parallel to one side, which disjoints the *square* from itself we need a Hamiltonian which has slope $\sim \sqrt{\pi}$ and is increasing along lines orthogonal to the above side. By cutting this Hamiltonian off smoothly we see that we can separate the square from itself by a Hamiltonian H with $\|H\| \sim \pi$. Hence, a simple corollary:

$$d(\mathbf{B}^2(1) \times \mathbf{B}^{2n-2}(R)) \leq \pi \quad \forall R > 0$$

$$D(\mathbf{B}^2(1) \times \mathbf{R}^{2n-2}) \leq \pi.$$

Implementing the above construction for a disjoint union of small cubes via a suitable highly oscillatory Hamiltonian we can prove the following.

Proposition 3.1 *Assume U and V are open bounded sets in $\mathbf{R}^{2n}$ with smooth boundary and equal volume. Given $\varepsilon > 0$ there exists $H_\varepsilon \in C$ with $\|H_\varepsilon\| < \varepsilon$ and a subset $M_\varepsilon \subset U$ with meas $(M_\varepsilon) \leq \varepsilon$ such that $\Psi := \Psi_{H_\varepsilon}$ satisfies*

$$\Psi(U \setminus M_\varepsilon) \subset V.$$

So with other words we can displace a set S from itself modulo sets of small measure with arbitrary small energy. However, the surprising fact is

Theorem 3.2

$$d(\mathbf{B}^{2n}(1)) = d(Z^{2n}(1)) = \pi.$$

In particular the above construction is the best one can do! Clearly, the map d can be used to define a capacity $c : Op(\mathbf{R}^{2n}, \sigma) \to \Gamma$ such that in addition to the axioms (N) and (C) we have

$$c(\mathbf{B}^{2n}(1), \sigma) = c(Z^{2n}(1), \sigma) = \pi.$$

This capacity for example implies Gromov's Squeezing Theorem. It also has some very interesting consequences for the topological group $\mathcal{D}$. Define the energy of a map $\Psi \in \mathcal{D}$ by

$$E : \mathcal{D} \to \mathbf{R} : E(\Psi) = inf\ \{\|H\| \mid \Psi_H = \Psi\}$$

One easily verifies, [7], that

$$E(\Psi) = E(\Psi^{-1})$$

$$E(\Psi\Phi) \leq E(\Psi) + E(\Phi).$$

Assume now $\Psi \neq id$. Then

$$\Psi(\bar{B}_\varepsilon(x_0)) \cap \bar{B}_\varepsilon(x_0) = \phi$$

for some point $x_0 \in \mathbf{R}^{2n}$ and $\varepsilon > 0$. Hence

$$E(\Psi) \geq d\left(\bar{B}_\varepsilon(x_0)\right) = \pi\varepsilon^2 > 0.$$

So we find, summing up

Theorem 3.3 *The map $\rho : \mathcal{D} \times \mathcal{D} \to \mathbf{R}$ defined by*

$$\rho(\Psi, \Phi) = E(\Psi^{-1} \circ \Phi)$$

defines a bi-invariant metric on $\mathcal{D}$. Moreover the map

$$\mathcal{C} \ni H \to \Psi_H \in \mathbf{D}$$

is Lipschitz continuous with Lipschitz constant equal to 1:

$$\rho(\Psi_H, \Psi_K) \leq \|H - K\|.$$

Remark 3.4 *The map $H \to \Psi_H$ extends to the completions $\tilde{\mathcal{C}} \to \tilde{\mathcal{D}}$. Since for $k \geq 2$ the map $H \to \Psi_H$ can be considered as a map $C^k \to C^{k-1}$, the above theorem shows philosophically that this map extends to a map $C^\circ \to C^{-1}$. So in some sense ϱ defines a C^{-1}-topology.*

Remark 3.5 *The topology on $\mathcal{D}$ is very weak. It does for example not even imply pointwise convergence. For example we can have*

$$\Psi_k \to id \qquad in\ (\mathcal{D}, \rho)\ and \qquad \int_{\mathbf{R}^{2n}} | \Psi_k(x) - x |^p \, dx \to +\infty$$

as $k \to \infty$ for $1 \le p \le \infty$. It would be interesting to have a good model for the completion $\widetilde{\mathcal{D}}$.

Remark 3.6 *The displacement capacity can be considered – to a certain extent – as a nonlinear extension of the 2–dimensional Lebesgue measure to higher dimensions. To see this, note that bounded simply connected domain S with smooth boundary is symplectomorphic to $B^2(R)$ for some R. Hence*

$$d(S) = d(B^2(R)) = \pi R^2 = meas\,(B^2(R)) = meas\,(S).$$

Next we consider whether there are other invariants of a set $S \subset \mathbf{R}^{2n}$ which are related to the displacement capacity. Let us assume S is a connected compact smooth hypersurface in $\mathbf{R}^{2n}$. The canonical line boundle $\mathcal{L}_S \to S$ is defined by

$$\mathcal{L}_S = \{(x, \xi) \in TS \mid \omega(\xi, \eta) = 0, \quad \text{for all } \eta \in T_x S\}.$$

We denote by $L_S(x)$ the leaf through $x \in S$ and call it a characteristic on S. Of particular interest are the closed characteristics. We denote the set of closed characteristics of S by $\mathcal{D}(S)$. We give $\mathcal{L}_S$ an orientation by choosing a section ξ of $\mathcal{L}_S \to S$ such that $\omega(\xi(x), n(x)) > 0$, where $x \to n(x)$ is the outward pointing normal vector field. Recall that by Alexander duality $\mathbf{R}^{2n} \setminus S$ has precisely two components.

It is an open problem if $\mathcal{D}(S) \neq \phi$ for every compact hypersurface. By a result of C. Viterbo, [8], we have that $\mathcal{D}(S) \neq \phi$ if S is of contact type, [9].

Now let S be a compact connected smooth hypersurface with the orientation of $\mathcal{L}_S \to S$ just defined. If $P \in \mathcal{D}(S)$ we have

$$TP = \mathcal{L}_{S|P},$$

and therefore P inherits on orientation from $\mathcal{L}_S$. Define the action of P by

$$A(P) = \int_D \sigma$$

where D is a disk in $\mathbf{R}^{2n}$, with $\partial D = P$ and compatible orientation. Clearly $A(P)$ is well defined.

We have the following strong squeezing theorem which follows immediately from results in [10].

Theorem 3.7 *(Strong Squeezing) Assume S is a smooth compact hypersurface bounding a convex domain C. Then*

$$sup\ \{\pi r^2 \mid \Psi(\mathbf{B}^{2n}(r)) \subset C \quad for\ some\ \Psi \in \mathcal{D}\}$$

$$\leq \inf\ \{A(P) \mid P \in \mathcal{D}(S)\}$$

$$\leq \inf\ \{\pi r^2 \mid \Psi(C) \subset Z^{2n}(r) \quad for\ some\ \Psi \in \mathcal{D}\}$$

For $C = \mathbf{B}^{2n}(1)$ we recover Gromov's squeezing theorem.

It might be possible that we have in fact equality for the three expressions. This is still an open problem.

The following theorem which is a consequence of results in [7], gives a nice a priori estimate for the displacement energy.

Theorem 3.8 *Assume S is a smooth compact hypersurface bounding a convex domain C. Then*

$$d(C) \geq \inf\ \{A(P) \mid P \in \mathcal{D}(S)\}.$$

Note that for ellipsoids we have equality. That symplectic capacities are related to closed characteristics on hypersurfaces is no accident and has been detected in [2]. To make it plausible consider the following:

Let C be a subset of $\mathbf{R}^{2n}$ and define its capacity as the infimum of the outer measure of the projections of all symplectic images of C onto the $(z_1 = (x_1, y_1))$–coordinate–plane.

This defines a symplectic capacity, as an easy consequence of Gromov's squeezing theorem. Assume $S = \partial C$ is a compact hypersurface. Suppose also that the projection onto the z_1–coordinate has minimal area. In order to decrease this projection–area further (which is impossible by our assumption) one would like to take a Hamiltonian H which is increasing on the leaves of S, because the associated Hamiltonian vector field pushes S into C. Of course such an H does not exist globally, but on subregions of S. If some region contains a closed characteristic it cannot be the domain of a Hamiltonian which is increasing on leaves. So closed characteristics occur as obstructions against decreasing the measure of certain projections by pushing certain subregions of S into C.

We already motivated the occurence of periodic Hamiltonian trajectories in the study of symplectic capacities. The next construction makes extensive use of periodic solutions of Hamiltonian Systems and gives a whole sequence of independent capacities.

Periodic Hamiltonian Trajectories and Capacities

In [11] I. Ekeland and the author studied periodic solutions of Hamiltonian Systems on a prescribed energy surface. In [11] they associated to convex energy surfaces S symplectic invariants and to their periodic trajectories three independent invariants and showed that these three quantities are not independent. A little bit later C. Viterbo, [8], proved the $\mathbf{R}^{2n}$–Weinstein conjecture, [9]. Viterbo's key ingredient in his proof, meanwhile called Viterbo's trick, was simplified and exploited by E. Zehnder and the author, [11], in the study of periodic Hamiltonian trajectories on a prescribed energy surface. Roughly speaking one associates to a bounded open set $U \subset \mathbf{R}^{2n}$ a variational problem in the loop–space of $\mathbf{R}^{2n}$, which gives useful information about the Hamiltonian flow on the energy surface ∂U. These developments were the starting point for the following construction of symplectic capacities.

We define a particular class of Hamiltonian systems on $\mathbf{R}^{2n}$, which we denote by $\mathcal{H}$, as follows:

A Hamiltonian $H : \mathbf{R}^{2n} \to \mathbf{R}$ belongs to $\mathcal{H}$ if

H1 There exists an open nonempty set U such that $H \mid U \equiv 0$. Moreover $H(z) \geq 0$ for all $z \in \mathbf{R}^{2n}$.

H2 There exist numbers $\rho, a > 0$, such that $a > \pi$, $a \notin \pi \cdot \mathbf{Z}$, and $H(z) = a \mid z \mid^2$ for all $\mid z \mid \geq \rho$.

Denote by $\hat{F}$ the loop space of $\mathbf{R}^{2n}$, i.e. the space of all smooth maps $\mathbf{R}/\mathbf{Z} \to \mathbf{R}^{2n}$. We put $S^1 := \mathbf{R}/\mathbf{Z}$ and define $\Phi_H : \hat{F} \to \mathbf{R}$ for $H \in \mathcal{H}$ by

$$\Phi_H(z) = \frac{1}{2}\int_0^1 < -J\dot{z}, z > dt - \int_0^1 H(z)\, dt.$$

Here J is given by

$$J = \begin{pmatrix} \begin{pmatrix} 0 & -1 \\ 1 & 0 \end{pmatrix} & \cdots & 0 \\ \vdots & \ddots & \vdots \\ 0 & 0 & \begin{pmatrix} 0 & -1 \\ 1 & 0 \end{pmatrix} \end{pmatrix}$$

The critical points of Φ_H are precisely the 1–periodic solutions z of the differential equation $-J\dot{z} = H'(z)$, where H' is the gradient of H with respect to the standard inner product $< \cdot, \cdot >$. Alternatively we have of course $\dot{z} = X_H(z)$, since $X_H = JH'$ is the Hamiltonian vector field associated to H. Our aim is to find infinitely many universal minmax–characterisations for critical points of Φ_H. For this we take the completion F of $\hat{F}$ with respect to the $H^{\frac{1}{2}}$–norm on $\hat{F}$. This is defined as follows: Every element in $\hat{F}$ can be written as

$$z = \sum_{k \in \mathbf{Z}} exp\,(2\pi t k J) z_k, \quad z_k \in \mathbf{R}^{2n}$$

such that $\sum | k |^s | z_k |^2 < \infty$ for every $s \geq 0$. We define a norm by

$$\|z\|^2 = | z_0 |^2 + 2\pi \sum_{k \in \mathbf{Z}} | k || z_k |^2,$$

and denote the completion of $(\hat{F}, \| \cdot \|)$ by F. F has an orthogonal splitting $F = F^- \oplus F^0 \oplus F^+$ by cutting a Fourier series into pieces $k < 0$, $k = 0$ and $k > 0$. The map $z \mapsto \frac{1}{2} \int_0^1 \langle J\dot{z}, z \rangle \, dt$ extends to a quadratic form $z \mapsto a(z)$ on F, with a gradient a' given by

$$a'(z) = -z^- + z^+, \quad z = z^- + z^0 + z^+ \in F.$$

Moreover one can show that F embeds compactly into every $L^P([0,1]; \mathbf{R}^{2n})$ for $1 \leq p < \infty$. We have a natural S^1–action on F by phase-shift, i.e. $\tau \in \mathbf{R}/\mathbf{Z}$ acts via

$$(\tau * z)(t) = z(t + \tau).$$

We define a distinguished subgroup $\mathcal{B}$ of $homeo(F)$ the homeomorphism group of F by saying $h \in \mathcal{B}$ if $h : F \to F$ is a homeomorphism admitting the representation

$$h(z) = exp\,(\gamma^-(z)) z^- + z^0 + exp\,(\gamma^+(z)) z^+ + K(z),$$

where $\gamma^+, \gamma^- : F \to \mathbf{R}$ are continuous and S^1–invariant, mapping bounded sets in F into bounded sets in $\mathbf{R}$. Moreover $K : F \to F$ is continuous and S^1–equivariant and maps bounded sets into precompact sets. It is easily verified that usual composition turns $\mathcal{B}$ into a group. Next we need a pseudoindex theory in the sense of Benci, [13], associated to the Fadell-Rabinowitz index, [14], in order to *measure* the size of S^1–invariant sets. This goes as follows: Given a paracompact S^1–space X we build a free S^1–space $X \times S^\infty$ by letting S^1 act through the diagonal action. Here $S^\infty = \bigcup S^{2n-1}$, with $S^{2n-1} \subset \mathbf{R}^{2n} \cong \mathbf{C}^n$. Taking the quotient with respect to S^1 we obtain a principal S^1–bundle

$$X \times S^\infty \longrightarrow (X \times S^\infty)/S^1.$$

The classifying map

$$f : (X \times S^\infty)/S' \to \mathbf{C}P^\infty$$

induces a homomorphism

$$f^* : \bar{H}(\mathbf{C}P^\infty) \to \bar{H}_{S^1}(X) := \bar{H}((X \times S^\infty)/S^1).$$

in Alexander-Spanier-Cohomology with rational coefficients. Here $\bar{H}_{S^1}$ is the well-known Borel construction of a S^1–equivariant cohomology theory. We know that

$\bar{H}(\mathbf{C}P^\infty) = \mathbf{Q}[t]$, the generator t being of degree 2. We define the index of X, denoted by $\alpha(X)$ to be the largest number k such that

$$f^*(t^{k-1}) \neq 0.$$

We put $\alpha(X) = \infty$ if $f^*(t^{k-1}) \neq 0$ for all k and $\alpha(\emptyset) = 0$. The α–index is the well–known Fadell-Rabinowitz construction of a S^1–index theory. Next we define for an S^1–invariant set $\xi \subset F$ an index $\text{ind}(\xi)$ by

$$\text{ind}(\xi) = \inf \left\{ \alpha(h(\xi) \cap S^+) \mid h \in \mathcal{B} \right\}$$

where S^+ is the unit sphere in F^+. The key topological result is the following:

Lemma 3.9 *Let X be an S^1–invariant linear subspace of F^+ of dimension $2k$. Then*

$$ind\,(F^- \oplus F^0 \oplus X) = k.$$

For a proof see [3].

Finally we define for $H \in \mathcal{H}$

$$c_{k,H} = \inf \{ \sup \Phi_H(\xi) \mid \xi \subset F \text{ is } S^1\text{–invariant and ind } (\xi) \geq k \}.$$

By construction, it is clear that

$$0 < c_{1,H} \leq c_{2,H} \leq \ldots \leq +\infty.$$

For a bounded set $S \subset \mathbf{R}^{2n}$ we put

$$\mathcal{H}(S) = \{ H \in \mathcal{H} \mid H \mid U(\overline{S}) \equiv 0 \}.$$

where $U(\bar{S})$ denotes an open neighbourhood of $\bar{S}$ depending on H. For a bounded set S let

$$c_k(S) = \inf_{H \in \mathcal{H}(S)} c_{k,H}$$

For an unbounded set T let

$$c_k(T) = sup\{ c_k(S) \mid S \subset T \quad \text{bounded} \}.$$

The first result proved in [3] is

Theorem 3.10 *For every $k \in \{1, 2, \ldots\}$ the map c_k defines a symplectic capacity. Moreover*

$$c_1 \leq c_2 \leq \ldots.$$

The c_k have been computed in [3] for certain sets. For $r = (r_1, \dots, r_n)$ with $0 < r_1 \le r_2 \dots \le r_n$, define an ellipsoid $E(r)$ by

$$E(r) = \left\{ z \,\middle|\, \sum_{i=1}^{n} \left| \frac{z_i}{r_i} \right|^2 < 1 \right\}.$$

For such a r define with $\mathbf{N}^* = \{1, 2, \dots\}$:

$$d_k(r) = \inf \left\{ \tau > 0 \mid card \left\{ (l, j) \mid l \in \mathbf{N}^*,\ j \in \{1, \dots, n\},\ \pi r_j^2 l \le \tau \right\} \ge k \right\}.$$

For example for $r = (1, 1, \dots, 1)$ we have

$$d_k(r) = \left[\frac{n + k - 1}{n} \right] \pi$$

where $[*]$ denotes the integer part. Moreover observe that

$$\lim_{n \to \infty} d_k((1, n, \dots, n)) = k\pi.$$

It has been proved in [3] that

Theorem 3.11 *The following equations hold:*

i $c_k(E(r)) = d_k(r)$

ii $c_k(\mathbf{B}^2(1) \times \mathbf{B}^2(R) \dots \times \mathbf{B}^2(R)) = k\pi \quad$ *for* $R \ge 1$

Here all sets are equipped with the standard structure σ. The capacities c_k enjoy a useful representation property explained as follows. Call a compact connected smooth hypersurface S to be of restricted contact type provided there exists a vector field η on $\mathbf{R}^{2n}$, such that

$$\eta \quad \text{is transversal to } S, \quad L_\eta \sigma = \sigma.$$

We have, denoting by B_S the bounded component of $\mathbf{R}^{2n} \setminus S$:

Theorem 3.12 *Let S be as just described. Then there exists a sequence $(P_k) \subset \mathcal{D}(S)$ and a sequence $n_k \in \mathbf{N}^*$ such that*

$$c_k(B_S) = n_k A(P_k).$$

Finally let us give an application to an embedding problem:

Theorem 3.13 *Assume $B^2(1) \times \dots \times B^2(1)$ admits a symplectic embedding into $\mathbf{B}^{2n}(r)$. Then $r \ge \sqrt{n}$.*

PROOF : Let $\Psi : B^2(1) \times \ldots \times B^2(1) \hookrightarrow \mathbf{B}^{2n}(r)$ be the symplectic embedding. We find, see [2], a symplectic map $\widehat{\Psi} \in \mathcal{D}$, such that

$$\widehat{\Psi}\left(B^2(\delta) \times \ldots \times B^2(\delta)\right) \subset B^{2n}(r)$$

where $\widehat{\Psi} = \widehat{\Psi}_\delta$ and $0 < \delta < 1$ was given. Hence, taking the n-th capacity we obtain

$$n\pi\delta^2 \leq \pi r^2.$$

Since $\delta < 1$ was arbitrary we deduce $r \geq \sqrt{n}$. We had to introduce the map $\widehat{\Psi}$ since the c_k are by construction only invariant under symplectomorphisms in $\mathcal{D}$. ■

4 A capacity on Symp^{2n} and the Weinstein conjecture

Here we present a more general construction for a symplectic capacity on arbitrary symplectic manifolds. Assume $(M, \omega) \in \mathrm{Symp}^{2n}$. We denote by $\mathcal{H}(M)$ the set of all autonomous Hamiltonian Systems $H : M \to \mathbf{R}$ such that

H1 There exists a non–empty open set $U \subset M \setminus \partial M$ such that $H \mid U \equiv 0$.

H2 There exists a compact subset $K \subset M \setminus \partial M$ and a constant $m(H) > 0$ such that

$$0 \leq H(x) \leq m(H) \quad \text{for } x \in M$$

and

$$H(x) = m(H) \quad \text{for } x \in M \setminus K.$$

Given a symplectic embedding $\Psi : M \hookrightarrow N$ we define a morphism

$$\Psi_* : \mathcal{H}(M) \to \mathcal{H}(N)$$

by

$$(\Psi_* H)(x) = \begin{cases} H \circ \Psi^{-1}(x) & \text{if } x \in im\,(\Psi) \\ m(H) & \text{if } x \in N \setminus im\,(\Psi) \end{cases}$$

We call a Hamiltonian $H \in \mathcal{H}(M)$ admissible provided every T–periodic solution of $\dot{x} = X_H(x)$ for $T \in [0,1]$ is constant. We denote by $\mathcal{H}_a(M)$ the collection of all admissible Hamiltonians in $\mathcal{H}(M)$. Clearly $\Psi_*\mathcal{H}_a(M) \subset \mathcal{H}_a(N)$ and $\mathcal{H}_a(M) \neq \phi$ as one easily verifies. We define $c : \mathrm{Symp}^{2n} \to \Gamma$ by

$$c\,(M, \omega) = sup\{m(H) \mid H \in \mathcal{H}_a(M, \omega)\}.$$

Clearly c is a covariant functor and $c\,(M, \alpha \cdot \omega) = \mid \alpha \mid \cdot c\,(M, \omega)$ for $\alpha \neq 0$. The difficulty lies in proving axiom (N) which is done in [10]. Some other results listed below are taken from [15,16].

Theorem 4.1 $c : Symp^{2n} \to \Gamma$ *is a symplectic capacity and*

$$c\,(\mathbf{B}^{2n}(1), \sigma) = c\,(\mathbf{Z}^{2n}(1), \sigma) = \pi.$$

Moreover for every every convex bounded domain C with smooth boundary $S = \partial C$ we have

$$c\,(C, \sigma) = inf\{A(P) \mid P \in \mathcal{D}(S)\}$$

and additionally for every compact symplectic manifold (N, ω) with $\omega \mid \pi_2(N) = 0$

$$c\,(\mathbf{N} \times B^{2k}(r), \omega \oplus \sigma) = \pi r^2$$

provided $k \geq 1$. Also for the standard $\mathbf{CP}^n$:

$$c\,(\mathbf{CP}^n, \omega) = \int_{\mathbf{CP}^1} \omega.$$

The above capacity is very useful in dealing with the Weinstein conjecture, [9]. Let us recall some notations. Consider a compact $(2n-1)$-dimensional manifold $\mathbf{S}$ equipped with a 1-form λ such that $\lambda \wedge (d\lambda)^{n-1}$ is a volume. One calls λ a contact form.

The manifold S carries a natural line bundle defined by

$$\mathcal{L}_S = \{(x, \xi) \in TS \mid d\lambda(x)(\xi, \eta) = 0 \ forall\ \eta \in T_x S\}.$$

Hence $\mathcal{L}_S \subset TS$ and since $\lambda \wedge (d\lambda)^{n-1}$ is a volume $\lambda(x, \xi) \neq 0$ for $(x, \xi) \in \mathcal{L}_S$, if $\xi \neq 0$. Denote by $\mathcal{D}(S)$ the closed integral curves for the (integrable) distribution $\mathcal{L}_S \to S$. Weinstein conjectured that $\mathcal{D}(S) \neq \phi$ at least if $H^1(S; \mathbf{R}) = 0$. Given (S, λ) we define a germ of a symplectic manifold structure on $(-\varepsilon, \varepsilon) \times S$ by $\omega = d(e^t \lambda)$ where $(t, x) \in (-\varepsilon, \varepsilon) \times S$. We consider S as the subset $\{0\} \times S$ in $(-\varepsilon, \varepsilon) \times S$. First we have the following trivial observation.

Proposition 4.2 *Assume the previously defined capacity of $(-\varepsilon, \varepsilon) \times S$ for some $\varepsilon > 0$ is finite. Then $\mathcal{D}(S) \neq \phi$.*

PROOF : Fix a smooth map $\varphi : (-\varepsilon, \varepsilon) \to \mathbf{R}$ such that $\varphi \equiv 0$ near zero and $\varphi(S) = m > 0$ for S close to ε or $-\varepsilon$. Define a Hamiltonian

$$H \in \mathcal{H}\left((-\varepsilon, \varepsilon) \times S, d\left(e^t \lambda\right)\right)$$

by

$$H(x) = \varphi(\delta) \; for \; x \in \{S\} \times S.$$

We pick m in such a way that $m > c((-\varepsilon, \varepsilon) \times S)$.

Hence H is not admissible and has a non-constant periodic solution x of period $T \in [0,1]$. We have

$$x(\mathbf{R}) \subset \{\delta\} \times S$$

for some δ and

$$t \rightarrow \frac{d}{dt}(pr_S \circ x(t))$$

is a non-vanishing section of $\mathcal{L}_S \rightarrow S$. Hence it parametrises a closed characteristic. ∎

Let us say that a compact smooth hypersurface S in a symplectic manifold (M, ω) is of contact type provided there exists a 1-form λ on S such that

$$d\lambda = i^*\omega$$

where $i : S \hookrightarrow M$ is the inclusion and $\lambda(x, \xi) \neq 0$ for non-zero elements in $\mathcal{L}_S$. It is not difficult to show that $((-\varepsilon, \varepsilon) \times S, d(e^t\lambda))$ admits for some small $\varepsilon > 0$ a symplectic embedding into (M, ω) mapping $\{0\} \times S$ into S.

Definition 4.3 *Let (S, λ) be a compact $(2n-1)$-dimensional manifold equipped with a contact form. We call (S, λ) embeddable into the symplectic manifold $(M, \omega) \in Symp^{2n}$ provided there exists an embedding $\phi : S \hookrightarrow M$ such that*

$$\phi^*\omega = d\lambda.$$

As a corollary we obtain

Theorem 4.4 *Let (S, λ) be described as before and assume (S, λ) is embeddable into $N \times B^{2k}(R)$ or $\mathbf{C}P^n$, where (N, ω) is a compact symplectic manifold with $\omega \mid \pi_2(N) = 0$. Then $\mathcal{D}(S) \neq \phi$. (Here of course we talk about codimension 1-embeddings).*

The Weinstein conjecture is one of the key conjectures in the Existence theory for periodic solutions of Hamiltonian Systems. For more details we refer the reader to [8,9,10,12].

5 Capacities and Instanton Homology

In this section we describe some results presented at the symplectic year at MSRI, Berkeley. We restrict ourselves to the case of open sets in $\mathbf{R}^{2n}$. The construction can be done in much greater generality. For this we refer the reader to a forthcoming paper, [17], in which a theory, called symplectology, is developed (Symplectology: Symplectic Homology). For the following we require some familarity with Floer–homology for example as described in [18,19,20]. We consider as in 3.3 the family of Hamiltonians $\mathcal{H}$. For $H \in \mathcal{H}$ we have the map $\Phi_H : \hat{F} \to \mathbf{R}$ as introduced in 3.3. Φ_H induces a map, still denoted by Φ_H,

$$\Phi_H : (\hat{F} \times S^\infty)/S^1 \to \mathbf{R} : [z, \gamma] \to \Phi_H(z)$$

Roughly speaking we study the L^2–gradient flow on $(\hat{F} \times S^\infty)/S^1$. In practice we have to replace S^∞ by a sufficiently large compact approximation S^{2n-1} and to show that the following construction stabilizes. From now on we shall simply ignore any technical difficulty in order to present the idea. For a generic $H \in \mathcal{H}$ the map Φ_H has finitely many critical points. For numbers $c \leq d$ we consider the free abelian group $\wedge_c^d(H)$ defined by

$$\wedge_c^d(H) = \oplus_{x \in G} \mathbf{Z}x$$

where G is the set of all x with:

$$\begin{cases} \Phi'_H(x) = 0 \\ \Phi_H(x) \in (c, d]. \end{cases}$$

We can associate to critical points of Φ_H a relative Morse index $\mu \in \mathbf{Z}$, which can be understood in terms of a Maslov class for example as in [21]. So $\wedge_c^d(H)$ has a natural $\mathbf{Z}$–grading. We define a boundary operator $\partial : \wedge_c^d(H) \to \wedge_c^d(H)$ by

$$\partial y = \sum_{\mu(x) = \mu(y) - 1} \# < x, y > x$$

where $\# < x, y >$ denotes the number of trajectories of the *flow equation* $\gamma' = \Phi'_H(\gamma)$ running from x to y. Here these orbits carry natural orientations so that we actually count their *signed* number. We have $\partial^2 = 0$ and define a homology group $I_c^d(H)$. The choice of an inner product and a positive almost complex structure with respect to the symplectic form is involved in studying $\gamma' = \Phi'_H(\gamma)$. However it turns out that $I_c^d(H \circ \Psi) = I_c^d(H)$ for all $\Psi \in \mathcal{D}$, where $\mathcal{D}$ consists of all symplectomorphisms obtained as time–1–maps for compactly supported time–dependent Hamiltonians. We define an ordering on $\mathcal{H}$ by

$$H \geq K \iff \begin{cases} \text{There exists } \Psi \in \mathcal{D} \text{ such that} \\ H(z) \geq K \circ \Psi(z) \text{ for all } z \in \mathbf{R}^{2n}. \end{cases}$$

Taking a homotopy from $K \circ \Psi$ to H which is increasing, say L_s such that $L_s = K \circ \Psi$ for $s \leq -1$ and $L_s = H$ for $s \geq 1$ and $\frac{\partial}{\partial s} L_s(x) \geq O$ we can study the flow equation

$$\gamma' = \Phi'_{Hs}(\gamma),$$

where one looks for solutions running from critical points of $\Phi_{K\circ\Psi}$ to critical points of Φ_H. Studying this problem for critical points of the same Morse index one obtains an induced morphism

$$I_c^d(K) \to I_c^d(H).$$

One defines for a bounded subset S of $\mathbf{R}^{2n}$ using $\mathcal{H}(S)$ as defined in 3.3

$$I_c^d(S) = \lim_{\mapsto} I_c^d(H)$$

where the direct limit is taken over all H with $H \in \mathcal{H}(S)$. Recall that $\mathcal{H}(S)$ is partially ordered. We also note that we have for $c \leq d \leq e$ morphisms

$$I_c^d(H) \to I_c^e(H),$$

which induce

$$I_c^d(S) \to I_c^e(S).$$

We define a *terminal object* $\ominus$ by

$$\ominus := \lim I_0^{+\infty}(H) \text{with} H \in \mathcal{H}(*).$$

It turns out that $\ominus := \mathbf{Z}[t]$ is the polynominal ring in one variable of degree 2 (recall that we have a $\mathbf{Z}$–grading).

Assume $\Psi(S) \subset T$ for some $\Psi \in \mathcal{D}$, where S and T are bounded sets. Then we obtain a natural map

$$I_c^d(T) \to I_c^d(S)$$

We define for an unbounded set U the group $I_c^d(U)$ to be the inverse limit of the $I_c^d(S)$ for S running over the bounded subsets of U. The family I_c^d, for $c \leq d$, of functors is the symplectology. Using our terminal object $\ominus$ we have induced maps

$$I_0^d \to \ominus,$$

such that if $\Psi(S) \subset T$, $\Psi \in \mathcal{D}$ and $c \leq d$

$$\begin{array}{ccc} I_0^d(T) & \to & I_0^d(S) \\ & \searrow \quad \swarrow & \\ & \ominus & \end{array}$$

$$\begin{array}{ccc} I_0^c(T) & \to & I_0^d(T) \\ & \searrow \quad \swarrow & \\ & \ominus & \end{array}$$

Taking the second diagram we see that the image of the map $I_0^d(T) \to \ominus$ is increasing in d. Studying the images of the maps $I_0^d \longrightarrow \ominus$ we define a non–decreasing sequence

$\{d_k\}$ as follows. The sequence $\{d_k\}$ consists of all points where the rank of the image changes. Here we repeat a point of discontinuity according to its multiplicity, namely the net change of the rank. The first diagram shows that the numbers d_k are monotonic invariants, i.e. if $\Psi(S) \subset T$ then $d_k(S) \leq d_k(T)$. This Instanton–homology approach to symplectic capacities shows that symplectic capacities can be understood as numbers, where a certain classifying map into $\mathbf{Z}[t]$ changes its range.

References

[1] M. Gromov: *Pseudoholomorphic curves in symplectic manifolds*, Inv. Math. , 1985 , **82**, 307–347.

[2] I. Ekeland, H. Hofer: *Symplectic topology and Hamiltonian Dynamics*, Math. Zeit. , 1989, **200**, 355–378.

[3] I. Ekeland, H. Hofer: *Symplectic topology and Hamiltonian Dynamics II*, Math. Zeit. , 1990, to appear.

[4] M. Gromov: *Partial differential relations*, Springer, Ergebnisse der Mathematik, 1986.

[5] Y. Eliashberg: *A Theorem on the structure of Wave Fronts*, Funct. Anal. Appl., 1987, **21**, 65–72.

[6] M. Gromov: *Soft and Hard Symplectic Geometry*, Proc. of the ICM at Berkeley 1986, 1987, 81–98.

[7] H. Hofer: *On the topological properties of symplectic maps*, Proc. Royal Soc. of Edinburgh, special volume on the occasion of J. Hale's 60th birthday.

[8] C. Viterbo: *A proof of the Weinstein conjecture in* $\mathbf{R}^n$. , Ann. Inst. Henri Poincaré, Analyse non lineare, 1987, **4**, 337–357.

[9] A. Weinstein: *On the hypotheses of Rabinowitz's periodic orbit theorems*, J. Diff. Eq. , 1979 , **33**, 353–358.

[10] H. Hofer, E. Zehnder: *A new Capacity for symplectic Manifolds*, to appear in the proceedings of a conference on the occasion of J.Moser's 60th birthday.

[11] I. Ekeland, H. Hofer: *Convex Hamiltonian Energy Surfaces and their periodic Trajectories*, Comm. Math. Phy. , 1987 , **113**, 419–469.

[12] H. Hofer, E. Zehnder: *Periodic Solutions on Hypersurfaces and a result by C. Viterbo*, Inv. Math. , 1987 , **90**, 1–9.

[13] V. Benci : *On the critical Point Theory for indefinite Functionals in the Presence of Symmetries*, Transactions Am. Math. Soc. , 1982 , **274** , 533–572.

[14] E. Fadell, P. Rabinowitz: *Generalized cohomological index theories for Lie Group Actions with an Application to Bifurcation Questions for Hamiltonian systems*, Inv. Math. , 1978 , **45**, 139–173.

[15] A. Floer, H. Hofer, C. Viterbo: *The Weinstein Conjecture in* $\mathbf{P} \times \mathbf{C}^l$, Math. Zeit., 1990, to appear.

[16] H. Hofer, C. Viterbo : *The Weinstein Conjecture in the Presence of holomorphic Spheres*, in preparation.

[17] A. Floer, H. Hofer: in preparation.

[18] D. Salamon : *Morse theory, the Conley Index and the Floer Homology*, Bull. of the London Math. Soc., to appear.

[19] D. McDuff : *Elliptic Methods in symplectic geometry*, Lecture notes.

[20] A. Floer: *Morse theory for Lagrangian intersection theory*, J. Diff. Geom., 1988, **28**, 513–547.

[21] D. Salamon, E. Zehnder : *Floer Homology, the Maslov Index and periodic orbits of Hamiltonian Equations*, preprint , Warwick , 1989 .

[22] Y. Eliashberg, H. Hofer : *Towards the definition of a symplectic boundary*, in preparation.

The Nonlinear Maslov index

A.B. GIVENTAL

Lenin Institute for Physics and Chemistry, Moscow

I will present here a nonlinear generalisation of the Maslov-Arnold index concept [**1**],and use it to deduce the following theorem

THEOREM (GIVENTAL).
Let $\mathbf{RP}^{n-1} \subset \mathbf{CP}^{n-1}$ *be the fixed-point set of the standard anti-holomorphic involution of* $\mathbf{CP}^{n-1}$. *Then if* $f : \mathbf{CP}^{n-1} \to \mathbf{CP}^{n-1}$ *is a map which can be deformed to the identity through a Hamiltonian isotopy, the image* $f(\mathbf{RP}^{n-1})$ *intersects* $\mathbf{RP}^{n-1}$ *in at least* n *points.*

For a simple example, take $n = 2$; it is evident that the equatorial circle in the 2-sphere meets any area bisecting circle at least twice.
This theorem is a typical fact of symplectic topology ; similar to results proved by Conley-Zehnder and Floer. We shall see, I hope, that the nonlinear Maslov index provides a natural and convenient language to formulate "Arnold- type" conjectures on symplectic fixed points , or Langrangian intersections (see [**2**]).

The linear Maslov index.
By the linear Maslov index we mean the only homotopy invariant of loops in the Lagrange-Grassman manifold Λ^n, the space of n-dimensional Lagrangian linear subspaces of $\mathbf{R}^{2n}$. Before generalising this notion it is convenient to projectivise it.
Let $\mathbf{C}^n$ be complex $n-$space , endowed with its standard symplectic structure. The real projectivization, $\mathbf{RP}^{2n-1}$ has a standard contact structure. A point p in $\mathbf{RP}^{2n-1}$ is a real line in $\mathbf{C}^n$; its' skew-orthogonal complement is a hyperplane containing this line. In projective space we get a hyperplane through p , and the tangent space of this defines the element of the contact structure at p. With this contact structure, the Legendrian projective subspaces of $\mathbf{RP}^{n-1}$ are exactly the projectivisations of Lagrangian subspaces of $\mathbf{C}^n$ (if a Lagrangian subspace contains a line then it is contained in the skew-orthogonal complement of the line, i.e. it is tangent everywhere to the contact structure). Thus Λ^n is the manifold of projective Legendrian subspaces in $\mathbf{RP}^{2n-1}$. It is a homogeneous space of the the group $G_n = Sp(2n, \mathbf{R})/\pm 1$, and its' subgroup $H_n = U(n)/\pm 1$. A linear Maslov index $m(\gamma) \in \mathbf{Z}$ of a loop γ , in any of these three spaces, is just its' homotopy class under the canonical identification ([**1**])

$$\pi_1(H_n) \cong \pi_1(G_n) \cong \pi_1(\Lambda_n) \cong \mathbf{Z}.$$

Legendre-Grassmann manifolds.

We shall deal with the following infinite-dimensional manifolds:

(1) $\mathfrak{G}_n$- the identity component of the contactomorphism group of $\mathbf{RP}^{2n-1}$.

(2) $\mathfrak{L}_n$- the space of all embedded Legendrian *submanifolds* of the contact manifold $\mathbf{RP}^{2n-1}$ which can be obtained from the standard $\mathbf{RP}^{n-1} \subset \mathbf{RP}^{2n-1}$ by a Legendrian isotopy. We call $\mathfrak{L}_n$ the *Legendre-Grassmann manifold.* It is a homogeneous space of $\mathfrak{G}_n$.

(3) $\mathfrak{H}_n$ the identity component of the subgroup of $\mathfrak{G}_n$ consisting of transformations which preserve the standard ($U(n)$-invariant) contact 1-form α_0 on $\mathbf{RP}^{2n-1}$.

One may consider the contact form α_0 as a pre-quantisation connection on the squared Hopf bundle $\mathbf{RP}^{2n-1} \to \mathbf{CP}^{n-1}$, having structural group the circle $T = \{e^{it}\}/\pm 1$. Thus $\mathfrak{H}_n$ is a central T-extension of the identity component of the symplectomorphism group of $\mathbf{CP}^{n-1}$. We call $\mathfrak{H}_n$ a *quantomorphism group* because it realises, at the group level, the Poisson bracket extension of the Lie algebra of Hamiltonian vector fields:

$$0 \to \mathbf{R} \to C^\infty(\mathbf{CP}^{n-1}) \to sym(\mathbf{CP}^{n-1}) \to 0.$$

The finite dimensional manifold Λ_n is naturally contained in the infinite dimensional space $\mathfrak{G}_n$. Similarly, $\mathfrak{G}_n$ contains G_n and $\mathfrak{H}_n$ contains H_n. The weak statement about these spaces is that the linear Maslov index can be extended to the loops in these infinite dimensional spaces, that is we have a commutative diagram of homomorphisms:

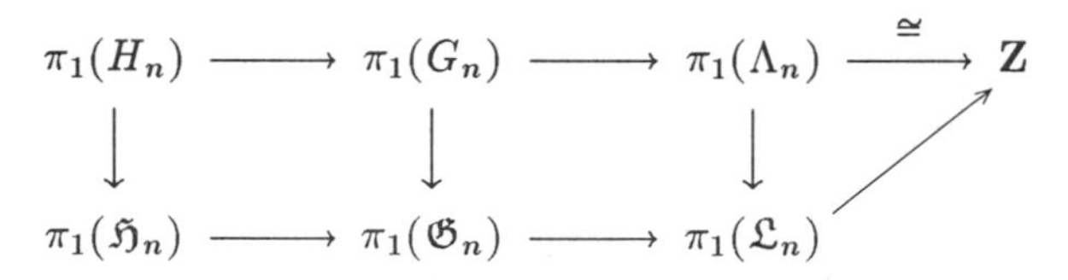

The stronger statement is that Arnold's geometrical definition ([**1**]) of the linear Maslov index can also be extended to these infinite dimensional spaces.

Discriminants.

To extend Arnold's geometrical definition we define a *discriminant*, a subspace $\Delta \subset \mathfrak{L}_n$. Let us mark a point in $\mathfrak{L}_n$, for example the projectivized imaginary subspace $i\mathbf{R}^n \subset \mathbf{C}^n$. The discriminant Δ consists of all Legendrian submanifolds which intersect this marked one. Two Legendrian subspaces are , in general, linked so Δ is a hypersurface in $\mathfrak{L}_n$ (with singularities). We also define a discriminant in $\mathfrak{G}_n$ to consist of contactomorphisms g which fix a point of $\mathbf{RP}^{2n-1}$, and fix the contact 1-form at that point ; i.e. $g(x) = x$ and $g^*(\alpha_0)|_x = \alpha_0|_x$, for some x in $\mathbf{RP}^{2n-1}$. Analogously we define the discriminant in $\mathfrak{H}_n$ to consist of quantomorphisms which have fixed points (in fact, circles of fixed points, fibres of the bundle $\mathbf{RP}^{2n-1} \to \mathbf{CP}^{n-1}$).

THEOREM 1.

Each of these three discriminants admits a co-orientation such that their intersection numbers with oriented loops define commuting homomorphisms

$$\begin{array}{ccccccc}
\pi_1(H_n) & \longrightarrow & \pi_1(G_n) & \longrightarrow & \pi_1(\Lambda_n) & \xrightarrow{\cong} & \mathbf{Z} \\
\downarrow & & \downarrow & & \downarrow & & \\
\pi_1(\mathfrak{H}_n) & \longrightarrow & \pi_1(\mathfrak{G}_n) & \longrightarrow & \pi_1(\mathfrak{L}_n) & & \\
{\scriptstyle m=\text{int. index}}\downarrow & & {\scriptstyle m}\downarrow & & {\scriptstyle m}\downarrow & & \\
\mathbf{Z} & & \mathbf{Z} & & \mathbf{Z} & &
\end{array}$$

extending the linear Maslov indices.

This theorem yields two corollaries, as we shall explain below.

COROLLARY 1.
Let P_1, P_2 be two points in $\mathfrak{L}_n$, i.e. embedded Legendrian submanifolds diffeomorphic to $\mathbf{RP}^{n-1}$. The projections of P_1 and P_2 to $\mathbf{CP}^{n-1}$ have at least n intersection points in $\mathbf{CP}^{n-1}$.

COROLLARY 2, (KLEINER-OH).
The standard $\mathbf{RP}^{n-1} \subset \mathbf{CP}^{n-1}$ has the least volume among all its images under Hamiltonian isotopies (and, more generally, among all projections of Legendrian submanifolds in $\mathfrak{L}_n$).

Here the volume is measured with respect to the standard $U(n)$-invariant Riemannian metric on $\mathbf{CP}^{n-1}$. A simple illustration of Corollary 2 is furnished by the theorem of Poincaré which asserts that the equator in the 2-sphere has least length among all area-bisecting curves. The proof of Corollary 2 is based on integral geometry. The volume of a Lagrangian submanifold is proportional to the average number of its intersections with the translates of the standard $\mathbf{RP}^{n-1}$ by $U(n)$. This number is not less than n, by Corollary 1, and equals n if the Lagrangian submanifold is standard.

The Morse Inequality.
Each of the manifolds $\mathfrak{L}_n, \mathfrak{G}_n, \mathfrak{H}_n$ is modelled on a space of smooth functions. For example, a neighbourhood of any Legendrian submanifold L in a contact manifold is contactomorphic to the 1-jet space J^1L of functions on L (that is, $J^1L \equiv \mathbf{R} \times T^*L$, with the contact structure $du = pdq$). A Legendrian submanifold C^1-close to L is represented by a Legendrian section of J^1L, i.e. by the graph of a smooth function f on L , with $u = f(q)$, $p = d_q f$. The section meets the zero section L at a critical point of f with zero critical value. Thus the discriminant Δ , near the marked point, looks like the hypersurface in $C^\infty(L)$ of functions with *singular zero level.* Consider, for example the space of polynomials in one variable in place of $C^\infty(L)$ in which the analogue of Δ is the subset of polynomials with multiple roots. For polynomials of degree 4 this subvariety has the form of the "swallow tail" singularity

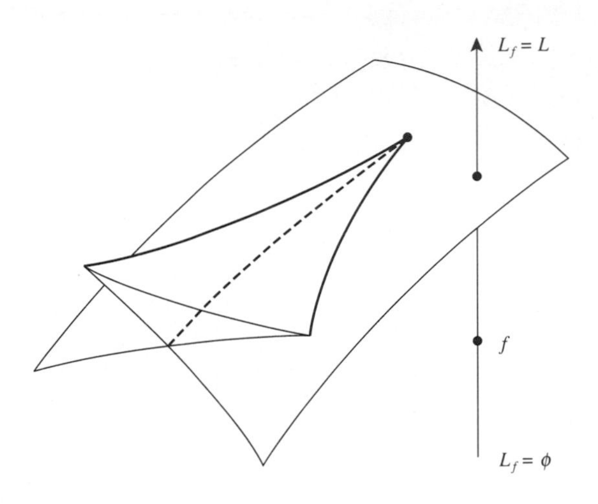

Diagram 1

in 3-space, depicted below, and this is in fact the general model for codimension-one singularities in the discriminants.
To co-orient the discriminant $\Delta \subset C^\infty(L)$ we introduce, for every C^∞ function f on L, a topological space

$$L_f = \{q \in L | f(q) \geq 0\},$$

and an integer $b_f = b_*(L_f, \mathbf{Z}/2)$ – the sum of the Betti numbers of L_f, with $\mathbf{Z}/2$ coefficients. As f moves in the complement of Δ the boundary $f^{-1}(0)$ remains smooth , and b_f is unchanged. When f crosses Δ at a nonsingular point a Morse bifurcation ocurs: a new cell is glued to L_f and b_f changes by either $+1$ or -1. We co-orient Δ in the direction in which b_f increases.
Now we give an interpretation of the total Morse inequality in our intersection-index terms. Let us consider the flow $f \mapsto f + t$ in $C^\infty(L)$. The *number*, $\sharp$, of intersections with Δ of the orbit of a Morse function f on L is, by the definition of Δ , equal to the number of critical levels of f. On the other hand the intersection *index* of the orbit with Δ is equal to

$$b_{f+\infty} - b_{f-\infty} = b_*(L : \mathbf{Z}/2).$$

Thus the total Morse inequality,$\sharp \geq b_*(L; \mathbf{Z}/2)$, follows from the fact that an intersection index is never more than an intersection number.
An analogous argument is used to deduce Corollary 1 from Theorem 1, at least in the case when the intersections are transverse. Intersections of Δ with a T-orbit through $L \in \mathfrak{L}_n$ corespond to the intersections of the image of L in $\mathbf{CP}^{n-1}$ with the standard $\mathbf{RP}^{n-1}$, just by the definition of Δ. But the intersection index, being homotopy invariant, is equal to the linear Maslov index of T- orbits in Λ_n , that is, to n.

The Calabi-Weinstein Invariant.
The Calabi-Weinstein invariant([9]) is the homomorphism

$$w : \pi_1(\mathfrak{H}_n) \to \mathbf{R}$$

which is defined by the Lie algebra homomorphism

$$W : C^\infty(\mathbf{CP}^{n-1}) \to \mathbf{R} \quad : \quad h \mapsto \int_{\mathbf{CP}^{n-1}} h d\mu,$$

– integration with respect to normalised Liouville measure. In fact such a homomorphism w is defined for any quantomorphism group.

PROPOSITION 1. *The Calabi-Weinstein invariant is proportional to the Maslov index: more precisely* $m = (n/\pi)w$.

COROLLARY. *The homomorphism* w *takes values in* $(\pi/n)\mathbf{Z} \subset \mathbf{R}$.

If $\mathfrak{H}_n$ were compact one could prove Proposition 1 by integral geometry arguments. There exists a (C^0, locally exact, adjoint invariant) 1-form M on $\mathfrak{H}_n$ which represents the cohomology class m. The value of M on a tangent vector $h \in T_q\mathfrak{H}_n$ is

$$M_q(h) = \sum_{\hat{q}(x)=x} \pm h(x),$$

where $\hat{q}$ denotes the underlying symplectomorphism, and signs are defined by the co-orientation of Δ at its' intersection points with the T-coset through q. If we could average this form over all translations of M we would obtain a left and right invariant form, which should clearly be proportional to W.

Generating functions.
To prove the existence of the nonlinear Maslov index we mark in $\mathbf{RP}^{2n-1}$ another Lagrangian subspace $L_0 = \text{Proj}(\mathbf{R}^n) \subset \mathbf{C}^n$. Given an ambient Hamiltonian isotopy h^t of $\mathbf{RP}^{2n-1}$ ($0 \leq t \leq 1$) , we factor h^1 into a large number N of small isotopies, and construct a function

$$f : \mathbf{RP}^{2D-1} \to \mathbf{R}\ ,\ D = nN.$$

This function is a kind of finite approximation to the action function of the isotopy, lifted to $\mathbf{C}^n$ homogeneously. It is chosen to have the property that

$$f^{-1}(0) \text{ is non-singular } \Leftrightarrow h^1(L_0) \in \Delta.$$

Then we define the *relative Maslov index* of the path γ in $\mathfrak{L}_n$ formed by $L_t = h^t(L_0)$ to be $m(\gamma) = b_*(f^{-1}(\mathbf{R}^+); \mathbf{Z}/2) - D$, and co-orient Δ in the increasing direction of the relative index when the end of the path crosses Δ. A crucial point in the proof of the validity of this definition is the *additivity* of the relative index. If γ_0 and γ_1 are, respectively, a loop and a path in $\mathfrak{L}_n$ which are subdivided into N_0 and N_1 parts , then $m(\gamma_0\gamma_1) = m(\gamma_0) + m(\gamma_1)$. Moreover, if f is the function associated to the composite $\gamma_0\gamma)_1$, and f_1 is associated to γ_1, then the space $f^{-1}(\mathbf{R}^+)$ is cohomologically equivalent to the Thom space of the sum of $D_0 = nN_0$ copies of the Mobius line bundle over $f^{-1}(\mathbf{R}^+)$. In particular, the two spaces $f_0^{-1}(\mathbf{R}^{\pm})$ are cohomologically equivalent to $\mathbf{RP}^{d_{\pm}-1}$, where the "inertia indices" $d_{\pm} = b_*(f_0^{-1}(\mathbf{R}^{\pm}); \mathbf{Z}/2)$ are complementary $(d_+ + d_- = 2D_0\ )$, and differ from the middle dimension by twice the Maslov index of the loop:

$$m(\gamma_0) = d_+ - D_0 = D_0 - d_-.$$

A similar method can be used to co-orient the discriminant in $\mathfrak{G}_n$. We decribe the generating function for this case explicitly. Let $h^1 = h_N \circ \cdots \circ h_1$ be a decomposition of the isotopy h^1into N small parts – lifted into $\mathbf{C}^n$ homogeneously. Then we take

$$f = \text{Proj } (Q - H) : (\mathbf{C}^n)^N \to \mathbf{R}\ ,$$

where Q is the quadratic generating function of the cyclic permutation in $(\mathbf{C}^(n))^N$ which maps $(x_1, \ldots, x_N)$ to $(x_2, \ldots, x_N, x_1)$, and $H = \bigoplus_{i=1}^N H_i$ is the generating function of the "vector" transformation which maps a point $(x_1, \ldots, x_N)$ to $(h_1x_1, \ldots, h_Nx_N)$. The singularities of $f^{-1}(0)$ correspond to fixed lines of h^1.

For the group $\mathfrak{H}_n$, the lifted isotopy preserves the unit ball in $\mathbf{C}^n$ and the generating function, being e^{it}-invariant, is naturally defined on a complex projective space:

$$f : \mathbf{CP}^{2D-1} \to \mathbf{R}\ ,\ D = n[N/2].$$

A topological comparison of all of these generating functions proves the consistency statements of Theorem 1. Besides, the "signature" (d_+, d_-) of the generating function of a loop (in $\mathfrak{G}_n$ for example) is actually a signature of the generating function at the critical points. The change in the signature of the second variation of the action functional along a path of Hamiltonian isotopies is known as the Conley-Zehnder index ([**3**]), so this last observation yields the corollary:

PROPOSITION 2.
The Maslov index of a loop in $\mathfrak{G}_n$ is equal to the Conley-Zehnder index of the corresponding trajectory of Hamiltonian isotopies of $\mathbf{C}^n$.

According to the "Sturm theory " the variation in the signature is equal to the linear Maslov index of the Hamiltonian flow , linearised along its trajectory. This gives

PROPOSITION 3.
The nonlinear Maslov index of a loop (in $\mathfrak{G}_n$) is equal to the linear Maslov index of loops (in G_n) formed by linearising flows along the the trajectories in $\mathbf{C}^n$.

We use this result, applied to the subgroup $\mathfrak{H}_n$, to prove Proposition 1. Averaging over all trajectories, one finds that the nonlinear Maslov index is equal to the average value over the ball of the Laplacian *divgrad*H_t of the Hamiltonian. For homogeneous H_t this is equal to an average of H_t over the unit sphere, which proves Proposition 1.

Tails.
To complete the proof of our Lagrange intersection theorem for the non- transverse case one needs to modify the definition of the discriminant. Let us define the cohomological *length*, $l(X)$, of a subset X of a real projective space to be the least degree in which the restriction of the generator of the cohomology of the projective space vanishes on X. The length is *monotone* with respect to inclusion. Let us denote by Γ , and call a *tail* the subset of the discriminant Δ where the length of the positive subset $f^{-1}(\mathbf{R}^+)$ changes – rather than the Betti sum we considered before. The tail is a hypersurface with no topological singularities in any finite codimension (see the diagram). Its' co-orientation by the length-increasing direction coincides with the positive T-co-orientation (monotonicity) and makes Γ into a 1-cocycle cohomologous to Δ (additivity). Thus the inequality between the intersection *number* and *index*, applied to Γ and T-orbits, gives estimates for *geometrically different* Lagrangian intersections, or fixed points of symplectomorphisms.

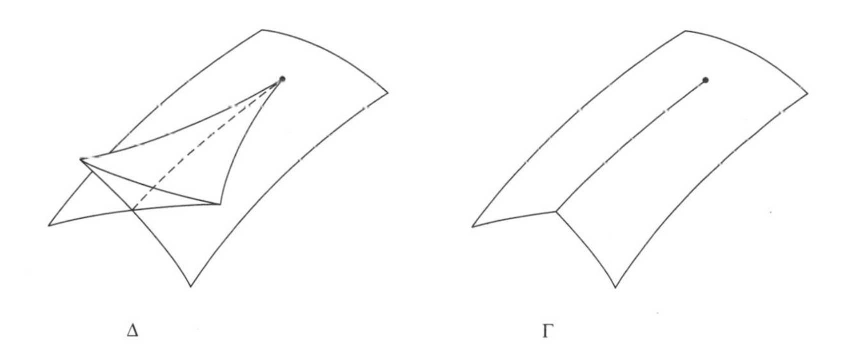

Diagram 2

COROLLARY 3 (FORTUNE-WEINSTEIN [5]).
Any Hamiltonian transformation of $\mathbf{CP}^{n-1}$ has at least n different fixed points.

The asymptotical Maslov index.

Let α be a contact 1-form for the standard contact structure on $\mathbf{RP}^{2n-1}$; so $\alpha = h\alpha_0$ for some positive function h. The *characteristic vector field* v_α is defined by

$$v \in \ker d\alpha \ , \ \alpha(v) = 1,$$

and this vector field generates the *characteristic flow* of contactomorphisms. Let us fix on such a flow g_t or, more generally, a nonautonomous flow of a *time periodic* characteristic field. Let $L \in \mathfrak{L}_n$ and let $m_{L_\tau}(t)$ (respectively $m(t)$) be the intersection index of the path $(g^\tau L)$ (respectively (g^τ) , $0 \le \tau \le t$, with the tail $\Gamma \subset \mathfrak{L}_n$ (respectively $\mathfrak{G}_n$).

THEOREM 2.

The following limits exist and are equal :

$$m = \lim m(t)/t \ = \ \lim m_L(t)/t.$$

The asymptotical Maslov index m is monotone and homogenerous of degree −1 on the space of contaxct 1-forms; it is continuous in C^0-norm and does not change under conjugations in $\mathfrak{G}_n$.

For example , a positive quadratic Hamiltonian in $\mathbf{C}^n$ generates a projective characteristic flow on $\mathbf{RP}^{2n-1}$. Its nonlinear asymptotical Maslov index is equal to the linear one; the sum of frequencies $m = \sum \omega_i$. For example α_0 generates the Hopf flow with $m = n/\pi$.

COROLLARY 1.

$$n/(\pi \text{max. } h) \le m \le n/(\pi \text{min. } h).$$

COROLLARY 2 (VITERBO). *Any closed 1-form on $\mathbf{RP}^{2n-1}$ admits a closed characteristic.*

COROLLARY 3 (GIVENTAL-GINZBURG).

The number $\sharp$ of chords (i.e. characteristics starting and finishing on the same Legendrian submanifold L) of length less than or equal to t grows at least linearly with t :

$$\sharp(L,t) \ge mt - const.(L) \ge n[t/\pi \max h] - const.(L).$$

Remarks.

(1) For the quantomorphism group $\mathfrak{H}_n$ one can also define the asymptotical Calabi-Weinstein index. If a characteristic flow of a contact 1-form $h\alpha_0$ (where h is a positive function on $\mathbf{CP}^{n-1}$) is *precompact* in $\mathfrak{H}_n$ then its Calabi-Weinstein index coincides with its asymptotical Maslov index, in accordance with Proposition 1 :

$$m = (n/\pi) \int h^{-1}.$$

(2) In general the asymptotical Maslov index is equal, just by the definition, to the density of the length spectrum of closed characteristics and seems to coincide with the density of the *capacity spectrum* of the domain $\{r^2 h \le 1\} \subset \mathbf{C}^n$ which has been

defined by Ekeland-Hofer and Floer-Hofer ([7]). When applied to a precompact flow in $\mathfrak{H}_n$ this turns (i) into the *Duistermaat- Heckmann integration formula* :

$$n \int h^{-1} = \sum_{dh=0} h^{-1}(x).$$

(see [9]).

(3) Our proof of Theorem 2 is mainly based on the monotonicity of cohomological length and the *quasiadditivity* of generating functions : a homogeneous generating function for the flow $\{g^\tau\}$, $\tau \leq t_1 + t_2$, coincides with the direct sum of such functions for $\tau \leq t_1$ and $\tau \leq t_2$ *after* their restriction to a subspace of low codimension (that is, low compared with large values of t_1 and t_2).

Let us note finally that our intersection-theoretic interpretation of Arnold's fixed point and Lagrange intersection problems can be extended to some other symplectic and contact manifolds using direct methods of Floer homology ([4]), instead of singular homology and Morse theory.

References

1. Arnold,V.I., *On a characteristic class entering the quantisation conditions*, Funct. Anal. Appl. 1 (1967), 1-6.
2. Arnold,V.I., *First steps in symplectic topology*, Russian Math. Surveys 41:6 (1986), 1-21.
3. Conley,C. and Zehnder,E., *Morse type index theory for flows and periodic solutions of Hamiltonian equations*, Comm.Pure Appl. Math. (1984), 207-253.
4. Floer,A., *Morse Theory for Lagrangian intersections*, J. Differential Geometry 28 (1988), 513-547.
5. Fortune,B. and Weinstein,A., *A symplectic fixed point theorem for complex projective space*, Bull.Amer. Math. Soc.12 (1985).
6. Givental,A., *Periodic maps in symplectic topology*, Funct. Anal. Appl. 23:4 (Russsian) (1989), 37-52.
7. Hofer,H., In this volume (1989).
8. Viterbo,C., *Intersections de sous variétés Lagrangiennes, functionelles d'action et indice des systemes Hamiltoniens*, Bull.Math. Soc France 115 (1987), 361-90.
9. Weinstein,A., *Cohomology of symplectomorphism groups and critical values of Hamiltonians*, Preprint, Berke (1988).

Filling by Holomorphic Discs and its Applications

YAKOV ELIASHBERG

Stanford University

The survey is devoted to application of the technique of filling by holomorphic discs to different symplectic and complex analytic problems.

1. COMPLEX AND SYMPLECTIC RECOLLECTIONS

1.1 J-Convexity

Let X, J be an almost complex manifold of the real dimension 4 and Σ be an oriented hypersurface in X of the real codimension 1. Each tangent plane $T_x(\Sigma)$, $x \in \Sigma$, contains a unique complex line $\xi_x \subset T_x(\Sigma)$ which we will call a *complex tangency* to Σ at x. The complex tangency is canonically oriented and, therefore, cooriented. Hence the tangent plane distribution ξ on Σ can be defined by an equation $\alpha = 0$ where the 1-form α is unique up to multiplication by a positive function. The 2-form $d\alpha\mid_\xi$ is defined up to the multiplication by the same positive factor. We say that Σ is *J-convex* (or pseudo-convex) if $d\alpha(T, JT) > 0$ for any non-zero vector $T \in \xi_x, x \in \Sigma$. We use the word "pseudo-convex" when the almost complex structure J is not specified.

An important property of a J-convex hypersurface Σ is that it cannot be touched inside (according to the canonical coorientation of Σ) by a J-holomorphic curve. In particular, if Ω is a domain in X bounded by a smooth J-convex boundary $\partial\Omega$ then all interior points of a J-holomorphic curve $C \subset X$ with $\partial C \subset \partial\Omega$ belong to IntΩ. Moreover, C is transversal to $\partial\Omega$ in all regular points of its boundary ∂C.

A hypersurface Σ is called *Levi-flat* if $d\alpha\mid_\xi = 0$. According to Frobenius' theorem this means that ξ integrates to a foliation of Σ by J-holomorphic curves.

A function φ on X is called *J-convex* or pseudoconvex if it is strictly subharmonic on any J-holomorphic curve in X. All level-sets $\{\varphi = c\}$ of a J-convex function φ are J-convex if we oriente them as boudaries of domains $\{\varphi \leq c\}$. Conversely, any function with this property can be made J-convex by a reparametrization: $\tilde{\varphi} = h \circ \varphi$ for a diffeomorphism $h : \mathbb{R} \to \mathbb{R}$.

1.2 Contact Structure on a J-Convex Hypersurface

The distribution ξ on a J-convex $\Sigma \subset X$ is completely nonintegrable (indeed if α is as in 1.1 then $\alpha \wedge d\alpha > 0$) and, by the definition, defines a *contact structure* on Σ. There are no local invariants of contact stuctures where the word "local" may be refered to the manifold as well as to the space of all contact structures on a given manifold. In particular, a deformation of J which leaves the hypersurface J-convex does not change the underlying contact structure ξ.

A curve $\gamma \subset \Sigma$ is called *Legendrian* if it is tangent to ξ. An oriented transversal $\gamma \subset \Sigma$ to ξ is called *positive* or *negative* according to the sign of the 1-form $\alpha \mid_{T(\gamma)}$ (recall that we have chosen the 1-form α with $\{\alpha = 0\} = \xi$ to define the canonical coorientation of the distribution ξ). Any Legendrian curve can be C^∞-approximated by a positive as well as negative transversal to ξ.

1.3 Invariants of Legendrian Curves

Let γ be a Legendrian curve in a J-convex hypersurface $\Sigma \subset X$. Suppose that γ is homological to 0 in Σ and fix a relative homology class $\beta \in H_2(\Sigma, \gamma)$. Pushing γ along a vector field transversal to $\xi \mid_\gamma$ in Σ one can compute the intersection number of the perturbed curve γ' with β. This number $tb(\gamma \mid \beta)$ we will call the *Thurston-Bennequin's invariant.* Note that in many cases (for instance, when Σ is a homology sphere) $tb(\gamma \mid \beta)$ does not depend on β. We will write $tb(\gamma)$ in these cases or when the choice of β is clear from the context.

Now let $M \subset \Sigma$ be any surface which is bounded by γ and represents the class $\beta \in H_2(\Sigma, \gamma)$. Take any trivialization of the bundle $\xi \mid_M$. Suppose that γ is oriented and let τ be a tangent to the γ vector field which defines the orientation. Then the degree of τ with respect to the chosen trivialization of $\xi \mid_M$ depends only on γ and β. We will denote it by $r(\gamma \mid \beta)$ and will call the *rotation* of γ with respect to β. Note that in contrast to $tb(\gamma \mid \beta)$ the rotation number changes the sign with the change of the orientation of γ. As above we will write $r(\gamma)$ when the choice of β is clear or irrelevant.

1.4 Complex Points of a Real Surface in an Almost Complex 4-Manifold

Let M be a real 2-surface in an almost complex 4-manifold X, J. A point $p \in M$ is called *complex* if the tangent plane $T_p(M)$ is a J-complex line in $T_p(X)$. A generic surface $M \subset X$ has isolated complex points. If the surface M is oriented then we call a complex point *positive* if the orientation of $T_p(M)$ coincides with its complex orientation and *negative* in the other case. A surface without complex points is called *totally real.*

If the surface M is contained in a J-convex hypersurface $\Sigma \subset X$ then complex points of M are points where M is tangent to the distribution ξ of complex tangencies to Σ. Intersections of ξ with tangent planes to M form an (orientable) line field M_ξ on the totally real part of M. In the generic case the index of this field at complex points is equal to ± 1. We will call a complex point $p \in M$ *elliptic* if the index of M_ξ at M is equal to $+1$ and *hyperbolic* in the other case. The field M_ξ integrates to a 1-dimensional foliation (which we will still denote by M_ξ) on M with singularities at complex points of M. The foliation M_ξ is called the *characteristic foliation.* Leaves of the characteristic foliation are, by the definition, tangent to ξ and, therefore, Legendrian. The foliation M_ξ has a focus type singularity near an elliptic point and the standard hyperbolic singularity near a hyperbolic one.

REMARK. The notion of ellipticity and hyperbolicity of a complex point can be extended to a generic complex point of an arbitrary surface in X (and not necessarily one which is contained in a pseudoconvex hypersurface). To do that, consider a fibration $Gr_2(X) \to X$ over X whose fiber at a point $x \in X$ is the Grassmanian of oriented 2-planes in $T_x(X)$. Let $\mathbb{C}P_+(X)$ and $\mathbb{C}P_-(X)$ be subbundles of $Gr_2(X) \to X$ which consist of complex lines with, respectively, complex or anticomplex orientation. Note that $\mathbb{C}P_+(X)$ and $\mathbb{C}P_-(X)$ have codimension 2 in $Gr_2(X)$. For a surface $M \subset X$ the inclusion map can be lifted to a Gaussian map $g : M \to Gr_2(X)$. Points from $g^{-1}(\mathbb{C}P_\pm(X))$ are exactly positive and negative complex points of M. Properly fixing a coorientation of $\mathbb{C}P_+(X) \cup \mathbb{C}P_-(X)$ in $Gr_2(X)$ (to agree with the special case which we considered above) we say that a complex point $p \in M$ is elliptic (hyperbolic) if g is transversal to $\mathbb{C}P_{\pm}(X)$ at p and the intersection index of $\mathbb{C}P_{\pm}(X)$ with $g(M)$ at the point $g(p)$ is equal to $+1(-1)$.

For an oriented surface $M \subset X$ (possibly with boundary) we denote by $e_{\pm}(M)$ and $h_{\pm}(M)$ numbers of positive or negative elliptic or hyperbolic *interior* points of M. Let $d_{\pm}(M) = e_{\pm}(M) - h_{\pm}(M)$.

If M is closed we denote by $c(M)$ the value of the first Chern class $c_1(X)$ of X on M and by $\nu(M)$ the normal Euler number of M in X. If $\partial M \neq \emptyset$ let τ be a vector field along ∂M which is tangent to ∂M and defines on ∂M the orientation induced by the orientation of M. Let n be a vector field tangent to M and outward transversal to ∂M. Suppose that M has no complex points at ∂M. Then vector fields τ and n are linearly independent over $\mathbb{C}$ and $J\tau$ is transversal to M in X. We denote by $c(M)$ the obstruction to the extension of τ and n on M as linearly independent over $\mathbb{C}$ vector fields and by $\nu(M)$ the obstruction to the extension of $J\tau$ on M as a transversal to M vector field.

The proof of the following formula which relates invariants $d_{\pm}(M)$, $c(M)$, $\nu(M)$ and the Euler characteristic $\chi(M)$ of M is straightforward (see [HE] for the discussion).

PROPOSITION 1.4.1. *Let $M \subset X$ be either closed or does not have complex points on ∂M. Then*

$$d_{\pm}(M) = \frac{1}{2}(\chi(M) + \nu(M) \pm c(M)) .$$

If M is contained in a J-convex hypersurface $\Sigma \subset X$ then the absence of complex points at ∂M can be guaranteed by the condition that ∂M is transversal to the distribution ξ. In this case $\nu(M) = 0$ and the formula 1.4.1 takes a simpler form

$$d_{\pm}(M) = \frac{1}{2}(\chi(M) \pm c(M)) .$$

We will need also an analog of 1.4.1 for a surface $M \subset \partial X$ bounded by a Legendrian curve ∂M. If, in addition, M has to be contained in a pseudo-convex hypersurface Σ it is impossible, in general, to avoid complex points at ∂M. Instead we want to standardize them in the following sense. There exists an isotopy of M in Σ which is fixed at ∂M and such that all complex points at the boundary of the resulting surface M' are elliptic and their signs alternates along $\partial M' = \partial M$. We say in this case that M' is in *normal form* near its Legendrian boundary $\partial M'$. It is easy to see that if M is in normal form near ∂M then the number of complex points on ∂M is equal to $2|tb(\partial M)|$, where the Thurston-Bennequin's invariant $tb(\partial M)$ is calculated with respect to the homology class of M in $H_2(\Sigma, \partial M)$.

PROPOSITION 1.4.2. *Suppose that a surface $M \subset \Sigma \subset X$ is in normal form near its Legendrian boundary ∂M. Then*

$$d_{\pm}(M) = \frac{1}{2}(\chi(M) + tb(M) \pm r(M))$$

where $tb(M)$ and $r(M)$ are calculated with respect to the homology class of M in $H_2(\Sigma, \partial M)$.

To prove the equality one should approximate M by surfaces M_+ and M_- bounded by positive and negative transversals, observe that $c(M_\pm) = tb(\partial M) \pm r(M)$ (comp. [Be]) and apply 1.4.1.

1.5 Tame Almost Complex Manifolds

Following Gromov (see [Gr]) we say that an almost complex manifold is *tame* if there exists a symplectic structure ω on X such that the form $\omega(T, JT)$, $T \in T(X)$ is positive definite.

Note that any Stein (i.e. affine) complex manifold, Kahlerian manifold or (genuine) complex manifold with pseudoconvex boundary is tame. Any symplectic manifold admits an almost complex structure tamed by the symplectic structure.

An almost complex manifold X, J is called *holomorphically aspherical* if there are no holomorphic embeddings $\mathbb{C}P^1 \to (X, J)$.

2. FILLING BY HOLOMORPHIC DISCS

In 1969 E. Bishop (see [Bi]) discovered that a punctured neighborhood of an elliptic point admits a (unique!) foliation by circles spanning holomorphic discs. By simple topological reasons these discs have to be disjoined and fill a piece of a Levi-flat

3-manifold. Many attempts were made to globalize Bishop's result: *How far from the elliptic point can the family of circles spanning holomorphic discs be developed?*

The first serious success in this direction was obtained by E. Bedford and B. Gaveau in 1982 (see [BG]). They considered the following problem: Given a 2-sphere S embedded in $\mathbb{C}^2$ with exactly two complex points (which are in this case necessarily elliptic), does there exist a Levi-flat 3-ball in $\mathbb{C}^2$ bounded by the sphere S? Or equivalently, does the complement of complex points in S admit a foliation by circles spanning holomorphic discs?

Bedford and Gaveau gave a positive answer to this question under two additional assumptions:

—S is contained in the pseudoconvex boundary of a domain $\Omega \subset \mathbb{C}^2$;

—orthogonal projection of S onto $\mathbb{R}^3 \subset \mathbb{C}^2$ is an embedding.

While the first assumption looks very important and cannot be completely removed (see Section 9 below) the second is absolutely unnecessary and even the original Bedford-Gaveau's proof can be easily modified to work without this assumption (see [El2]). The first complete proof of the result belongs to M. Gromov (see [Gr]). The key point in Bedford-Gaveau's proof is Lipshitz estimates for boundaries of holomorphic discs while Gromov uses the more general compactness theorem which he proves in the same paper.

In our joint paper with V. M. Harlamov we related the technique of filling by holomorphic discs with topological problems and, in particular, with Bennequin's theorem [Be] which was just proved. We showed that if one could prove a filling result for spheres which do not belong to a pseudo-convex boundary, that would imply very strong corollaries in low-dimensional topology (see Problem 10.7 below). Unfortunately (or fortunately?) the result in that generality is false (see Section 9 below) but the problem is far from being understood.

Since that time there were two major breakthroughs which sufficiently extended possibilities of the method. In 1985 Gromov wrote his famous "Pseudoholomorphic curves in symplectic manifolds" where he showed that holomorphic curves and, in particular, holomorphic discs can be successfully used not only in complex but also in *tame* almost complex manifolds. Recently E. Bedford and W. Klingenberg (see [BK]) found a teccnique for developing the family of holomorphic discs in the presence of *hyperbolic* complex points. That allowed them to prove the result about the filling of a 2-sphere in $\mathbb{C}^2$ without any additional assumptions (besides pseudo-convexity).

In the next section I formulate without proofs a summary of results concerning regularity and compactness properties for families of holomorphic discs. The main results (Theorems 3.4.1 and 3.4.2) can be proved within the ideology of [BG], [BK] and [Gr] but are not straightforward corollaries of results of these papers. The generalization of [BK] for an almost complex case requires a fine analysis of singularities of J-holomorphic curves which was recently done by D. McDuff (see [McD2]).

3. COMPACTNESS AND REGULARITY PROPERTIES FOR FAMILIES OF HOLOMORPHIC DISCS

3.1 φ-admissible sets

By a Morse function we mean a function with nondegenerate critical points and pairwise different critical values. By an almost Morse function we mean a function which has either non-degenerate or birth-death type critical points and at most double critical values.

Let K be a 3-manifold with the boundary $\partial K = M$ and $\varphi : K \to \mathbb{R}$ be a function whose restriction on M is almost Morse. For a closed subset $A \subset K$ we denote by $\partial' A$ the intersection $A \cap \partial K$ and by $\partial'' A$ the rest of its boundary: $\frac{\partial'' A =}{\partial A \setminus \partial' A}$.

A closed subset $A \subset K$ is called φ-admissible if the following conditions A1–A5 are satisfied:

A1. The function φ does not have critical points in A.

A2. Int$A$ is a union of some components of sets $\{x \in \text{Int}\, K,\ c < \varphi(x) < c'\}$ where c, c' are critical values of $\varphi\,|_M$. in particular, φ is constant on components of $\partial'' A$.

A3. For any $c \in \mathbb{R}$ all components of the level set $A_c = \{x \in \text{Int}\, A,\ \varphi(x) = c\}$ are simply connected.

A4. For any critical value c of $\varphi\,|_M$ the intesection of closures in K of sets $\{\varphi(x) < c,\ x \in A\}$ and $\{\varphi(x) > c,\ x \in A\}$ is an open-closed subset of $\{\varphi = c,\ x \in A\}$.

A5. Any component C of $\partial'' A$ contains a critical point of the function $\varphi|_M$ and it is either not simply connected or it does not cover the whole component of the level set $\{x \in K,\ \varphi(x) = \varphi|_C\}$.

If $A \subset K$ is φ-admissible then one can form a graph $G_{A,\varphi}$ identifying each component of level sets $\{x \in A,\ \varphi(x) = c\}$, $c \in \mathbb{R}$, to a point. Vertices of G correspond to local maxima and minima of $\varphi|_M$ and to components of $\partial'' A$.

Let π be a projection $A \to G_{A,\varphi}$. For a subgraph $\Gamma \subset G_{A,\varphi}$ we will denote by A_Γ the inverse image $\pi^{-1}(\Gamma)$ and by M_Γ the intersection $A_\Gamma \cap M = \partial' A_\Gamma$. Note that if the set A_Γ is φ-admissible then $G_{A_\Gamma,\varphi} = \Gamma$. We will say in this case that Γ is an *admissible* subgraph of G.

The function φ factorizes to a function $\tilde{\varphi} : G_{A,\varphi} \to \mathbb{R}$ with $\varphi = \pi \circ \tilde{\varphi}$. The function $\tilde{\varphi}$ is non-degenerate on each edge of $G_{A,\varphi}$ and, therefore, defines an orientation of the graph $G_{A,\varphi}$.

3.2 Partial and maximal fillings

Let M now be an oriented closed 2-surface in an almost complex manifold X, J.

By a *partial filling of M by holomorphic discs* we mean a quintuple $\zeta = (K, \varphi, A, \mathcal{H}, F)$ which consists of:
—a 3-manifold K bounded by M,
—a function $\varphi : K \to \mathbb{R}$ with an almost Morse restriction $\varphi|_M$,
—a φ-admissible set A,
—a family $\mathcal{H}$ of complex structures on levels $A_c = \{x \in \operatorname{Int} A,\ \varphi(x) = c\}$ which depends smoothly on $c \in \mathbb{R}$,
—a smooth map $F : A \to X$ which is holomorphic on levels A_c, $c \in \mathbb{R}$, and coincides with the inclusion $\partial' A \hookrightarrow X$ on $\partial' A = A \cap M$.

Note that for any component Δ of A_c, $c \in \mathbb{R}$, the image $F(\Delta)$ is a holomorphic disc in X and its boundary is contained in M. For any partial filling ζ all critical points of $\varphi|_M$ from A are, necessarily, complex points of M and elliptic points correspond to local maxima and minima of $\varphi|_M$. We will require, *in addition*, that local maxima and minima of $\varphi|_M$ correspond to negative and positive elliptic points, respectively. With this choice of the orientation of M the orientation of any curve $\{\varphi(x) = c,\ x \in \partial' A = M \cap A\}$ as the boundary of the set $\{\varphi(x) \leq c,\ x \in \partial' A\}$, $c \in \mathbb{R}$, coincides with its orientation as the boundary of a corresponding holomorphic disc $F(\Delta)$.

A partial filling is called *maximal* if *all* elliptic points of M belong to A and are, therefore, local maxima and minima of $\varphi\,|_M$. A partial filling is called *generic* if $\varphi\,|_M$ is a Morse function.

We say that M *can be filled by holomorphic discs* if it admits a maximal filling with $A = K$. Note that in this case K must be a handlebody bounded by the surface M. In particular, if M is diffeomorphic to S^2 and can be filled by holomorphic discs then it bounds a 3-ball in X.

3.4 Existence of the maximal filling

THEOREM 3.4.1. *Let X, J be a tame almost complex manifold, $\Omega \subset X$ be a domain with a smooth J-convex boundary $\partial\Omega$ and $M \subset \partial\Omega$ be a closed 2-surface. Then M can be C^2-approximated by a surface $\widetilde{M}$ which admits a unique generic maximal filling by holomorphic discs.*

THEOREM 3.4.2. *Let X, J and Ω be as above. Suppose that, in addition, Ω is holomorphically aspherical. Let M be a closed 2-surface and Φ be an embedding $M \times I \to \partial\Omega$. Then the embedding Φ can be C^2-perturbed in such a way that each surface $\Phi(M \times t)$, $t \in I$, admits a maximal filling by holomorphic discs. If each surface $\Phi(M \times t)$, $t \in I$, can be filled by holomorphic discs then there exist a handlebody K and an embedding $\widetilde{\Phi} : K \times I \to \Omega$ such that $\widetilde{\Phi}$ coincides with Φ on $M \times I$ and for any $t \in I$ the image $\Phi(M \times t) \subset \Omega$ is Levi-flat.*

3.5 Graph of a filling

With any partial filling $\zeta = (K, \varphi, A, \mathcal{H}, F)$ we can associate the graph $G_{A,\varphi}$ (see 3.1). If ζ is a maximal filling of M then we will write G_M instead of $G_{A,\varphi}$. Vertices of G_M correspond to all elliptic and some hyperbolic points of M. Elliptic points are end-points of G_M. The orientation of G_M has been chosen in such a way that edges start at negative elliptic points and end at positive ones. If M is filled by holomorphic discs then all end-points of G_M correspond to elliptic points of M. If this is not so then any connected component of G_M must contain an end-point which is a hyperbolic point of M. Note that edges can start at positive hyperbolic points and end at negative ones.

3.6 Cutting A out of M

Let $\zeta = (K, \varphi, A, \mathcal{H}, F)$ be a partial filling of M and $G = G_{A,\varphi}$ be the graph of the filling. The boundary of A is a piecewise smooth surface which consists of two parts: $\partial A = \partial' A \cup \partial'' A$. It can be smoothed in such a way that the new surface M_A is filled by holomorphic discs. The graph G_{M_A} is isomorphic to G. End-points of G which correspond to hyperbolic points of M are replaced in G_{M_A} by end-points which correspond to ellipic points of M_A.

4. SURFACES WHICH CAN BE FILLED BY HOLOMORPHIC DISCS

Through the rest of the paper X, J denotes a tame almost complex manifold, Ω denotes a domain in X with a smooth J-convex boundary and M denotes a connected 2-surface (sometimes closed, sometimes with a boundary) which is contained in $\partial\Omega$.

For purposes of our applications nothing changes with a small C^2-perturbation. So we will speak about the filling of a surface M meaning that it can be done after, probably, some small C^2-perturbation.

THEOREM 4.1. *Any closed surface $M \subset \partial\Omega$ different from S^2 satisfies the inequality*

$$\chi(M) \leq -|c(M)| \ .$$

If M is diffeomorphic to S^2 then it can be filled by holomorphic discs. In particular, it bounds an embedded ball $B^3 \subset \Omega$.

Proof. If M is not S^2 but can be filled by holomorphic discs then $c(M) = 0$ and the inequality is automatically satisfied. Hence it is enough to consider the case when M cannot be filled by holomorphic discs. Let $\zeta = (K, \varphi, A, \mathcal{H}, F)$ be a maximal filling of M. Then $\partial' A$ contains no closed components. Cutting A out of M we get a surface $M' = M_A$ which can be filled by holomorphic discs. The graph $G_{M'}$ of this filling coincides with G_M. Let us denote by k_+ and k_- numbers of positive and negative hyperbolic points of M which have been substituted by elliptic points of M_A. Then

$$e_\pm(M') = e_\pm(M) + k_\mp$$

$$h_\pm(M) \geq h_\pm(M') + k_\pm$$

Note that if $b_0 = b_0(M')$ is the number of components of M' then $k_+ + k_- \geq b_0$.

Now we have $c(M') = 0$ because M' is filled by holomorphic discs. Therefore, equalities 1.4.1 imply that

$$d_+(M') = d_-(M') = \frac{1}{2}\chi(M') \leq b_0 .$$

Hence,

$$e_\pm(M') - h_\pm(M') = e_\pm(M) + k_\mp - h_\pm(M') \leq b_0 .$$

Then

$$\begin{aligned} d_\pm(M) &= e_\pm(M) - h_\pm(M) \\ &\leq b_0 - k_\mp + h_\pm(M') - h_\pm(M) \\ &\leq b_0 - k_\mp - k_\pm = b - (k_- + k_+) \leq 0 \end{aligned} .$$

But

$$d_\pm(M) = \frac{1}{2}(\chi(M) \pm c(M))$$

which implies the required inequality

$$\chi(M) \leq -|c(M)|$$

Q.E.D.

Note that Theorem 4.1 provides an obstruction for a 4-manifold to carry a structure of a tame almost complex manifold with the J-convex boundary. For example,

COROLLARY 4.2 *$S^2 \times D^2$ has no tame almost complex structure with the J-convex boundary.*

Indeed, a sphere $S^2 \times p, p \in \partial D^2$ does not bound any ball in $S^2 \times D^2$ which contradicts 4.1.

A contact structure on a 3-manifold V is called *fillable* if it can arise as the distribution ξ of complex tangencies to the J-convex boundary $V = \partial\Omega$ of a domain Ω in a tame almost complex manifold X, J. Note that the first Chern class $c_1(X)$ restricted to V coincides with the Euler class $e(\xi)$ of the bundle ξ. Let us denote by $Fill(V)$ the subset of $H^2(V)$ which consists of classes $e(\xi)$ for all *fillable* contact structures ξ on V.

COROLLARY 4.3 *The set $Fill(V)$ is finite.*

Proof. Any homology class $\mu \in H^2(V)$ can be realized by a smooth surface $M \subset V$. According to 4.1 we have $|e(\xi)[\mu]| = |c(M)| \leq -\chi(M)$. Thus $e(\xi)$ can take only finite many values on any homology class from $H^2(V)$.

Q.E.D.

Now we consider the relative case. Let $M \subset \partial\Omega$ be a 2-surface with the boundary ∂M transversal to the distribution ξ of complex tangencies to $\partial\Omega$. Let us choose the orientation of M which induces the positive orientation of ∂M (see 1.2).

THEOREM 4.4. *If M is as above then $d_- \leq 0$ or, equivalently, $\chi(M) \leq c(M)$.*

This is a generalized Bennequin's theorem. In [Be] he proved the inequality for the case when Ω is a round ball in $X = \mathbb{C}^2$.

Proof. Let us double M: consider a closed surface $\widehat{M}$ which arises as the boundary of a tubular neighborhood of M in $\partial\Omega$. We can consider the oriented surface M to be contained in M as a half of $\widehat{M}$. Let $\widehat{G} = G_{\widehat{M}}$ be the graph of the maximal filling of $\widehat{M}$. Let V_0 be the set of those vertices of $\widehat{G}$ which correspond to negative elliptic points of $\widehat{M}$ which belong to M. Let G_0 be the least admissible (see 3.1) subgraph of $\widehat{G}$ which contains V_0. Let $A_0 = A_{G_0}$ and $M_0 = \widehat{M}_{G_0}$ be parts of A and $\widehat{M}$ which correspond to G_0 (see 3.1). Note that M_0 is contained in $\operatorname{Int} M$. Indeed, if it is not so then there exists a holomorphic disc Δ whose boundary $\partial\Delta \subset M$ is tangent to ∂M at a point $p \in \partial M$. By our choice of orientation of M the orientation of $\partial\Delta$ (as the boundary of the holomorphic disc Δ) at p coincides with the orientation of ∂M as the boundary of M. But the complex orientation of $\partial\Delta$ is opposite to the orientation of $\partial\Delta$ induced by the orientation of the set $\{\varphi \leq c\}$ (see 3.2). This contradiction proves that $M_0 \subset \operatorname{Int} M$. Cutting A_0 out of $\widehat{M}$ we get a surface $\widehat{M_0} = \widehat{M}_{A_0}$ (see 3.6). Then, arguing as in 4.1, we get

$$0 = c(\widehat{M_0}) = d_+(\widehat{M_0}) - d_-(\widehat{M_0}), \quad d_+(M_0) + d_-(\widehat{M_0}) = \chi(\widehat{M_0})$$
$$e_-(\widehat{M_0}) = e_-(M)$$
$$h_-(\widehat{M_0}) \leq h_-(M) - b_0(\widehat{M_0})$$

$$e_-(M) = e_-(\widehat{M_0}) = \frac{1}{2}\chi(\widehat{M_0}) + h_-(\widehat{M_0})$$
$$\leq \frac{1}{2}\chi(\widehat{M_0}) + h_-(M) - b_0(M_0)$$
$$= h_-(M) - \frac{b_1(\widehat{M_0})}{2} \leq h_-(M)$$

which means that $d_-(M) = e_-(M) - h_-(M) \leq 0$. Q.E.D.

COROLLARY 4.5. *Let $M \subset \partial\Omega$ be a surface with a Legendrian boundary $\partial M = L$ and μ be the homology class of M in $H_2(\partial\Omega; L)$. Then*

$$\chi(M) \leq tb(L|\mu) - |r(L|\mu)|.$$

To prove 4.5 we can consequently approximate M by a surface whose boundary is a negative or positive transversal to ξ and apply 4.4 and 1.4.2 (see[Be] for similar arguments).

5. MANIFOLDS BOUNDED BY A J-CONVEX SPHERE

The theorem of this section indicates that the topology of the J-convex boundary imposes very strong restrictions on the topology of the domain.

THEOREM 5.1. *let X and Ω be as above and suppose, in addition, that Ω is holomorphically aspherical. If $\partial\Omega$ is diffeomorphic to S^3 then Ω is diffeomorphic to the ball B^4. Moreover, Ω admits a J-convex function which is constant on $\partial\Omega$ and has exactly one critical point—the minimum—in Ω.*

Proof. Take a Morse function $h: \partial\Omega \to \mathbb{R}$ with exactly two critical points. All non-critical levels of h are diffeomorphic to S^2. By a small C^2-perturbation of ∂W near critical points of h we can arrange that level-sets of h which are close to critical points are filled by holomorphic discs. This defines a foliation of a neighborhood of critical points in Ω by Levi-flat 3-balls. By an additional C^2-perturbation of $\partial\Omega$ we can arrange (via 3.4.2) that all other levels also admit maximal filling by holomorphic discs. But all these levels are diffeomorphic to S^2 and, therefore, according to Theorem 4.1 they are actually filled by holomorphic discs. Hence each 2-sphere $\{h = c\} \subset \partial\Omega$ bounds a Levi-flat 3-ball B_c. The family of these balls depends smoothly on $c \in \mathbb{R}$ (see 3.4.2) and are pairwise disjoint. Indeed, if $B_c \cap B_{c'} \neq \emptyset$ for $c \neq c'$ then there exist holomorphic discs Δ_c and $\Delta_{c'}$ with $\partial\Delta_c \in \partial B_c$, $\partial\Delta_{c'} \in \partial B_{c'}$ and $\Delta_c \cap \Delta_{c'} \neq \emptyset$. The intersection index of two holomorphic discs Δ_c and $\Delta_{c'}$ is positive (see [McD2] for the non-integrable case) but it has to be zero because their boundaries $\partial\Delta_c \subset \partial B_c$ and $\partial\Delta_{c'} \subset \partial B_{c'}$ bound discs in disjoint spheres ∂B_c and $\partial B_{c'}$. Hence the domain Ω is foliated by Levi-flat balls B_c and, therefore, diffeomorphic to B^4. To prove the second part of the theorem, denote by H the function on Ω which extends h and which has balls H_c as its level-sets. There exists a diffeomorphism $\varphi: \Omega \to \Omega$ which is fixed at the boundary $\partial\Omega$ and such that all level-sets of the function $H_1 = H \circ \varphi$ are J-convex. As it was explained in 1.1 there exists a function $\gamma: \mathbb{R} \to \mathbb{R}$ such that the composition $\gamma \circ H_1$ is a J-convex function. Let $M = \max_\Omega H_2$, $m = \min_\Omega H_2$. Let H_3 be a J-convex function which is defined in a tubular neighborhood $U \supset \partial\Omega$ and which is constant on $\partial\Omega$ and ∂U. We can arrange that $H_3 |_{\partial\Omega} > M$ and $H_3 |_{\partial U \cap \Omega} < m$. Now extend somehow H_3 on $\Omega \setminus U$ with $\max H_3 |_{\Omega \setminus U} < m$ and let $H_4 = \max(H_3, H_2)$. Smoothing H_4 we get the required J-convex function $H_5 : \Omega \to \mathbb{R}$.

Q.E.D.

D. McDuff's study of blowing down in tame almost complex manifolds allows us to understand what is going on when Ω is not holomorphically aspherical.

THEOREM 5.2. *Let X, J be a tame symplectic manifold and $\Omega \subset X$ be a domain bounded by a J-convex 3-sphere. Then for an almost complex structure J' which is C^∞-close to J the manifold (Ω, J') is a 4-ball up to blowing up a few points. In particular, Ω is diffeomorphic to $B^4 \# k\overline{\mathbb{C}P^2}$.*

Proof. According to [McD1] there exists an almost complex structure J' which is C^∞-close to J and such that all embedded J-holomorphic spheres in Ω are disjoint, non-singular and have self-intersection number (-1). These spheres can be blown down to give a non-singular holomorphically aspherical almost complex manifold Ω' with the same boundary $\partial\Omega$. Thus we can apply 5.1 to conclude that Ω' is diffeomorphic to B^4 and, therefore, Ω is diffeomorphic to $B^4 \# k\overline{CP^2}$.

COROLLARY 5.3. *Let X, J and Ω be as in 5.2. Then the contact structure ξ on $\partial\Omega$ (formed by complex tangencies to $\partial\Omega$) is isomorphic to the standard contact structure on S^3 (i.e. structure induced on a unit sphere in $\mathbb{C}^2$.)*

Proof. Perturbation of J does not change the contact structure. So using 5.2 and 5.1 we can get a ball Ω' bounded by $\partial\Omega$ and a J-convex function H on Ω which is constant on $\partial\Omega = \partial\Omega'$ and has exactly one (non-degenerate) critical point: the minimum. All non-singular level sets of H are J-convex and diffeomorphic to S^3. Hence all of them have contact structure and, therefore, all these contact structures are isomorphic (see 1.2). But near the critical point the almost complex structure can be made integrable and level sets can be made biholomorphically equivalent to the round sphere in $\mathbb{C}^2$ and, therefore, the contact structure on all level-sets is the standard one.

6. KILLING OF ELLIPTIC COMPLEX POINTS

If a surface $M \subset X$ has hyperbolic and elliptic points $p, q \in M$ of the same sign then one can easily C^0-perturb M along a path connecting p and q to kill these 2 points (see [HE]). If M is contained in a J-convex boundary $\partial\Omega$ of a domain $\Omega \subset X$ then it would be important to be able to kill p and q by an isotopy of M *inside* $\partial\Omega$. This is impossible to do in a neighborhood of a path connecting p and q (it would contradict the symplectic rigidity) but as it follows from Theorem 6.1 below this is possible when enough room is provided.

We start with the following definition. Let G be the graph of a maximal filling of M. An *elementary contraction* C_l of G is the removal of an edge l with ends which correspond to elliptic and hyperbolic points of the same sign. The following theorem which will be proved in [El3] shows that any elementary contraction can be realized by an isotopy of M inside $\partial\Omega$.

THEOREM 6.1. *Let M, $M \subset \partial\Omega$, be a closed surface, G be a graph of a generic partial filling and l be an edge of G with ends which correspond to elliptic and hyperbolic points of the same sign. Let M_l be a part of M which is covered by boundaries of holomorphic discs which correspond to points of l. Then there exists a diffeotopy $h_t : \partial\Omega \to \partial\Omega$, $t \in [0,1]$, such that $h_0 = id$, h_t is fixed outside a* *neighborhood of M_t for all $t \in [0,1]$ and the surface $h_1(M)$ admits a maximal filling whose graph is the result of the contraction C_l of G. In particular $h_1(M)$ has two complex points less than M.*

Now we apply 6.1 to prove

THEOREM 6.2. *Let M be a closed surface in $\partial\Omega$. Suppose that $\chi(M) = -|c(M)|$ (comp. 5.1). Then M either can be filled by holomorphic discs (and it has to be diffeomorphic to the torus T^2 in this case) or is isotopic in $\partial\Omega$ to a surface M' without elliptic points. In particular if M is a torus then either it bounds a Levi-flat solid torus in Ω or it is isotopic in $\partial\Omega$ to a totally real torus.*

Proof. Suppose that M cannot be filled by holomorphic discs and let G_M be the graph of a maximal filling $\zeta = (K, \varphi, A, \mathcal{H}, F)$. In view of 6.1 it is enough to prove that the graph G can be exhausted by a sequence of elementary contractions. In other words, that we can always find an edge whose ends correspond to elliptic and hyperbolic points of the same sign. Let M_A be a result of cutting A out of M (see 3.6). As in the proof of 4.1, let us denote by k_+ and k_- numbers of positive and negative hyperbolic points of M which have been substituted by elliptic points of M_A. Then we have

$$e_\pm(M_A) = e_\pm(M) + k_\mp,\ h_\pm(M) \geq h_\pm(M_A) + k_\pm,\ k_+ + k_- \geq b_0(M_A) \qquad (*)$$

Let us choose the orientation of M such that $c(M) \geq 0$. Then $\chi(M) = -c(M)$ and, therefore, $d_+(M) = \frac{1}{2}(\chi(M) + c(M)) = 0$, $d_-(M) = \frac{1}{2}(\chi(M) - c(M)) = \chi(M)$ or $e_+(M) = h_+(M)$, $e_-(M) = h_-(M) + \chi(M)$. Suppose that G_M admits *no* elementary contractions. This means that there is no edge whose ends are elliptic and hyperbolic points of the same sign which implies inequalities

$$h_\pm(M_A) \geq e_\mp(M)\ ,$$

Then

$$h_+(M_A) \geq \chi(M) + h_-(M)\ ,$$
$$h_-(M_A) \geq h_+(M) \qquad \text{and}$$
$$h_+(M_A) + h_-(M_A) \geq \chi(M) + h_+(M) + h_-(M)$$

or

$$k_+ + k_- \leq (h_+(M) - h_+(M_A)) + (h_-(M) - h_-(M_A)) \leq -\chi(M) . \qquad (**)$$

On the other hand, the equality $c(M_A) = 0$ implies that

$$e_+(M_A) - h_+(M_A) = e_-(M_A) - h_-(M_A) = \frac{1}{2}\chi(M_A) .$$

Now we get

$$h_+(M) + k_- - h_+(M_A) = \frac{1}{2}\chi(M_A) ,$$

$$h_-(M) + \chi(M) + k_+ - h_-(M_A) = \frac{1}{2}\chi(M_A) .$$

The first equality and $(*)$ give

$$b_0(M_A) \leq k_+ + k_- \leq \frac{1}{2}\chi(M_A) \leq b_0(M_A)$$

and therefore,

$$b_0(M_A) = k_+ + k_- = \frac{1}{2}\chi(M_A) .$$

But then the second equality and $(**)$ imply that

$$k_+ + k_- = b_0(M_A) - \chi(M) \leq -\chi(M)$$

and, therefore, $b_0(M_A) = 0$. But then M_A is empty which is impossible if M has elliptic points.

Q.E.D.

THEOREM 6.3. *Let M be a 2-surface in $\partial\Omega$ with the Legendrian boundary $L = \partial M$ and μ be the homology class of M in $H_2(\partial\Omega, L)$. Suppose that M is in normal form near ∂M and either $M = D^2$ or*

$$\chi(M) = -tb(L|\mu) - |r(L|\mu)|.$$

Then there exists an isotopy of M in $\partial\Omega$ which is fixed near L and moves M to a surface $\widehat{M}$ without interior elliptic points.

Proof. Let M' be any closed surface in $\partial\Omega$ which contains M and let $\zeta = (K, \varphi, A, \mathcal{H}, F)$ be its maximal filling by holomorphic discs. Let Γ be the minimal admissible (see 3.1) subgraph of the graph of ζ which contains all interior elliptic points of M.

Let $M_\Gamma = \partial' A_\Gamma = M' \cap A_\Gamma$ be the corresponding part of M'. I claim that M_Γ is contained in Int$M$. Indeed, if it is not then there must exist a holomorphic disc Δ whose boundary $\partial\Delta$ is tangent to $L = \partial M$. But that is impossible either in the totally real part of L or in a complex elliptic point which belong to L. The rest of the proof repeats the proof of 6.1 where, of course we need to use the formula from 1.4.2 instead of 1.4.1.

Q.E.D.

7. LEGENDRIAN KNOTS

Let S^3 be a boundary of a unit ball in $\mathbb{C}^2$ and ξ_0 be the standard contact structure on S^3 (formed by complex tangencies). By a standard Legendrian curve in S^3 we will mean an intersection L_0 of S^3 with a Lagrangian plane P in $\mathbb{C}^2$ or *any* Legendrian curve which is Legendrian isotopic to this intersection. Note that if L_0 is a standard Legendrian curve in S^3 then $r(L_0) = 0$ and $tb(L_0) = -1$. Any contact 3-manifold M, ξ is locally isomorphic to $S^3 \setminus p$ (Darboux chart) with the standard contact structure. We will call a Legendrian curve $L \subset M$ *standard* if it is Legendrian isotopic to a Legendrian curve which is standard in a Darboux chart.

The curve L_0 bounds in S^3 a hemisphere $\mathcal{D}_0$ which is the intersection of S^3 with a hyperplane which contains the Lagrangian plane P. The hemisphere $\mathcal{D}_0$ has no interior complex points and has exactly two elliptic points on the boundary L_0. That implies that leaves of the characteristic foliation $(\mathcal{D}_0)_{\xi_0}$ are Legendrian arcs which connect the two elliptic points on the boundary. Note that using these arcs one can easily construct a Legendrian isotopy in an arbitrary small neighborhood of $\mathcal{D}_0$ into a small standard Legendrian curve.

THEOREM 7.1. *let X, J and Ω be as in 6.3 and $\mathcal{D} \subset \partial\Omega$ be an embedded disc bounded by a Legendrian curve $\partial\mathcal{D} = L$. Suppose that $r(\partial\mathcal{D}) = 0$ and $tb(\partial\mathcal{D}) = -1$. Then the Legendrian curve L is standard.*

Proof. The disc $\mathcal{D}$ can be deformed (without changing the boundary) to a disc $\mathcal{D}'$ which is in normal form near L. The equality $1 = \chi(\mathcal{D}') = -tb(L) - r(L)$ and Theorem 6.3 guarantee a possibility of finding a disc $\mathcal{D}''$ which coincides with $\mathcal{D}'$ near the boundary and which does not have interior elliptic points. But according to 1.4.2 we have $d_\pm = 0$. Therefore, $\mathcal{D}''_\xi$ has no interior complex points at all and has two elliptic complex points p and q at the boundary. Then the characteristic foliation $\xi_{\mathcal{D}''}$ on $\mathcal{D}''$ is a foliation by Legendrian arcs connecting p and q. But this implies that the pair $\mathcal{D}'', \partial\mathcal{D}''$ is contactomorphic to $\mathcal{D}_0, L_0$ and, therefore, L can be contracted to a small standard Legendrian curve through a Legendrian isotopy in an arbitrary small neighborhood of $\mathcal{D}''$.

Q.E.D.

8. Prime decomposition of manifolds with a pseudo-convex boundary

In this section we consider tame almost complex manifolds Ω, J with a J-convex boundary $\partial\Omega$ to be defined up to a deformation of $\partial\Omega$ which keeps it J-convex.

Given a Levi-flat embedding $h : (B, \partial B) \to (\Omega, \partial\Omega)$ of a 3- ball B one can extend it to an embedding $H : B \times [-1,1] \to \Omega$ with $H|_{B\times 0} = h$, $H(\partial B \times [-1,1]) \subset \partial\Omega$ and such that balls $H(B \times 1)$ and $H(B \times (-1)$ are J-concave (with the orientation wich is induced on the boundary of the domain $H(B \times [-1,1])$. The boundary of the complement domain $\Omega \setminus H(B\times]-1,1[)$ can be canonically smoothed to become J-convex. We say that the new manifold Ω' is the result of *cutting Ω along a Levi-flat ball $h(B)$*.

Given two tame almost complex manifolds Ω_1, J_1 and Ω_2, J_2 with pseudo-convex boundaries one can canonically define the boundary connected sum $\Omega_1\#\Omega_2, J_1\#J_2$ as a tame almost complex manifold with the pseudo-convex boundary $\partial\Omega_1\#\partial\Omega_2$. The new manifold can be characterized by the property that cutting it along a Levi-flat ball we can get the disjoint union of original manifolds Ω_1, J_1 and Ω_2, J_2.

Let us call a manifold Ω, J *prime* if it cannot be presented as the boundary connected sum $\Omega_1\#\Omega_2, J_1\#J_2$ of two tame almost complex manifolds with pseudo-convex boundaries where Ω_1 and Ω_2 are not diffeomorphic to the 3-ball B. We say that Ω, J is *almost prime* if Ω_1 and Ω_2 are not diffeomorphic to $B\#k\overline{\mathbb{C}P^2}$.

THEOREM 8.1. *A manifold Ω, J with a J-convex boundary is almost prime if and only if its boundary $\partial\Omega$ is a prime 3-manifold.*

Proof. If $\partial\Omega$ is prime then for any decomposition $(\Omega, J) = (\Omega_1, J_1)\#(\Omega_2, J_2)$ one of manifolds Ω_1 or Ω_2 is bounded by the 3-sphere and, therefore, it is diffeomorphic to $B\#k\overline{\mathbb{C}P^2}$ according to Theorem 5.1. Conversely, if $\partial\Omega$ is not prime then there exists a 2-sphere $S \subset \partial\Omega$ which divides $\partial\Omega$ into two parts different from the 3-ball. According to Theorem 4.1 the sphere S bounds a Levi-flat ball B in Ω. Cutting Ω along B we decompose Ω, J into the connected sum $(\Omega_1, J_1)\#(\Omega_2, J_2)$. Boundaries $\partial\Omega_1$ and $\partial\Omega_2$ are not spheres and, therefore, Ω_1 and Ω_2 themselves are not diffeomorphic to $B\#k\overline{\mathbb{C}P^2}$. Hence, Ω, J is not almost prime.

Q.E.D.

9. EXAMPLE OF A FAILURE OF THE FILLING PROPERTY

We give here an example of a failure of the filling property for a surface which is not contained in a J-convex boundary. Namely, we give an example of a 2-sphere in $\mathbb{C}^2$ with exactly 2 complex points (which are necessarily elliptic and of opposite signs) which cannot be filled by holomorphic discs. The example was worked out jointly with V. Harlamov.

Let us take a hyperplane $\mathbb{R}^3 \subset \mathbb{C}^2$. Then $\mathbb{R}^3$ is foliated by complex lines. Let $h : \mathbb{R}^3 \to \mathbb{R}$ be a height-function whose level-sets $P_t = h^{-1}(t), t \in \mathbb{R}$ are these complex lines. Let i be the inclusion $S^2 \hookrightarrow \mathbb{R}^3$ as the boundary of the unit ball $B^3 \subset \mathbb{R}^3$. We normalize h in such a way that $h \circ i(S^2) = [-1, 1]$. Take an immersion $\varphi : S^2 \to \mathbb{R}^3$ such that the function $h \circ \varphi$ coincides with $h \circ i$. Then the immersed sphere $\varphi(S^2)$ has exactly two (elliptic) complex points at critical points p and q of the function $h \circ i$. Denote by S_t, $t \in [-1, 1]$, level sets of the function $h \circ i$. Suppose that φ coincides with i near p and q and that the restriction $\varphi \mid_{S_0}$ considered as an immersion of S_0 into the complex line $P_0 = h^{-1}(0)$ cannot be extended to an immersion of a 2-disc into P_0. Let now $\varphi_n : S^2 \to \mathbb{C}^2$, $n = 1, \ldots$, be a sequence of embeddings which C^∞-converges to φ. I claim that, at least for large n, the sphere $\varphi_n(S^2) \subset \mathbb{C}^2$ cannot be filled by holomorphic discs. More precisely, there is no smooth map $F_n : B^3 \to \mathbb{C}^2$ such that:

$F_n \mid_{S^2}$ coincides with φ_n near poles p and q;

F_n restricted to open discs $\mathcal{D}_t = P_t \cap \operatorname{Int} B^3$, $t \in]-1, 1[$, are holomorphic;

$F_n(S^2) = \varphi_n(S^2)$.

Suppose that F_n with these properties exists for all $n = 1, \ldots$. First of all the analysis of the local behavior of holomorphic discs near a given disc (see, for example, [Gr]) shows that $F_n |_{S^2}: S^2 \to \varphi_n(S^2)$ has to be an immersion and, therefore, an embedding. Then the boundary of a holomorphic disc $F_n(\mathcal{D}_t)$, $t \in]-1, 1[$, divides the sphere $\varphi_n(S^2)$ into two hemispheres $S^+_{n,t}$ and $S^-_{n,t}$. Let ω be the standard symplectic form in $\mathbb{C}^2$. Then, according to Stokes' theorem,

$$\text{Area } F_n(\mathcal{D}_t) = \int_{F_n(\mathcal{D}_t)} \omega = |\int_{S^+_{n,t}} \omega| \leq \text{Area } S^+_{n,t} \leq \text{Area } \varphi_n(S^2) .$$

This means that areas of all holomorphic discs are uniformly bounded and according to Gromov's compactness theorem (see [Gr]) for each $t \in]-1, 1[$ the sequence $F_n |_{\mathcal{D}_t}$, $n = 1, \ldots$, has a converging subsequence. (It is easy to check that cusp-degenerations cannot occur.) But then each point of the immersed sphere $\varphi(S^2) \subset \mathbb{R}^3 \subset \mathbb{C}^2$ has to belong to a holomorphic disc. But it is not so for points from $\varphi(\mathcal{D}_0) \subset \varphi(S^2)$. Q.E.D.

10. OPEN QUESTIONS AND CONJECTURES

10.1 *Which prime 3-manifolds are pseudo-convex boundaries of compact domains in tame almost complex manifolds?*

We will call such 3-manifolds *fillable.* No one example of a nonfillable manifold is known. Note that without the condition "tame" all 3-manifolds are, obviously, fillable.

10.1.A *If one drops the word "almost" in 10.1 does it change the class of fillable 3-manifolds?*

Note that results of [El1] raise the hope that the clas of fillable (in both senses) 3-manifolds is pretty large.

10.2 *How many different (up to blowing up) 4-manifolds can have the same 3-manifold as a pseudo-convex boundary?*

The sphere S^3 has the unique filling (see 5.1) but it is possible that the same M^3 can be filled by a few different 4-manifolds (for example, the lense space $L(4, 1)$ can be filled at least by two different manifolds, see [McD1]).

10.3 *How many different fillable structures can arise on the same 3-manifolds?*

Conjecture: *A fillable contact structure on a 3-manifold is uniquely determined by its Euler class* (see Section 4). Together with 4.3 it would imply the existence of only finite different fillable contact structures on a given 3-manifold. Theorem 5.3 proves the conjecture for $M = S^3$.

10.4 Let Ω be a compact domain with the J-convex boundary $\partial\Omega$ in a tame almost complex manifold X, J. *Does Ω admit a J-convex function $h : \Omega \to \mathbb{R}$ which is constant on $\partial\Omega$?* The answer is known to be negative if one does not allow certain degeneration of h.

Conjecture: *There exists a subset $\Sigma \subset \operatorname{Int}\Omega$ which is a union of J-holomorphic curves and Levi-flat hypersurfaces such that Ω admits a J-convex function $\varphi : \Omega \to \mathbb{R}$ which is constant on Σ and J-convex in $\Omega \setminus \Sigma$.* If J is integrable than this is true (even without Levi-flat components of Σ) according to a Grauert-Rossi's theorem.

10.5 Let L be a Legendrian knot in the standard contact S^3. Let $top(L)$ be L itself considered as a topological knot, $r(L)$ and $tb(L)$ be rotation number and Thurston-Bennequin's invariant of L (see 1.3). *Does the triple $(top(L), r(L), tb(L))$ define L up to a Legendrian isotopy?*

I think that the answer is positive when top(L) is trivial (compare 7.1) and negative in the general case.

10.6 **J-convex h-cobordism problem.** Let Ω be a domain in a tame almost complex (X, J) which is diffeomorphic to $M \times I$ for a 3-manifold M. Let $h : \Omega \to I$ be a J-convex function such that $h\,|_{M\times 0} = 0$, $h\,|_{M\times 1} = 1$. Does Ω admit a *nondegenerate* J-convex function with this property?

10.7 *Prove the conjecture from* [HE] *which gives 4-dimensional generalization of Bennequin's inequality* (see 4.2). Let Ω be a domain with J-convex boundary in a tame (almost) complex (X, J) and M be a surface in Ω whose boundary is contained in $\partial\Omega$ and is transversal to the contact structure ξ on $\partial\Omega$. Let us orient M in such a way that it induces the positive (see 1.2) orientation on the transversal ∂M. *Then $d_-(M) \leq 0$ or, equivalently, $\chi(M) \leq c(M)$.*

As it is shown in section 9, the procedure of filling by holomorphic discs which proves the inequality for the case $M \subset \partial\Omega$ (see 4.2) fails in the general case but one can hope that the result is, nevertheless, true.

REFERENCES

[Be] D. Bennequin, Entrelacements et equations de Pfaff, *Astérique* **107–108** (1983), 83–161.

[Bi] E. Bishop, Differentiable manifolds in complex Euclidean space, *Duke Math. J.*, **32** (1965), 1–22.

[BG] E. Bedford and B. Gaveau, Envelopes of holomorphy of certain 2-spheres in $\mathbb{C}^2$, *Amer. Journal of Math.*, **105** (1983), 975–1009.

[BK] E. Bedford and W. Klingehberg, On the envelope of holomorphy of a 2-sphere in $\mathbb{C}^2$, preprint, 1989.

[El1] Ya. Eliashberg, Topological characterization of Stein manifolds of dimension > 2, preprint, 1989, to appear in *International Journal of Math.*

[El2] Ya. Eliashberg, Three lectures on symplectic topology, *Proceedings of the Conference on Diff. Geometry in Cala Gonone*, 1988.

[El3] Ya. Eliashberg, Killing of elliptic complex points and the Legendrian isotopy, in preparation.

[Gr] M. Gromov, Pseudo-holomorphic curves in symplectic manifolds, *Inv. Math.*, **82** (1985), 307–347.

[HE] V. M. Harlamov and Ya. Eliashberg, On the number of complex points of a real surface in a compex surface, *Proceedings of Leningrad Int. Topology Conference*, 1982, 143–148.

[McD1] D. McDuff, The structure of rational and ruled symplectic manifolds, preprint, 1989.

[McD2] D. Mcduff, The local behavior of holomorphic curves in almost complex 4-manifolds, preprint, 1989.

PART 2

JONES/WITTEN THEORY

The discovery by Vaughan Jones of a new polynomial invariant of links in the 3-sphere was an important breakthrough which has lead to the introduction of a whole range of new techniques in 3-dimensional topology. The original Jones invariant, a Laurent polynomial in one variable, was obtained via a "braid" description of a link, utilising the remarkable properties of some representations of the braid group which arose in the theory of Von Neumann algebras. Early developments were largely combinatorial, leading to alternative definitions of the invariant and to generalisations, including a 2-variable polynomial which specialises to both the Jones polynomial and the classical Alexander polynomial after appropriate substitutions. The new invariants are comparatively easy to calculate and have had many concrete applications but for some time no really satisfactory conceptual definition of the invariants was known: one not relying on special combinatorial presentations of a link . It was not clear, for example, whether such invariants could be defined for links in other 3-manifolds. While there were many intriguing connections between the Jones theory and statistical mechanics, for example through the Yang-Baxter equations and the newly developed theory of quantum groups, it was a major problem to find the correct geometrical setting for the Jones theory. We refer to [L] for a survey of this phase of the theory.

In his lecture at the International Congress of Mathematical Physicists in Swansea, July 1988, Witten proposed a scheme which largely resolved this problem. He showed that the invariants should be obtained from a quantum field theory with Lagrangian involving the Chern-Simons invariant of connections. This scheme provided a truly natural definition of the invariants, and indeed allowed considerable generalisations, to links in arbitrary 3-manifolds. Taking, in particular, the empty link, Witten obtained new invariants of closed 3-manifolds. The challenge in this approach arose from the notorious difficulties of quantum field theory, in attaching real meaning to the functional integrals over the space of connections which were involved. It is, to a mathematician, a striking fact that, despite these serious foundational difficulties, Witten's theory made concrete predictions which could be verified on a more elementary level. The new invariants appear to have great potential, and it seems possible that they might even be able to detect counter examples to the Poincaré conjecture, should any exist.

To develop the quantum field theory of the Chern-Simons function on a more rigorous basis, Witten has also given a Hamiltonian approach, which unites these ideas with conformal field theory in two dimensions. This approach starts by quantising a certain symplectic manifold naturally associated to a closed surface Σ, the moduli space of representations of $\pi_1(\Sigma)$ in a given compact Lie group. This leads to the construction of a projectively flat connection on a certain bundle over the moduli space of complex structures on Σ, whose monodromy defines a representation of the mapping class group. More generally, one should take a Riemann surface with a set of marked points (corresponding to a link in a 3-manifold), and in the case when the surface is the Riemann sphere one gets representations of the braid group.

Meanwhile Reshetikhin and Turaev have defined new invariants, from a more combinatorial and algebraic point of view, which it seems should agree with the Witten invariants. They use a Dehn surgery description of a 3-manifold, and obtain a manifold invariant from a link invariant of Jones type. This approach by-passes the analytical difficulties in the quantum field theory, although the conceptual geometric picture is less clear.

We were fortunate to have a number of lectures in Durham on these recent developments, leading to the four articles below. The paper of Hitchin and Jeffrey's notes on Witten's three lectures concentrate on the geometric quantisation of the representation spaces for surfaces; the paper of Kirby and Melvin gives the definition of the Reshetikhin-Turaev invariants and evaluates them in some special cases. Jeffrey's notes also give Witten's explanation of how the Reshetikhin-Turaev formulae should be obtained in the Chern-Simons picture. The lecture of Atiyah lies closer to the earlier work in Jones theory, involving the Hecke algebras; the paper describes new topological constructions of braid group representations, due to Lawrence.

[L] Lickorish, W.B.R. *Polynomials for links* Bull. London Math. Soc. **20** (1988) 558-588

New Results in Chern-Simons Theory

Lectures by

Edward Witten

Institute for Advanced Study, Princeton, NJ, USA 08540

Notes by

Lisa Jeffrey[1]

Mathematical Institute, 24-29 St. Giles, Oxford OX1 3LB, UK

In these lectures, I will describe some aspects of an approach to the Jones polynomial of knots, and its generalizations, that is based on a three-dimensional quantum Yang-Mills theory in which the usual Yang-Mills Lagrangian is replaced by a Chern-Simons action. This approach gives a manifestly three-dimensional approach to the subject, but some of the key aspects of the story, involving the Feynman path integral, are somewhat beyond the reach of present rigorous understanding. In the first two lectures, I will describe aspects of the subject that can be developed rigorously at present. This basically consists of a gauge theory approach to the Jones representations of the braid group and their generalizations. In the last lecture, I will describe the more ambitious Feynman path integral approach, which is an essential part of the way that physicists actually think about problems such as this one, and which gives the most far-reaching results.

The first two lectures describe joint work with S. Della Pietra and S. Axelrod. In this work, methods of symplectic geometry and canonical quantization are used to associate a vector space $\mathcal{H}_\Sigma$ to every oriented surface Σ. In this approach, we will have to pick a complex structure J on Σ as an auxiliary tool in constructing the $\mathcal{H}_\Sigma$, and much of the effort will go into constructing a projectively flat connection whose

[1]Supported in part by an NSF Graduate Fellowship.

existence shows that in a suitable sense, the $\mathcal{H}_\Sigma$ are independent of J. There are several other approaches to this connection, due to Tsuchiya, Ueno, and Yamada, Segal, and Hitchin; Hitchin's approach, which is based on algebraic geometry, has been presented at this meeting.

Once one has understood the association $\Sigma \to \mathcal{H}_\Sigma$, one can use it to construct topological invariants of three-manifolds (and, more generally, of links in three-manifolds). To give a simple example, since the association $\Sigma \to \mathcal{H}_\Sigma$ is natural, a diffeomorphism f of Σ gives rise to an endomorphism Φ_f of $\mathcal{H}_\Sigma$. If one forms a three-manifold M_f by taking the mapping cylinder of f (that is, by multiplying Σ by a unit interval and gluing the ends using f), then the invariant one associates to M_f is the trace of Φ_f.

The first lecture will be devoted to background about canonical quantization. In the second lecture, we will focus more specifically on the problem of actual interest – quantization of the moduli space $\mathcal{M}$ of representations of the fundamental group of a surface in a compact Lie group G. The third lecture describes more concretely how these constructions lead to link invariants. A detailed description of these results can be found in [2] and [15], along with more extensive references.

1. GEOMETRIC QUANTIZATION

(a) The prototype case: affine space

We begin with a standard affine symplectic space $\mathbb{A} = \mathbb{R}^{2n}$ with coordinates $x^1, \ldots,$ x^{2n} and a symplectic form $\omega = \omega_{ij}\, dx^i \wedge dx^j$ ($\omega_{ij} = -\omega_{ji}$ is a real antisymmetric matrix such that ω is nondegenerate.) We may regard ω as a transformation $T \to T^*$ (T is the tangent space to $\mathbb{A}$) by sending a tangent vector to its interior product with ω. Then the inverse $\omega^{-1} : T^* \to T$ will be denoted ω^{ij}: thus we have

$$\omega \frac{\partial}{\partial x^k} = \omega_{kj} dx^j,$$

$$\omega^{-1} dx^k = \omega^{kj} \frac{\partial}{\partial x^j}.$$

We examine the *Heisenberg Lie algebra* (essentially just given by the Poisson bracket) given by $\{x^1, \ldots x^{2n}\}$ with the commutation relations

$$[x^i, x^j] = -i\omega^{ij} \tag{1}$$

The quantization procedure for $\mathbb{A}$ constructs an irreducible unitary Hilbert space representation of this algebra. In fact it gives a representation of the Heisenberg *group*, an extension by $U(1)$ of the group of affine translations of $\mathbb{R}^{2n}$. By a theorem of Stone and von Neumann, there is a unique such representation up to isomorphism: moreover, the isomorphism between any two such representations is unique up to multiplication by $S^1 \subset \mathbb{C}$. So any two such representations are canonically isomorphic up to a projective multiplier.

To construct representations of the Heisenberg group (first a rather uninteresting reducible one and then an irreducible subrepresentation of it), we fix a unitary line bundle $\mathcal{L}$ over $\mathbb{A}$ with a connection D such that the curvature $D^2 = -i\omega$. (Such a bundle exists whenever $\omega/2\pi$ represents an integral cohomology class - here it it the trivial one!) The isomorphism classes of such bundles are given by $H^1(\mathbb{A}, U(1))$, which is 0; so $\mathcal{L}$ is unique up to isomorphism. The isomorphism between two such bundles is unique up to a constant gauge transformation, which must be of modulus 1 to preserve the metric.

We define $\mathcal{H}_0 = \Gamma_{L^2}(\mathbb{A}, \mathcal{L})$. The C^∞ functions on $\mathbb{A}$ act naturally on $\mathcal{H}_0$ as follows. Given $\phi \in C^\infty(\mathbb{A})$, the Hamiltonian function associated to it is $V_\phi = \omega^{-1}(d\phi)$. Then ϕ acts on $\mathcal{H}_0$ via an operator

$$U_\phi : s \mapsto iD_{V_\phi}s + \phi s. \qquad (2)$$

(This represents a lifting to $\mathcal{L}$ of the vector field V_ϕ, which preserves the symplectic structure: V_ϕ thus acts on the space of sections of $\mathcal{L}$.) The map $\phi \mapsto U_\phi$ is a homomorphism of algebras, i.e.,

$$[U_\phi, U_\psi] = U_{\{\phi,\psi\}};$$

so this gives a representation of the Lie algebra of all Hamiltonian vector fields, including those, the infinitesimal translations, comprising the Heisenberg algebra.

To get an irreducible subrepresentation of this, we pick a translationally invariant complex structure J on $\mathbb{A}$ (i.e., a constant endomorphism of T whose square is -1) such that:

(i) ω is compatible with J in the sense that $\omega(JX, JY) = \omega(X, Y)$, or in other words, in the complex structure J, ω a form of type (1,1).

(ii) ω is *positive* with respect to J (i.e., $\omega(X, JX) > 0$ if $X \neq 0$).

Thus there are holomorphic linear functions z^i such that $\omega = i\sum_i dz^i \wedge d\bar{z}^i$.

Let us now consider the line bundle $(\mathcal{L}, D)$ which was introduced and fixed before picking J. It acquires a new structure once J is picked. Indeed, as D has curvature of type (1,1) with respect to J, D and J combine to give $\mathcal{L}$ a holomorphic structure.

(One defines a $\bar{\partial}$ operator on $\mathcal{L}$ as $D^{0,1}$; the condition that this actually defines a holomorphic structure is $\bar{\partial}^2 = (DD)^{0,2} = \omega^{0,2} = 0$.)

The line bundle $\mathcal{L}$ can be given a particularly simple description once J is picked. Let $\mathcal{L}_0$ be the trivial holomorphic line bundle on A with the Hermitian metric (for $\psi \in C^\infty(\mathcal{L}_0)$)

$$|\psi|^2 = \psi^* \psi \; \exp(-h), \qquad\qquad h(z) = \sum z^i \bar{z}^i$$

and the connection compatible with this metric and the standard holomorphic structure. The curvature of this connection is

$$-\bar{\partial}\partial h = \sum dz^i d\bar{z}^i = -i\omega$$

This is the formula that characterizes $\mathcal{L}$, so $\mathcal{L}$ is isomorphic to $\mathcal{L}_0$, by an isomorphism that is unique up to a projective ambiguity.

Now that J has been picked, the Heisenberg algebra that we are seeking to represent can be written

$$\begin{aligned} [z^i, \bar{z}^j] &= -\delta^{ij} \\ [z^i, z^j] &= [\bar{z}^i, \bar{z}^j] = 0 \end{aligned} \tag{3}$$

We can represent this algebra on the subspace

$$\mathcal{H}_J = H^0(\mathsf{A}_J, \mathcal{L})$$

of our original Hilbert space consisting L^2 sections that are also holomorphic. This representation ρ is

$$\begin{aligned} \rho(z^i) \cdot \psi &= z^i \, \psi \\ \rho(\bar{z}^i) \cdot \psi &= \tfrac{\partial}{\partial z^i} \, \psi \end{aligned} \tag{4}$$

The commutation relations are preserved, as

$$\frac{\partial}{\partial z^i}(z^j \psi) = \delta^{ij} \psi + z^j \frac{\partial}{\partial z^i} \psi.$$

To verify unitarity, note that

$$(\rho(\bar{z}^i) \, \psi \, , \chi) = \int d^n z \, d^n \bar{z} \, \overline{\frac{\partial \psi}{\partial z^i}} \; \exp(-\sum_k z^k \bar{z}^k) \, \chi$$

But

$$\overline{\frac{\partial \psi}{\partial z^i}} = \frac{\partial \bar{\psi}}{\partial \bar{z}^i}$$

so upon integrating by parts and using the holomorphicity of ψ, we find that this equals

$$\int d^n z\, d^n \bar{z}\, \bar{\psi}\, \exp(-\sum_k z^k \bar{z}^k)\, z^i \chi = (\psi\,, \rho(z^i)\,\chi).$$

This representation of the Lie algebra in fact exponentiates to a representation of the (complexified) Heisenberg *group* (see for instance the discussion in [8], p. 188). It is irreducible, because holomorphic functions can be approximated by polynomials. Thus we have found the unique irreducible representation of the Heisenberg group described by the Stone-von Neumann theorem.

We now want to let J vary, fixing its key properties (translation invariance and the fact that ω is positive of type (1,1)). The space $\mathcal{T}$ of all complex structures on $\mathbb{A}$ with these properties is the *Siegel upper half plane* of complex symmetric $n \times n$ matrices with positive imaginary parts. Over $\mathcal{T}$ we get a Hilbert space bundle $\mathcal{H}$ whose fibre over $J \in \mathcal{T}$ is $\mathcal{H}_J$. The uniqueness theorem for representations of the Heisenberg group implies that there is a natural way to identify the fibres of this bundle, i.e., $\mathcal{H}$ has a natural projectively flat connection.

We would like to identify this connection explicitly. To this end, we observe that if $x^i \mapsto \hat{x}^i$ is any representation of (1), then the objects

$$D^{ij} = \hat{x}^i \hat{x}^j + \hat{x}^j \hat{x}^i$$

themselves obey a Lie algebra, which can be worked out explicitly by expanding the commutators, i.e.,

$$[D^{ij}, D^{kl}] = -i(D^{il}\omega^{jk} + D^{ik}\omega^{jl} + D^{lj}\omega^{ik} + D^{kj}\omega^{il})$$

This is just the Lie algebra of $Sp(2n, \mathbb{R})$, if we identify $\mathbb{A}$ with $\mathbb{R}^{2n}$ (it is given by the algebra of homogeneous quadratic polynomials under Poisson bracket). Thus, the Lie algebra of $Sp(2n, \mathbb{R})$ acts in any representation of the Heisenberg algebra.

Actually, the *group* $Sp(2n, \mathbb{R})$ acts on the Heisenberg algebra by outer automorphisms, and so conjugates the representation $\mathcal{H}_J$ of the Heisenberg group to another representation. By the uniqueness of this irreducible representation, $Sp(2n, \mathbb{R})$ must act at least projectively on $\mathcal{H}_J$. The D^{ij} give the action explicitly at the Lie algebra level.

The z^i act on $\mathcal{H}_J$ as zeroth order differential operators, while the $\bar{z}^i$ act as first order operators. The D^{ij} are thus at most second order differential operators. The D^{ij}, which generate vector fields that act transitively on $\mathcal{T}$, essentially define a connection on $\mathcal{H}$, and this connection is given by a second order differential operator since the D's have this property. (Actually, the full group of symplectic diffeomorphisms of $\mathbb{A}$ has a natural "prequantum" action on the big Hilbert space $\mathcal{H}_0$. This gives a representation of the Lie algebra of $Sp(2n,\mathbb{R})$ by first order operators D_{ij}'. The connection is really the difference between the D_{ij} and the D_{ij}'.)

The connection can be described explicitly as follows. Let δ be the trivial connection on bundle $\mathbb{A} \times \mathcal{H}_0$, regarded as a trivial Hilbert space bundle over $\mathbb{A}$. δ does not respect the subspace $\mathcal{H}_J \subset \mathcal{H}_0$. A connection which does respect it can be described by adding to the trivial connection a certain second-order differential operator. In describing it, note that the Siegel upper half plane has a natural complex structure, so the connection decomposes into (1,0) and (0,1) parts. Let T denote the tangent space to $\mathbb{A}$; then a tangent vector to $\mathcal{T}$ corresponds to a deformation of the complex structure of $\mathbb{A}$ which is given explicitly by an endomorphism $\delta J : T \to T$ (which obeys $J\,\delta J + \delta J\, J = 0$). Associated with such an endomorphism is the object $\delta J \circ \omega^{-1} \in T \otimes T$, i.e., $\delta J \circ \omega^{-1} : T^* \to T$, which in fact lies in $\mathrm{Sym}^2 T$; indeed, it lies in $\mathrm{Sym}^2 T^{1,0} \oplus \mathrm{Sym}^2 T^{0,1}$. The (2,0) part of this (which is the (1,0) part of δJ with respect to the complex structure on $\mathcal{T}$) will be denoted δJ^{ij}. Then the connection on $\mathcal{H}$ can be written:

$$\begin{aligned} \nabla^{0,1} &= \delta^{0,1} \\ \nabla^{1,0} &= \delta^{1,0} - \mathcal{O} \\ \mathcal{O} &= -\frac{1}{4}\sum_{i,j=1}^{n} \frac{D}{Dz^i}\,\delta J^{ij}\,\frac{D}{Dz^j} \end{aligned} \tag{5}$$

where $\frac{D}{Dz^i}$ is the covariant derivative acting on sections of $\mathcal{L}$. As promised, this connection is given by a second-order differential operator on sections of $\mathcal{L}$. In more invariant notation we have

$$\nabla^{1,0} s = \delta^{1,0} s + \frac{1}{4} D(\delta J \circ \omega^{-1} D^{1,0} s)$$

for $s \in \Gamma_{L^2}(\mathbb{A},\mathcal{L})$ and with the first D acting as a map $T \otimes \Gamma_{L^2}(\mathbb{A},\mathcal{L}) \to \Gamma_{L^2}(\mathbb{A},\mathcal{L})$.)

One may check that $[\bar{D}_J, \nabla] = 0$ on holomorphic sections, where $\bar{D}_J$ is the $\bar{\partial}$ operator of the holomorphic bundle $\mathcal{L}$ induced by J, i.e., $\bar{D}_J = \sum_k d\bar{z}^k \frac{D}{D\bar{z}^k} = \frac{1}{2}(1+iJ)D$. The trivial connection δ does not commute with $\bar{D}_J$, but neither does $\mathcal{O}$, and the two contributions cancel. The contribution from $\mathcal{O}$ arises because one meets $[\frac{D}{Dz^i}, \frac{D}{D\bar{z}^j}]$, which is just the curvature of $\mathcal{L}$, i.e., $-i\omega$. Both commutators are first order differential operators.

(b): The torus

We will now describe several variations of this construction, in increasing order of relevance. For our first example, we consider the quantization of a torus. (This precise situation actually arises in the Chern-Simons theory for the gauge group $U(1)$.)

We pick a lattice Λ of maximal dimension $2n$ in the group of affine translations of $\mathbb{A}$, integral in the sense that the action of Λ on $\mathbb{A}$ can be lifted to an action on $\mathcal{L}$, and we pick such a lift. The quotient of $\mathbb{A}$ by Λ is a torus T, which we wish to quantize. Using the Λ action on $\mathcal{L}$, we can push down $\mathcal{L}$ to a line bundle over T which we will also call $\mathcal{L}$.

If a complex structure J is picked on $\mathbb{A}$ as before (compatible with ω and giving a metric on $\mathbb{A}$) then the complex structure on $\mathbb{A}$ descends to a complex structure on T, which thus becomes a complex torus and indeed an *Abelian variety* T_J. (An Abelian variety is precisely a complex torus which admits a line bundle with the properties of $\mathcal{L}$.) The holomorphic sections $\mathcal{H}_{\Lambda,J} = H^0(T_J, \mathcal{L}_\Lambda) = (\mathcal{H}_J)^\Lambda$ form a vector bundle over $\mathcal{T}$ as before. $H^0(T_J, \mathcal{L}_\Lambda)$ is a space of theta functions, of some polarisation depending on Λ. The Λ invariant subspace of the bundle $\mathcal{H}$ over $\mathcal{T}$ – whose fiber is given by $\mathcal{H}_{\Lambda,J}$ – will be denoted as $\mathcal{H}_\Lambda$.

Because the action of Λ commutes with the connection ∇, ∇ restricts to a connection on the subbundle $\mathcal{H}_\Lambda$. In this case the connection can be made flat (not just projectively flat) by tensoring $\mathcal{H}_\Lambda$ with a suitable line bundle with connection over $\mathcal{T}$; the theta functions as conventionally defined by the classical formulas are covariant constant sections of $\mathcal{H}_\Lambda$. The condition of being covariant constant is the "heat equation" obeyed by the theta functions , which is first order in the complex structure J and second order in the variables z along the torus. This is thus the origin, from the point of view of symplectic geometry, of part of the classical theory of theta functions.

(c): The symplectic quotient

Somewhat closer to our interests is a situation in which the lattice Λ is replaced by a compact group $\mathcal{G}$ acting linearly and symplectically on $\mathbb{A}$, with a chosen lift of the $\mathcal{G}$ action to an action on $\mathcal{L}$ preserving D. In this situation, we restrict the general discussion of quantization of affine spaces to the $\mathcal{G}$ invariant subspaces. Thus, we let $\mathcal{T}_\mathcal{G}$ be the subspace of $\mathcal{T}$ consisting of $\mathcal{G}$ -invariant complex structures. Over $\mathcal{T}_\mathcal{G}$, we form the Hilbert space bundle $\mathcal{H}^\mathcal{G}$ whose fibre over J is the $\mathcal{G}$ invariant subspace $(\mathcal{H}_J)^\mathcal{G}$ of $\mathcal{H}_J$. Restricting to $\mathcal{T}_\mathcal{G}$ and to $(\mathcal{H}_J)^\mathcal{G}$, equation (5) gives a natural connection on $\mathcal{H}^\mathcal{G}$, which of course is still projectively flat.

In geometric invariant theory, once one picks a complex structure J, the symplectic action of the compact group $\mathcal{G}$ on $\mathbb{A}$ may be extended to an action of the complexification $\mathcal{G}_c$, which depends on J. (The vector fields generated by the imaginary part of the Lie algebra are orthogonal to the level sets of the moment map, in the metric determined by the complex structure and the symplectic form.) As a *symplectic* manifold, $\mathbb{A}_J/\mathcal{G}_c$ is independent of J; it may be identified with $\mathbb{A}//\mathcal{G}$, the symplectic or Marsden-Weinstein quotient of $\mathbb{A}$. This is defined as $F^{-1}(0)/\mathcal{G}$, where $F : \mathbb{A} \to \text{Lie }(\mathcal{G})^*$ is the moment map for the $\mathcal{G}$ -action. However, $\mathbb{A}//\mathcal{G}$ acquires a complex structure from its identification as $\mathbb{A}_J/\mathcal{G}_c$. The line bundle $\mathcal{L}$ over $F^{-1}(0)$ pushes down to a unitary line bundle with connection $\tilde{\mathcal{L}}$ over $\mathbb{A}//\mathcal{G}$; this line bundle is holomorphic in the complex structure that $\mathbb{A}//\mathcal{G}$ gets from its identification with $\mathbb{A}_J/\mathcal{G}_c$ (for any J). The $\mathcal{G}$ invariant space $(\mathcal{H}_J)^{\mathcal{G}}$ considered in the last paragraph can be identified with $\mathcal{H}_J = H^0(\mathbb{A}_J/\mathcal{G}_c, \tilde{\mathcal{L}})$. This identification is very natural from the point of view of geometric invariant theory. The $\mathcal{H}_J$ sit inside the fixed Hilbert space $\mathcal{H}_0 = \Gamma_{L^2}(\mathbb{A}//\mathcal{G}, \tilde{\mathcal{L}})$. The connection described in (5), restricted to the $\mathcal{G}$ invariant subspace and pushed down to an intrinsic expression on $\mathbb{A}//\mathcal{G}$, is still given by a second order differential equation, with the same leading symbol but more complicated lower order terms.

(d): The moduli space of representations

Suppose G is a compact Lie group with Lie algebra $\mathfrak{g}$, and Σ an oriented compact surface without boundary. We fix a principal G- bundle $P \to \Sigma$; $\mathcal{A}$, the space of connections on P, is an affine space modelled on $\Omega^1(\Sigma, \text{ad}(P))$. The gauge group $\mathcal{G}$ is $\Omega^0(\Sigma, \text{Ad}(P))$, and acts on $\mathcal{A}$ by $d_A \mapsto g\, d_A\, g^{-1}$.

If $\alpha, \beta \in \Omega^1(\Sigma, \text{ad}(P))$, we may form a pairing

$$\{\alpha, \beta\} \mapsto \frac{1}{4\pi}\int_\Sigma (\alpha, \wedge\beta)$$

where we have used a G-equivariant inner product $(\cdot,\cdot)$ on Lie(G). This skew-symmetric form on $\Omega^1(\Sigma, \text{ad}(P))$ defines a natural symplectic structure on $\mathcal{A}$. The normalisation of the inner product is chosen as follows. If F is the curvature of the universal G-bundle $EG \to BG$, then an invariant inner product $(\cdot,\cdot)$ on Lie(G) defines a class $(F, \wedge F) \in H^4(BG)$. If G is *simple*, all invariant inner products are related by scalar multiplication; we choose the *basic inner product* to be the one such that $(F, \wedge F)/(2\pi)$ is a generator of $H^4(BG, \mathbb{Z}) = \mathbb{Z}$. (It has the property that $(\alpha, \alpha) = 2$ where α is the highest root.)

We take this to define the basic symplectic form ω_0 on $\mathcal{A}$; it is integral in that it may be obtained as the curvature of a line bundle over $\mathcal{A}$. The action of G on $\mathcal{A}$ preserves the symplectic structure. (This situation was extensively treated in [1].) In general, we pick an integer k (which, as it turns out, corresponds to the "level" in the theory of representations of loop groups) and consider the symplectic structure $\omega = k\omega_0$.

If $A \in \mathcal{A}$ is a connection, its curvature is $F_A = dA + A \wedge A \in \Omega^2(\Sigma, \text{ad}(P))$. Now the Lie algebra of the gauge group is $\Omega^0(\Sigma, \text{ad}(P))$, so under the pairing given by the symplectic form, the dual Lie$(\mathcal{G})^* = \Omega^2(\Sigma, \text{ad}(P))$. The moment map for the $\mathcal{G}$

action is $A \mapsto F_A$: thus the symplectic quotient is

$$\begin{aligned} \mathcal{M} = \mathcal{A}//\mathcal{G} &= \{A : F_A = 0\}/\mathcal{G} \\ &= \text{Rep}(\pi_1(\Sigma) \to G)/\text{conjugation} \end{aligned} \tag{6}$$

$\mathcal{M}$ does not have a natural complex structure, but a complex structure can be picked as follows. Any choice of complex structure J on Σ decomposes $T_{\mathbb{C}}\mathcal{A}$ into:

$$\begin{aligned} T^{1,0}\mathcal{A} &= \Omega^{0,1}(\Sigma, \mathfrak{g}) \\ T^{0,1}\mathcal{A} &= \Omega^{1,0}(\Sigma, \mathfrak{g}) \end{aligned} \tag{7}$$

(This conventional but seemingly inverted choice is made so that the operator $\bar{\partial}_A$ will be a map Lie$(\mathcal{G}_c) \to T^{1,0}\mathcal{A}$ or equivalently so that the moduli space of holomorphic bundles varies holomorphically with the complex structure on Σ.) This complex structure on $\mathcal{A}$ is compatible with the symplectic form, as the symplectic form naturally pairs $(0,1)$- forms on Σ with $(1,0)$-forms.

The complex structure that $\mathcal{M}$ gets in this way has a very natural interpretation. According to the Narasimhan-Seshadri theorem (which was interpreted by Atiyah and Bott as an analogue for the infinite dimensional affine space of connections of considerations that we sketched earlier for symplectic quotients of finite dimensional affine spaces), once a complex structure J is picked on Σ, $\mathcal{M}$ has a natural identification with the moduli space of holomorphic principal $\mathcal{G}_c$ bundles on Σ. The latter has an evident complex structure, and this is the complex structure that $\mathcal{M}$ gets by pushing down the complex structure (7) on the space of connections.

In fact the complex structure on $\mathcal{M}$ depends on the complex structure J on Σ only up to isotopy. (The interpretation of $\mathcal{M}$ as a moduli space of representations of the fundamental group shows that diffeomorphisms isotopic to the identity act trivially on $\mathcal{M}$; and the Hodge decomposition that gives the complex structure of $\mathcal{M}$ is likewise invariant under isotopy.) Thus, we actually obtain a family of complex structures on $\mathcal{M}$ parametrized by *Teichmüller space*, which we will denote as $\mathcal{T}$. So in this case we will construct a projectively flat connection on a bundle over $\mathcal{T}$. Just as for symplectic quotients of finite dimensional symplectic manifolds, the connection form will be a second order differential operator.

2. THE GAUGE THEORY CASE

In this lecture, we will give more detail about the preceding discussion. To begin with, we will describe more precisely how to push down the basic formula (5) for the connection that arises in quantizing an affine space, to an analogous formula describing the quantization of a symplectic quotient.

The $\mathcal{G}$ action on A and the complex structure on the latter give natural maps

$$\begin{aligned} T &: \text{Lie}(\mathcal{G}) \to T(F^{-1}(0)) \\ T_c &: \text{Lie}(\mathcal{G}_c) \to T_{\mathbb{C}}\mathsf{A} \\ T_z &: \text{Lie}(\mathcal{G}_c) \to T^{1,0}\mathsf{A} \\ T_{\bar{z}} &: \text{Lie}(\mathcal{G}_c) \to T^{0,1}\mathsf{A} \end{aligned}$$

If we introduce an invariant metric on $\text{Lie}(\mathcal{G})$ and take its extension to $\text{Lie}(\mathcal{G}_c)$ (note A already has an invariant metric from the symplectic form), we may form an operator $T_z^{-1} : T^{1,0}\mathsf{A} \to \text{Lie}(\mathcal{G}_c)$, which is zero on the orthogonal complement to $\text{Im}(T_z)$ in $T^{1,0}\mathsf{A}$, and maps into the orthogonal complement of $\text{Ker}(T_z)$. We may also form $\mathcal{K}$, the operator that projects $T^{1,0}\mathsf{A}$ onto $T^{1,0}\mathcal{M}$, the orthocomplement of $\text{Im}(T_z)$ in $T^{1,0}\mathsf{A}$. (One sees this because

$$T\mathsf{A} = \text{Lie}(\mathcal{G}) \oplus J\text{Lie}(\mathcal{G}) \oplus T\mathcal{M},$$

since the codimension of $F^{-1}(0)$ in A is then $\dim \mathcal{G}$ and $J\text{Lie}(\mathcal{G})$ is orthogonal to $TF^{-1}(0)$. Thus $T_c\mathsf{A} = T_c\mathcal{M} \oplus \text{Lie}(\mathcal{G}_c)$ and the projection onto $(1,0)$ parts preserves this decomposition, so also $T^{1,0}\mathsf{A} = T^{1,0}\mathcal{M} \oplus \text{Im}(T_z)$.) From this, $\text{id} = \mathcal{K} + T_z T_z^{-1}$ on $T^{1,0}\mathsf{A}$.

If X is a vector field in the image of T, and s is a $\mathcal{G}$ invariant section, one has $D_X s = iF_X s$. This condition, which determines the derivatives of s in the gauge directions, permits one to express the connection form $\mathcal{O} = -(1/4)D_i \delta J^{ij} D_j$ acting on $\mathcal{G}$ -invariant holomorphic sections over $F^{-1}(0)$, in a manner that only involves derivatives along the $T^{1,0}\mathcal{M}$ directions and has the other directions projected out. The result is:

$$\begin{aligned} \mathcal{O} &= -\frac{1}{4} D_{(\mathcal{K}\frac{\partial}{\partial z^i})} \delta J^{ij} D_{(\mathcal{K}\frac{\partial}{\partial z^j})} - \frac{1}{4}\delta J^{ij}(D_i \mathcal{K}^k{}_j) D_{(\mathcal{K}\frac{\partial}{\partial z^k})} + \\ &+ \frac{i}{4} Tr(T_z^{-1} \delta J\, T_{\bar{z}}) \end{aligned} \tag{8}$$

where the indices denote bases for $T^{1,0}\mathcal{A}$ or $T^{1,0}\mathcal{M}$. It turns out that this formula can be expressed in a way that depends only on the intrinsic Kähler geometry of $\mathsf{A}_J/\mathcal{G}_c$ and the function

$$H = \det{}'\triangle,$$

where the operator $\triangle : \text{Lie}(\mathcal{G}_c) \to \text{Lie}(\mathcal{G}_c)$ is defined by

$$\triangle = T_z{}^* T_z = \frac{1}{2} T_c{}^* T_c.$$

(Here, "det " "denotes the product of the nonzero eigenvalues.) The final expression for the connection is:

$$\nabla^{1,0} = \delta^{1,0} - \mathcal{O}$$

$$\begin{aligned} \nabla^{0,1} &= \delta^{0,1} \\ \mathcal{O} &= -\frac{1}{4}\left\{D_i \delta J^{ij} D_j + \delta J^{ij}(D_i \log H) D_j\right\} \end{aligned} \tag{9}$$

where now the indices represent a basis of $T^{1,0}\mathcal{M}$.

The appearance of $\log H$ has a natural explanation: this object appears in the expression for the curvature of the canonical bundle of $\mathcal{M}$. Assume for simplicity $\operatorname{Ker} T_z = 0$. Then

$$\operatorname{Im}(T_z) \oplus T^{1,0}\mathcal{M} = T^{1,0}\mathbb{A},$$

so

$$\begin{aligned} (\Lambda^{\max}\operatorname{Lie}(\mathcal{G}_c))^* &\otimes \left(\Lambda^{\max} T^{1,0}\mathbb{A}\right) \\ &\cong (\Lambda^{\max}\operatorname{Lie}(\mathcal{G}_c))^* \otimes (\Lambda^{\max}\operatorname{Im}T_z) \otimes (\Lambda^{\max} T^{1,0}\mathcal{M}) \end{aligned}$$

Now $(\Lambda^{\max}\operatorname{Lie}(\mathcal{G}_c))^* \otimes (\Lambda^{\max}\operatorname{Im}T_z)$ is isomorphic to the trivial bundle, with the norm $\det T_z{}^* T_z$. If $\mathcal{G}$ acts trivially on $\Lambda^{\max} T^{1,0}\mathbb{A}$, then a "constant" section of the line bundle on the left hand side is $\mathcal{G}_c$ invariant and gives under this isomorphism a section s of $(\Lambda^{\max} T^{1,0}\mathcal{M})^*$ It then has norm $\text{const} \cdot \det(T_z{}^* T_z)$. In other words the *Ricci tensor* (the curvature of the dual of the canonical bundle $\mathcal{K}_{\mathcal{M}}$) is

$$R = -\bar{\partial}\partial \log H \tag{10}$$

where $H = \det(T_z{}^* T_z)$.

We now consider the gauge theory case, in which one is trying to quantize the finite dimensional symplectic quotient $\mathcal{M}$ of an infinite dimensional affine space $\mathcal{A}$, with the symplectic structure $\omega = k\omega_0$. Since we do not have a satisfactory theory of the quantization of the infinite dimensional affine space $\mathbb{A}$ which could *a priori* be pushed down to a quantization of $\mathcal{M}$, we simply take the final formula (9) that arises in the finite dimensional case and attempt to adapt it by hand to the gauge theory problem. The Kähler geometry of the quotient $\mathcal{M}$ exists in this situation, just as it would in a finite dimensional case, according to the Narasimhan-Seshadri theorem. Also, the pushing down of a trivial line bundle $\mathcal{L}$ on $\mathbb{A}$ to a line bundle $\tilde{\mathcal{L}}$ on $\mathcal{M}$, which is holomorphic in each of the complex structures of $\mathcal{M}$, can be carried out rigorously even in this infinite dimensional setting. We will explain this point in some detail. Start with a trivial line bundle $\mathcal{L} = \mathcal{A} \times \mathbb{C}$. We will describe a lift of the $\mathcal{G}$ action on $\mathbb{A}$ to $\mathcal{L}$; the required line bundle $\tilde{\mathcal{L}}$ over $\mathcal{M}$ is then the quotient of $\mathcal{L}$ under this action. Actually, we will lift not just $\mathcal{G}$, but its semidirect product with the mapping class group, in order to ensure that the action of the mapping class group on $\mathcal{M}$ lifts to an action on $\tilde{\mathcal{L}}$. We have exact sequences

$$\begin{array}{ccccccccc} 0 & \longrightarrow & \mathcal{G} & \longrightarrow & \mathrm{Aut}(P) & \longrightarrow & \Gamma & \longrightarrow & 0 \\ & & \| & & \cup & & \cup & & \\ 0 & \longrightarrow & \mathcal{G} & \longrightarrow & \mathrm{Aut}_0(P) & \longrightarrow & \Gamma_0 & \longrightarrow & 0 \end{array}$$

where Γ are the diffeomorphisms of Σ, and Γ_0 those isotopic to the identity. Thus the mapping class group $\Gamma/\Gamma_0 \cong \mathrm{Aut}(P)/\mathrm{Aut}_0(P)$. The action of any automorphism $\chi \in \mathrm{Aut}(P)$ covering $\phi \in \Gamma$ enables one to form a bundle over the mapping torus $\Sigma \times_\phi [0,1]$ by gluing the bundle P using χ. Given a connection A on P, one forms a connection on this new bundle by interpolating between $\chi^* A$ and A. The *Chern-Simons invariant* of this connection is an element of $U(1)$; thus one gets a map $\mathcal{A} \times \mathrm{Aut}(P) \to U(1)$, and one may use the $U(1)$ factor as a multiplier on $\mathcal{L} = \mathcal{A} \times \mathbb{C}$ to lift the action of $\mathrm{Aut}(P)$ on $\mathcal{A}$. (Restricting to $\mathcal{G} \subset \mathrm{Aut}(P)$, one obtains the moment map multiplier that is used to lift the $\mathcal{G}$ action to $\mathcal{L}$.) Thus the mapping class group action on $\mathcal{M}$ lifts to $\tilde{\mathcal{L}}$.

The only additional ingredient that we must define in order to adapt (9) to this situation is the determinant H. In this case the map $T_z : \mathrm{Lie}(\mathcal{G}_c) \to T^{1,0}\mathcal{A}$ is

$$\bar{\partial}_A : \Omega^0(\Sigma, \mathfrak{g}_c) \to \Omega^1(\Sigma, \mathfrak{g}_c)$$

so

$$\triangle = \bar{\partial}_A{}^* \bar{\partial}_A = \frac{1}{2} d_A{}^* d_A : \Omega^0(\Sigma, \mathfrak{g}_c) \to \Omega^0(\Sigma, \mathfrak{g}_c)$$

is the Laplacian on the Riemann surface Σ, twisted by the connection A. Following Ray and Singer, the determinant of the Laplacian can be defined by zeta functional regularization. With this choice, we can use the formula (9) to define a connection on the bundle over Teichmüller space whose fiber is $H^0(\mathcal{M}_J, \tilde{L})$. However, in contrast to the case of quantization of the symplectic quotient of a finite dimensional affine space, in which one knows *a priori* that this connection form commutes with the $\bar{D}$ operator and is projectively flat, in the gauge theory problem we must verify these properties.

In verifying the projective flatness of the connection (9), or more precisely of a slightly modified version thereof, the main ingredients required, apart from general facts about Kähler geometry, are formulas for the derivatives of H which are consequences of Quillen's local families index theorem. One important consequence of this theorem is the formula

$$R + \bar{\partial}\partial \log H = 2h(-i\omega_0) = -2ih\frac{\omega}{k}, \tag{11}$$

where R is the Ricci tensor of $\mathcal{M}$, h is the dual Coxeter number of the gauge group, and ω_0 is the fundamental quantizable symplectic form on $\mathcal{M}$.

The term proportional to h, which is absent in the analogous finite dimensional formula (10), is what physicists would call an "anomaly"; it is, indeed, a somewhat disguised form of the original Adler-Bell-Jackiw gauge theory anomaly, or more exactly of its two dimensional counterpart. Because of this term, when one tries to verify the desired properties of the connection, one runs into trouble, and it is necessary to modify the formula (9) in a slightly *ad hoc* way. The modified formula is

$$\begin{aligned}\nabla^{1,0} &= \delta^{1,0} - \frac{k}{k+h}\mathcal{O} \\ \nabla^{0,1} &= \delta^{0,1}\end{aligned} \tag{12}$$

where the new formula for $\mathcal{O}$ is the same as the old one (9). The identity (11) enters crucially into the proof that this connection satisfies $[\bar{D}_J, \nabla] = 0$ on holomorphic sections and thus preserves holomorphicity.

The formula (11) corresponds to the local index theorem formula for the $(\mathcal{M}, \mathcal{M})$ component of the curvature of the determinant line bundle, which is, however, defined over $\mathcal{M} \times \mathcal{T}$. The local index formula completely determines the curvature: explicitly,

$$\frac{2h}{k}(-i\omega) + c_1(\operatorname{Ind} T_z) = \bar{\partial}\partial \log H + R_{\mathcal{M},\mathcal{M}} + R_{\mathcal{M},\mathcal{T}} + R_{\mathcal{T},\mathcal{T}} \tag{13}$$

The R terms represent the curvature the determinant bundle would have had for the original metric without Quillen's correction factor H; $\partial, \bar{\partial}$ now refer to $\mathcal{M} \times \mathcal{T}$. The left hand side is the local index, and the right hand side the curvature computed from the Quillen metric. This identity enters in determining the curvature of the connection (12) over $\mathcal{T}$. One finds that the $(1,1)$ part of the curvature is central, with the explicit formula being

$$R^{1,1} = \frac{k}{2(k+h)} c_1(\operatorname{Ind} T_z). \tag{14}$$

The $(0,2)$ component of the curvature is trivially zero since $\nabla^{0,1} = \delta^{0,1}$, but to show that the $(2,0)$ component is zero using techniques of the sort I have been sketching requires a great deal of work. (There is also a simple global argument, of a very different flavor, which was explained in Hitchin's lecture.)

This is in contrast to the finite-dimensional case, where the vanishing of the $(2,0)$ component of the curvature would follow simply from the fact that the bundle $\mathcal{H}$ has a *unitary structure* that is preserved by ∇ (i.e. ∇ is the unique connection preserving the metric and the holomorphic structure on $\mathcal{H}$). In the gauge theory case, we do not have such a unitary structure rigorously. *Formally*, there is such a unitary structure: for $\tilde{\psi} \in \mathcal{H}_J = H^0(\mathcal{M}, \tilde{\mathcal{L}})_J$, we pull $\tilde{\psi}$ up to a $\mathcal{G}_c$ -invariant section $\psi \in (H^0)(\mathcal{A}, \mathcal{L})_J$ and define

$$\| \tilde{\psi} \|^2 = \int_{\mathcal{A}} d\mu \, |\psi|^2$$

where $d\mu$ is the formal "symplectic volume " on $\mathcal{A}$. (This is the formal analogue of the unitary structure in the case of a finite dimensional symplectic quotient.) In the finite dimensional case, one would integrate over the $\mathcal{G}_c$ orbits to get a measure on $\mathcal{M}$ rather than on $\mathcal{A}$. In the gauge theory case, Gawedzki and Kupiainen [5] have shown that it is miraculously possible to do this explicitly (though not quite rigorously) for the case when Σ is a torus; the result is

$$\| \tilde{\psi} \|^2 = \int_{\mathcal{M}} \frac{\omega^n}{n!} H^{1/2} \, |\tilde{\psi}|^2$$

where as before $H = \det T_z^* T_z$ is a factor from the "volume"of the gauge group orbit.

Their construction does not generalise to other Σ, for it uses the fact that $\pi_1(\Sigma)$ is abelian. However, one may construct an asymptotic expansion

$$\| \tilde{\psi} \|^2 = \int_{\mathcal{M}} \frac{\omega^n}{n!} H^{1/2} \, (\sum_k b_k k^{-n}) \, |\tilde{\psi}|^2$$

where $b_0 = 1$ and the higher b_k's are functions on $\mathcal{M} \times \mathcal{T}$ that can be computed by a perturbative evaluation of the integral over the $\mathcal{G}_c$ orbits. (The required techniques are similar to the methods that we will briefly indicated in the next lecture.) If one can establish unitarity to some order in $1/k$, then obviously $(\nabla^2)^{2,0}$ vanishes to the same order. But it is actually possible to show that unitarity to order $1/k^2$ is enough to imply that the $(2,0)$ curvature vanishes exactly. This is an interesting approach to proving that statement.

At this point, we can enjoy the fruits of our labors. The monodromy of the projectively flat connection that we have constructed gives a projective representation of the mapping class group. The representations so obtained are genus g analogues of the Jones representations of the braid group. The original Jones representations arise on setting $g = 0$ and generalizing the discussion to include marked points; the details of the latter have not been worked out rigorously from the point of view sketched here.

3. LINK INVARIANTS

In this lecture we aim to describe more concretely the way in which the theory constructed in the first two lectures gives rise to link invariants. Also, I want to tell you a little bit about how physicists actually think about the subject. First we shall put the theory in context. Symplectic manifolds arise in physics in a standard way, as phase spaces. For example, the trajectories of the classical mechanics problem

$$\ddot{x}^i = -\frac{\partial V}{\partial x^i} \qquad (i = 1, \ldots, n)$$

are determined by the values of x, $\dot{x}$ at $t = 0$. The above equation is the equation for the critical trajectories of the Lagrangian

$$\mathcal{L} = \int \left(\frac{1}{2}\dot{x}^2 - V(x)\right) dt \tag{15}$$

or the equivalent Lagrangian

$$\mathcal{L} = \int \left(p\dot{x} - (\frac{1}{2}p^2 + V)\right) dt \tag{16}$$

in which the momentum p (which equals $\dot{x}$ for classical orbits) has been introduced as an independent variable. Classical phase space is by definition the space of classical solutions of the equations of motion. In this case, a classical solution is determined by the initial values of x and $\dot{x}$ or in other words of x and p; so we can think of the phase space as the symplectic manifold $\mathbb{R}^{2n}$ with the symplectic form $\omega = dp \wedge dx$. The space of critical points of such a time dependent variational problem always has such a symplectic structure.

Our moduli space $\mathcal{M}$ is no exception. Consider an oriented three-manifold M. Let G be a compact Lie group, P the trivial G bundle on M and A a connection on P. The *Chern-Simons invariant* of A is

$$I = \frac{1}{4\pi} \int_M Tr(A \wedge dA + \frac{2}{3} A \wedge A \wedge A) \tag{17}$$

(It may also be defined as the integral of $p_1(F_A)$ over a bounding four manifold over which A has been extended.) The condition for critical points of this functional is $0 = F_A = dA + A \wedge A$. I is used as the Lagrangian of a quantum field theory whose fields are the connections A. There are two standard ingredients in understanding such a theory:

(a): Canonical quantization

We separate out a "time" direction by considering the manifold $M = \Sigma \times \mathbb{R}$; then we consider the moduli space of critical points of the Chern-Simons functional for this M. It is the space of equivalence classes of flat connections under gauge transformations, or of conjugacy classes of representations:

$$\mathcal{M} = \mathrm{Rep}(\pi_1(\Sigma \times \mathbb{R}), G)/\mathrm{conj} = \mathrm{Rep}(\pi_1(\Sigma), G)/\mathrm{conj}$$

In other words, we recover our earlier moduli space, but we have a new interpretation of the rationale for considering it: it is the phase space of critical points arising in a three dimensional variational problem. This change in point of view about where $\mathcal{M}$ comes from is the germ of the understanding of the three dimensional invariance of the Jones polynomial.

After constructing the classical phase space associated with some Lagrangian, the next task is to quantize it. Quantization means roughly passing from the symplectic manifold to a quantum vector space of "functions in half the variables" on the manifold (holomorphic sections of $\tilde{\mathcal{L}} \to \mathcal{M}$ in the Chern- Simons theory; "wave functions" $\psi(p)$, or $\psi(x)$, or holomorphic functions on $\mathbb{C}^n$ for some identification of $\mathbb{C}^n$ with $\mathbb{R}^{2n}$, for classical mechanics). Quantization of the phase space $\mathcal{M}$ of the Chern-Simons theory is precisely the problem that we have been discussing in the first two lectures.

Often one is interested in some group Γ of symplectic transformations of the classical phase space (the mapping class group for Chern-Simons; the symplectic group $Sp(2n;\mathbb{R})$ for classical mechanics). Under favorable conditions, quantization will then give rise to a projective representation of this group on the quantum vector space. (The classical mechanics version is the metaplectic representation of $Sp(2n;\mathbb{R})$.)

The Chern-Simons function does not depend on a metric on M (or a complex structure on Σ): thus the association of a vector space $\mathcal{H}_\Sigma$ to a surface Σ by quantization is purely topological, although in order to specify $\mathcal{H}_\Sigma$ we need to introduce a complex structure on Σ, as we have seen. This is analogous to trying to define the topological invariant $H^1(\Sigma,\mathbb{C})$ by picking J and identifying $H^1{}_J(\Sigma,\mathbb{C})$ as the space of meromorphic differential forms with zero residues modulo exact forms. Here, the analogue of our projectively flat connection is the Gauss-Manin connection: it enables one to identify the $H^1{}_J$ for different J, so that one recovers the topological invariance though it is not obvious in the definition. Of course there are other, more manifestly invariant ways to define $H^1(\Sigma,\mathbb{C})$! For Chern-Simons theory with nonabelian gauge group, however, we do not at present know any other definition.

(b): The Feynman integral approach

In canonical quantization, after constructing the "physical Hilbert space" of a theory, one wishes to compute the "transition amplitudes," and for this purpose the Feynman path integral is the most general tool. It is here that – in the case of the Chern-Simons theory – the three dimensionality will come into play.

We work over the space $\mathcal{W}$ of connections A on $P \to M$ (M being of course a three manifold). The Feynman path integral is the "integral over the space of all connections modulo gauge transformations"

$$Z(M) = \int_{\mathcal{W}} \mathcal{D}A \exp(ikI) \qquad (k \in \mathbb{Z}^+) \tag{18}$$

Here I is the Chern-Simons invariant of the connection A, and k is required to be an integer since I is gauge invariant only modulo 2π. (The comparison of the path integral and Hamiltonian approaches shows that the path integral (18) is related to quantization with the symplectic structure $\omega = k\omega_0$.) The path integral (18), which may at first come as a surprise to mathematicians but which is a very familiar sort of object to physicists, is the basic three manifold invariant in the Chern-Simons theory. To physicists, $Z(M)$ is known as the "partition function" of the theory. More generally, a physicist wishes to compute "expectation values of observables." These correspond to more general path integrals

$$Z_{\mathcal{O}}(M) = \int_{\mathcal{W}} \mathcal{D}A \exp(ikI)\mathcal{O}(A) \tag{19}$$

where $\mathcal{O}(A)$ is a suitable functional of the connection A.

The functional that is important in defining link invariants is the 'Wilson line'(which was introduced in the theory of strong interactions to treat quark confinement). If C is a loop in M, and R a representation of G , we define

$$\mathcal{O}_R(C) = Tr_R(\mathrm{Hol}_C(A)) \tag{20}$$

where Hol denotes the holonomy of the connection A around the loop C. A link in M is a collection of such loops C_i; we define a link invariant by assigning a representation ("colouring") R_i of G to each loop C_i, and taking the product $\prod_i \mathcal{O}_{R_i}(C_i)$. This is precisely the situation considered in R. Kirby's lecture at this conference, for $G = SU(2)$, and that is not accidental; the invariants he described are the ones obtained from the Feynman path integral, as we will discuss in more detail later.

In particular, the original Jones polynomial arises in this framework if one takes $M = S^3$ and one labels all links with the 2-dimensional representation of $SU(2)$; the HOMFLY polynomial arises from the N dimensional representation of $SU(N)$. Other representations yield generalised link invariants that have been obtained by considering quantum groups; however, the Chern-Simons construction gives a manifestly three-dimensional explanation for their origin. These invariants are link "polynomials" in the sense that for $M = S^3$ (but not arbitrary M) they can be shown, at least in the HOMFLY case, to be Laurent polynomials in

$$q = \exp\{2\pi i/(h+k)\}.$$

The Chern-Simons quantum field theory that we are discussing here is atypical in that it is exactly soluble. The arguments that give the exact solution (such as the rigorous treatment of the canonical quantization sketched in the last lecture) are somewhat atypical of what one is usually able to do in quantum field theory. To gain some intuition about what Feynman path integrals mean, it is essential to attempt a direct assault using general methods that are applicable regardless of the choice of a particular Lagrangian. The most basic such method is the construction of an asymptotic expansion for small values of the "coupling constant"$1/k$. In the Chern-Simons problem, even though it is exactly soluble by other methods, the asymptotic expansion gives results that are significant in their own right.

To understand the construction of the asymptotic expansion, we consider first an analogous problem for finite dimensional integrals. To evaluate the integral

$$\int \exp(ikf(x))d^n x \qquad (f : \mathbb{R}^n \to \mathbb{R})$$

for large k, one observes that the integrand will oscillate wildly and thus contribute very little, except near critical points p_i (in the sense of Morse theory) where $df(p_i) = 0$. The leading order contribution from such points is what we get by approximating f by a quadratic function near p_i and performing the Gaussian integral (suitably regularized):

$$\pi^{n/2} \sum_{p_i} \exp\{ikf(p_i)\} \frac{\exp\{\frac{i\pi}{4} \operatorname{sign} H_{p_i} f\}}{|\det H_{p_i} f|^{1/2}} \tag{21}$$

where $H_{p_i} f$ is the Hessian.

If we assume there are only finitely many gauge equivalence classes A_α of flat connections, the analogous expression in Chern-Simons gauge theory is:

$$Z(M) = \sum_\alpha e^{ikI_{CS}(A_\alpha)} \left[T_{RRS}(A_\alpha)\right]^{1/2} \cdot e^{i\pi\eta(0)/2} \cdot e^{ihI_{CS}(A_\alpha)} \tag{22}$$

Here, T_{RRS} is the *Reidemeister-Ray-Singer torsion* [11] of the flat connection A_α: it is a ratio of regularised determinants of Laplacians of d_A, and results from the formal analogue of the determinant of the Hessian. $\eta(0)$ is the *eta invariant* of the trivial connection: the eta invariant is a way to regularise the signature of a self-adjoint operator that is not positive. h is, as before, the dual Coxeter number.

At each A_α, this leading term is multiplied by an asymptotic expansion

$$1 + \sum_{n=1}^{\infty} \frac{b_n(\alpha)}{k^n}.$$

(In quantum field theory, such asymptotic expansions are usually not convergent.) Each $b_n(\alpha)$ is a topological invariant, capturing *global* information about M. The

$b_n(\alpha)$ are constructed from *Green's functions* , which are integral kernels giving the formal inverses of operators such as $*d_A : \Omega^1(M, \mathrm{ad}(P)) \to \Omega^1(M, \mathrm{ad}(P))$.

One may also expand the integral for $Z_{\mathcal{O}}(M)$ by this method. The leading term is the *Gauss linking number*: for $G = U(1)$, links indexed by a and representations indexed by integers n_a, this is

$$\exp\left(\frac{i}{2k}\sum_{a,b} n_a n_b \int_{\vec{x}\in C_a} d\vec{x} \cdot \int_{\vec{y}\in C_b} d\vec{y} \times \frac{(\vec{x}-\vec{y})}{|\vec{x}-\vec{y}|^3}\right) \tag{23}$$

Higher order terms in the asymptotic expansions give multiple integrals of the Gauss linking number. If the stabiliser of a flat connection A in the gauge group has virtual dimension m, one gets a contribution $\sim k^{m/2}$: for instance one sees this behaviour in the explicit formula for $Z(S^3)$ for $G = SU(2)$, which is

$$Z(S^3) = \sqrt{\frac{2}{k+2}} \sin\frac{\pi}{k+2} \sim k^{-3/2}. \tag{24}$$

The exponent reflects the fact that the trivial connection on S^3, since it has a three dimensional stabilizer, corresponds to a component of the moduli space of flat connections of virtual dimension -3.

(c): Putting these approaches together

We now discuss how to fit together the path integral and quantization approaches. If M is a manifold with boundary Σ, let A^Σ be a G connection on Σ. Define a functional Ψ on connections A^Σ by integrating over those connections on M that restrict to A^Σ:

$$\Psi(A^\Sigma) = \int_{A:A|_\Sigma = A^\Sigma} \mathcal{D}A \, \exp\{ikI(A)\} \tag{25}$$

Because of the behavior under gauge transformations on Σ, this integral does not define a *function* on the space of connections but a section of a line bundle. In fact, it defines a $\mathcal{G}(\Sigma)$ invariant section $\Psi(A^\Sigma) \in \Gamma(\mathcal{A}, \mathcal{L}^k)$. In fact this is a *holomorphic* section, i.e., $\Psi(A^\Sigma) \in \mathcal{H}_\Sigma$. (Holomorphicity may be formally proved by writing the path integral in the form

$$\Psi(A^\Sigma) = \lim_{T\to\infty} \exp\{-T\triangle_{\bar\partial}\}\Phi(A^\Sigma)$$

for some section Φ, with $\triangle_{\bar\partial}$ the $\bar\partial$ Laplacian on $\mathcal{M}$. As $T \to \infty$ this Laplacian obviously projects on holomorphic sections. See [10] or [4] for a derivation.) As Ψ is a holomorphic section over $\mathcal{M}$, it corresponds to a $\mathcal{G}_c$ invariant section of $\mathcal{L}$ over $\mathcal{A}$.

To calculate the invariants $Z_{\mathcal{O}}(M)$, we split the three-manifold M into two pieces M_L, M_R with a common boundary Σ. Then we split the path integral into path integrals over connections on the two halves:

$$\begin{aligned} Z(M) &= \int \mathcal{D}A \exp\{ikI(A)\} \\ &= \int_{A^\Sigma \in \mathcal{A}(\Sigma)} \mathcal{D}A^\Sigma \int_{A_L \in \mathcal{A}(M_L)} \mathcal{D}A_L \exp\{ikI(A_L)\} \cdot \\ &\qquad \cdot \int_{A_R \in \mathcal{A}(M_R)} \mathcal{D}A_R \exp\{ikI(A_R)\} \\ &= \int \mathcal{D}A^\Sigma \, \Psi_L{}^*(A^\Sigma) \, \Psi_R(A^\Sigma) \end{aligned}$$

(Since the boundary $\bar{\Sigma}$ of M_L has opposite orientation to that, Σ, of M_R, $\Psi_L{}^*(A^\Sigma)$ is an antiholomorphic section of $\tilde{\mathcal{L}} \to \mathcal{M}$.) The vector spaces $\mathcal{H}_\Sigma$ and $\mathcal{H}_{\bar{\Sigma}}$, being spaces of holomorphic and antiholomorphic sections of $\tilde{\mathcal{L}}$, are canonically dual; the integral over A^Σ formally defines this pairing, i.e.,

$$Z(M) = (\Psi_L, \Psi_R).$$

(This is similar to the way invariants are constructed in the topological quantum field theory for Floer-Donaldson theory. The difference between these situations is that the vector spaces in the Chern-Simons theory have a unitary structure, unlike those in the Floer-Donaldson theory. This reflects the more truly quantum-mechanical nature of the Chern-Simons theory.)

Finally we describe a way to explicitly evaluate the three manifold and link invariants by surgery. In this way, we will see how the formula which was adopted as the definition of these invariants in R. Kirby's lecture follows from the quantum field theory. More detail about the following can be found in ([15]).

Let C be any knot in a three manifold. The boundary of a tubular neighborhood of C is a torus Σ. If we label the link C with the representation R_i, the path integral for the Wilson line gives an element ψ_i of $\mathcal{H}_\Sigma$. The Hilbert space $\mathcal{H}_\Sigma$ is finite-dimensional, so a finite set of representations R_i will give a basis $\{\psi_i\}$. (For instance for $G = SU(2)$ in level k, we can find a basis with R_i the i dimensional representation, $i = 1, \ldots, k+1$.) We always include the trivial representation in such a basis as the element ψ_1.

Given a framing of the torus, the mapping class group gets identified with $SL(2, \mathbb{Z})$. Thus any element $u \in SL(2, \mathbb{Z})$ acts on $\mathcal{H}_\Sigma$, and is represented by some explicit matrix $M(u)$:

$$M(u)\psi_i = \sum_j M(u)_{ij} \psi_j.$$

For instance, for

$$T = \begin{pmatrix} 1 & 1 \\ 0 & 1 \end{pmatrix}$$

we have

$$M(T)_{ij} = \delta_{ij} \exp 2\pi i(h_i - c/24),$$

where the h_i are certain quadratic expressions in i, the *conformal weights* of the representations indexed by i, and the *central charge* c is a constant depending on k and G. For

$$S = \begin{pmatrix} 0 & -1 \\ 1 & 0 \end{pmatrix}$$

we have

$$M(S)_{ij} = \sqrt{\frac{2}{k+2}} \sin(\frac{\pi}{k+2})\, [ij]$$

(for $G = SU(2)$), in the notation used by Kirby. This formula can be obtained by explicitly integrating the flat connection that we constructed in the second lecture. (It was originally obtained, with different physical and mathematical interpretations, from the Weyl-Kac character formula for loop groups.) As is well known, the elements S and T generate $SL(2,\mathbb{Z})$, so the above formulas determine the representation.

We will now explain a formula describing how the quantum field theory behaves under surgery. We wish to compute the invariant $Z[M; [L, \{k\}])$ for a three manifold M containing a link L labeled by representations $\{k\}$. Let C be a knot in a three manifold M, disjoint from L. Let M_R be a tubular neighborhood of C (also disjoint from L), and M_L the complement of M_R in M. The path integral on M_L or M_R determines a vector Ψ_L or Ψ_R in the Hilbert space associated with quantization of Σ. The element Ψ_R is the path integral over the tubular neighbourhood with no Wilson lines, i.e., with the trivial representation labelling C: $\Psi_R = \psi_1$. So the quantum field theory invariant is

$$Z(M; [L, \{k\}]) = (\Psi_L, \psi_1).$$

Surgery on C corresponds to the action of some u in the mapping class group $SL(2,\mathbb{Z})$. We act on Σ_R by u and glue M_R back to form a new manifold M_u: in terms of the representation of the mapping class group this says

$$Z(M_u; [L, \{k\}]) = (\Psi_L, M(u)\psi_1) = \sum_j M(u)_{1j}(\Psi_L, \psi_j).$$

In other words, for the purpose of evaluating link invariants we may replace the surgery curve C by a Wilson line along C and sum over representation labels on C weighted by the representation of the mapping class group:

$$Z(M_u; [L, \{k\}] = \sum_j M(u)_{1j} Z(M; [L, \{k\}], [C, j]).$$

(Here $Z(M;[L,\{k\},[C,j])$ is the path integral for the three manifold M containing the link L labeled by $\{k\}$ and an additional circle C labeled by j.) One thus reduces to simpler manifolds (and eventually to S^3) by adding more links; so eventually one obtains the invariant of any manifold M as a sum over "colourings" of links in S^3. This is the formula that was to be explained.

References

[1] M.F. Atiyah and R. Bott, *The Yang-Mills Equations over Riemann Surfaces*, Phil. Trans. R. Soc. London **A308** (1982) 523.

[2] S. Axelrod, S. Della Pietra and E. Witten, *Geometric Quantization of Chern Simons Gauge Theory*, preprint IASSNS-HEP-89/57 (1989).

[3] J. Bismut and D. Freed, *The analysis of elliptic families I*, Commun. Math. Phys. **106** (1986) 159-176; D. Freed, *On determinant line bundles*, in *Mathematical aspects of string theory*, ed. S.T. Yau, World Scientific (1987) p. 189.

[4] P. Braam, *First Steps in Jones-Witten Theory*, Univ. of Utah lecture notes, 1989.

[5] K. Gawedzki and A. Kupiainen, *Coset Construction from functional integrals*, Nucl Phys. **B320**, 625-668 (1989).

[6] N. Hitchin, *Flat Connections And Geometric Quantization*, Oxford University preprint (1989).

[7] B. Kostant, *Quantization And Unitary Representations*, Lecture Notes in Math. **170** (Springer-Verlag, 1970) 87.

[8] A. Pressley and G. Segal, *Loop Groups*, Oxford University Press, Oxford, 1988.

[9] D. Quillen, *Determinants of Cauchy-Riemann operators over a Riemann surface*, Funct. Anal. Appl. **19** (1985) 31-34.

[10] T.R. Ramadas, I.M. Singer, and J. Weitsman, *Some comments on Chern-Simons gauge theory*, Commun. Math. Phys. **126**, 409-430 (1989).

[11] D. Ray and I. Singer, *R-Torsion and the Laplacian on Riemannian manifolds*, Adv. Math **7** (1971) 145-210.

[12] G. Segal, *Two Dimensional Conformal Field Theories And Modular Functors*, in *IXth International Congress on Mathematical Physics*, eds. B. Simon, A. Truman, and I. M. Davies (Adam Hilger, 1989) 22-37.

[13] J.-M. Souriau, *Quantification Geometrique,* Comm. Math. Phys. **1** (1966) 374.

[14] A. Tsuchiya and Kanie, *Vertex Operators In Conformal Field Theory on* $\mathbf{P}^1$ *And Monodromy Representations Of Braid Groups, Adv. Studies in Pure Math.* **16** (1988) 297.

[15] E. Witten, *Quantum field theory and the Jones polynomial*, Commun. Math. Phys. **121** (1989) 351-399.

Geometric quantization of spaces of connections

N.J. HITCHIN

Witten's three manifold invariants require, in the Hamiltonian approach, the *geometric quantization* of spaces of flat connections on a compact surface Σ of genus g. Recall that if G is a Lie group with a biinvariant metric, then the set of smooth points M^s of

$$M = \mathrm{Hom}(\pi_1(\Sigma); G)/G$$

acquires the structure of a *symplectic manifold*. This can be observed most clearly in the approach of Atiyah and Bott [1] which views M, the space of gauge equivalence classes of flat G-connections, as a symplectic quotient of the space of all connections.

The symplectic manifold M^s clearly depends only on the topology of Σ. To quantize it, we require a projective space with the same property. However, the method of geometric quantization requires first the choice of a polarization, the most tractable case of which is a *Kähler polarization*.

Briefly, if (M, ω) is a symplectic manifold with $\frac{1}{2\pi}[\omega] \in H^2(M; \mathbb{R})$ an integral class, then we can find a line bundle L with unitary connection whose curvature is ω. If we additionally choose a complex structure on M for which ω is a Kähler form (this is what a Kähler polarization means) then the $(0,1)$ part of the covariant derivative ∇ of this connection defines a holomorphic structure on L and we may consider the space of global holomorphic sections

$$V = H^0(M; L) = \ker \nabla^{0,1} : \Omega^0(L) \to \Omega^{0,1}(L).$$

The corresponding projective space $P(V)$ is the quantization relative to the polarization. What is required to make this a successful geometric quantization is to prove that $P(V)$ is, in a suitable sense, independent of the choice of polarization.

One way of approaching this question is to pass to the infinitesimal description of invariance. If X is a smooth family of Kähler polarizations of M, then an identification of $P(V_x)$ and $P(V_y)$ for $x, y \in X$ can be thought of as parallel translation of a *connection*, defined up to a projective factor, on a vector bundle $\mathbf{V}$ over X. To say that the identification is independent of the path between x and y is to say that the connection is *flat* (up to a scalar factor). One seeks this way a flat connection on the kernel of $\nabla^{0,1}$ as we vary the complex structure.

For a complex structure I, the space V consists of the solutions to the equation

$$(1 + \mathrm{i}I)\nabla s = 0.$$

Differentiating with respect to a parameter t (i.e. where X is one-dimensional) we obtain

$$\mathrm{i}\dot{I}\nabla s + (1 + \mathrm{i}I)\nabla \dot{s} = 0.$$

Here $\dot{I} \in \Omega^{0,1}(T)$ is a 1-form with values in the holomorphic tangent bundle and since $\nabla^{0,1}s = 0$ we can write this equation as

$$\mathrm{i}\dot{I}\nabla^{1,0}s + \nabla^{0,1}\dot{s} = 0.$$

A *connection* will be defined by a section $A(s, \dot{I}) \in \Omega^0(L)$, depending bilinearly on s and $\dot{I}$ and such that

$$\mathrm{i}\dot{I}\nabla^{1,0}s + \nabla^{0,1}A(s, \dot{I}) = 0. \qquad (*)$$

Now $\mathrm{i}\dot{I}\nabla^{1,0}$ is a $(1,0)$-form with values in $\mathcal{D}^1(L)$, the holomorphic vector bundle of first-order differential operators on L. The equation $(*)$ can be given a cohomological interpretation if we introduce the complex

$$C^p = \Omega^{0,p-1}(L) \oplus \Omega^{0,p}(\mathcal{D}^1(L))$$

and the differential

$$d_s(u, D) = (\bar{\partial}u + (-1)^{p-1}Ds, \bar{\partial}D).$$

From the integrability of the complex structure I and the compatibility with the fixed symplectic structure, the $(0,1)$-form $\mathrm{i}\dot{I}\nabla^{1,0} \in \Omega^{0,1}(\mathcal{D}^1(L))$ is $\bar{\partial}$. this, together with the fact that s is holomorphic, shows that a solution A to equation $(*)$ gives a 1-cochain in this complex:

$$d_s(A, \mathrm{i}\dot{I}\nabla^{1,0}) = 0$$

and hence a cohomology class.

Under the fairly mild hypothesis that there are no holomorphic vector fields preserving the line bundle L, we obtain the following:

Proposition: A connection on the projective bundle $P(H^0(M; L))$ over X is determined by a cohomology class in $\mathbf{H}^1_s(\mathcal{D}^1(L))$ — the first cohomology group of the complex above.

(This point of view was emphasised by Welters in his paper on abelian varieties [6].)

One way of obtaining such classes is to consider the sheaf sequence

$$\begin{array}{ccccccccc}
0 & \longrightarrow & \mathcal{D}^1(L) & \longrightarrow & \mathcal{D}^2(L) & \stackrel{\sigma}{\longrightarrow} & S^2T & \longrightarrow & 0 \\
 & & \downarrow s & & \downarrow s & & \downarrow & & \\
0 & \longrightarrow & L & \cong & L & \longrightarrow & 0 & \longrightarrow & 0
\end{array} \qquad (**)$$

where $\mathcal{D}^2(L)$ is the sheaf of second-order differential operators on L, σ is the symbol map and the vertical arrows consist of evaluation on the section s. The complex C^p defined above actually gives the Dolbeault version of the hypercohomology of the complex sheaves $\mathcal{D}^1(L) \xrightarrow{s} L$. In the exact cohomology sequence of (**), there is a coboundary map

$$\delta: H^0(S^2T) \to \mathbb{H}^1_s(\mathcal{D}^1(L))$$

so that we can obtain a class in the required cohomology group for each holomorphic *symmetry tensor* on M.

In fact, the spaces of flat connections we are considering have many such tensors. For this we have to introduce on M a Kähler polarization for each complex structure on the surface Σ. The theorem of Narasimhan and Seshadri [5] then describes the complex structure of m, the space of flat $U(n)$ connections — it is the moduli space of *stable holomorphic bundles* of rank n on the Riemann surface Σ.

The tangent space at a point represented by a holomorphic bundle E is the sheaf cohomology group $H^1(\Sigma; \text{End } E)$ and by Serre duality the cotangent bundle is $H^0(\Sigma; \text{End } E \oplus K)$, with K the canonical bundle of Σ. The cup product map

$$H^1(\Sigma; K^{-1}) \oplus H^0(\Sigma; \text{End } E \oplus K) \to H^1(\Sigma; \text{End } E) \qquad (***)$$

then defines for each vector in the $(3g-3)$-dimensional vector space $H^1(\Sigma; K^{-1})$ a global holomorphic section G of S^2T on M and hence a class $\delta(G)$ in $\mathbb{H}^1_s(\mathcal{D}^1(L))$.

This effectively defines a connection over the space of polarizations of M parametrized by the space of equivalence classes of complex structures on Σ — namely *Teichmüller space*. The minor (but universally occurring) feature in all of this is the fact that each symmetric tensor G arises in $(***)$ from a vector in $H^1(\Sigma; K^{-1})$ which is naturally the tangent space of Teichmüller space, but also from the exact sequence

$$0 \longrightarrow 0 \longrightarrow \mathcal{D}^1(L) \xrightarrow{\sigma} T \longrightarrow 0$$

the class $\delta(G)$ defines a cohomology class $\sigma\delta(G)$ in $H^1(T)$. This is the tangent space of Teichmüller space. The composition of these two maps is not however the identity but is the factor $1/2(k+l)$ where k is essentially the degree of L and l is a universal invariant of the Lie group.

The above procedure defines naturally a (holomorphic) connection. To see that it is flat, one spells out the cohomology and coboundary maps in explicit terms using a Cech covering $\{U_\alpha\}$ of M. Given the symmetric tensor G, we choose on each open set U_α a second-order differential operator Δ_α with symbol $(1/2(k+l))G$. On $U_\alpha \cap U_\beta$, $\Delta_\alpha - \Delta_\beta$ is first-order since $G_\alpha = G_\beta$ and this defines a class in $H^1(\mathcal{D}^1(L))$. On the other hand, if t is a deformation parameter, the Kodaira–Spencer class of the deformation has a similar representative. The vector field $\frac{\partial}{\partial t}\big|_\alpha - \frac{\partial}{\partial t}\big|_\beta$ is tangent to M and represents a class in $H^1(T)$ which lifts to one in $H^1(\mathcal{D}^1(L))$. Identifying this class from these two points of view gives a globally defined *heat operator* $\frac{\partial}{\partial t} - \Delta$. Covariant constant sections of the connection are then solutions to the heat equations

$$\frac{\partial s}{\partial t_A} - \Delta_A s = 0$$

where $A = 1, \ldots, 3g - 3$.

This point of view leads to the flatness of the connection, for

$$\left| \frac{\partial}{\partial t_A} - \Delta_A, \frac{\partial}{\partial t_B} - \Delta_B \right|$$

is a globally defined holomorphic differential operator on L. Considering it locally, it can be written as

$$-\frac{\partial \Delta_B}{\partial t_A} + \frac{\partial \Delta_A}{\partial t_B} + [\Delta_A, \Delta_B].$$

However, as shown in [3], the symbols of Δ_A and Δ_B *Poisson-commute* when considered as functions on the cotangent bundle of M. This means that $[\Delta_A, \Delta_B]$ is, like the two derivative terms, a *second*-order differential operator. When, as happened in our situation, the map $H^0(S^2T) \to H^1(\mathcal{D}^1(L))$ is injective, then the cohomology sequence of $(**)$ tells us that every second-order operator on L is first-order. On the other hand the hypothesis of the proposition, that no vector field preserves L, tells us that it must be zero-order and by compactness of M a constant.

The connection is thus *projectively flat* as required.

(The necessary hypothesis is satisfied for the Jacobian and automatically satisfied for non-abelian moduli spaces which have no global holomorphic vector fields [3].)

The details of the above outline of the connection may be found in [4]. The appearance of the heat equation in the context of symplectic quotients of affine spaces is treated by Axelrod, Della Pietra and Witten [2] where a direct computation of the curvature appears.

REFERENCES

[1] M.F. Atiyah and R. Bott, The Yang-Mills equation over Riemann surfaces, *Phil. Trans. R. Soc. Lond. A* **308** (1982), 523–615.

[2] S. Axelrod, S. Della Pietra and E. Witten, Geometric quantization of Chern-Simons gauge theory, preprint *IASSNS-HEP-89/57.*

[3] N.J. Hitchin, Stable bundles and integrable systems, *Duke Math. Journal* **54** (1987), 91–114.

[4] N.J. Hitchin, Flat connections and geometric quantization, preprint, Oxford, (1989).

Evaluations of the 3-Manifold Invariants of Witten and Reshetikhin–Turaev for $sl(2, \mathbf{C})$

ROBION KIRBY AND PAUL MELVIN

In 1988 Witten [**W**] defined new invariants of oriented 3-manifolds using the Chern–Simons action and path integrals. Shortly thereafter, Reshetikhin and Turaev [**RT1**] [**RT2**] defined closely related invariants using representations of certain Hopf algebras $\mathcal{A}$ associated to the Lie algebra $sl(2, \mathbf{C})$ and an rth root of unity, $q = e^{2\pi i m/r}$. We briefly describe here a variant τ_r of the Reshetikhin–Turaev version for $q = e^{2\pi i/r}$, giving a cabling formula, a symmetry principle, and evaluations at $r = 3$, 4 and 6; details will appear elsewhere.

Fix an integer $r > 1$. The 3-manifold invariant τ_r assigns a complex number $\tau_r(M)$ to each oriented, closed, connected 3-manifold M and satisfies:

(1) (multiplicativity) $\tau_r(M \# N) = \tau_r(M) \cdot \tau_r(N)$
(2) (orientation) $\tau_r(-M) = \overline{\tau_r(M)}$
(3) (normalization) $\tau_r(S^3) = 1$

$\tau_r(M)$ is defined as a weighted average of colored, framed link invariants $J_{L,\mathbf{k}}$ (defined in [**RT1**]) of a framed link L for M, where a coloring of L is an assignment of integers k_i, $0 < k_i < r$, to the components L_i of L. The k_i denote representations of $\mathcal{A}$ of dimension k_i, and $J_{L,\mathbf{k}}$ is a generalization of the Jones polynomial of L at q.

We adopt the notation $e(a) = e^{2\pi i a}$, $s = e(\frac{1}{2r})$, $t = e(\frac{1}{4r})$, (so that $q = s^2 = t^4$), and

$$[k] = \frac{s^{k_i} - \bar{s}^{k_i}}{s - \bar{s}} = \frac{\sin \frac{\pi k}{r}}{\sin \frac{\pi}{r}} .$$

DEFINITION: Let

$$\tau_r(M) = \alpha_L \sum_{\mathbf{k}} [\mathbf{k}] J_{L,\mathbf{k}} \tag{4}$$

where α_L is a constant that depends only on r, the number n of components of L, and the signature σ of the linking matrix of L, namely

$$\alpha_L = b^n c^\sigma \stackrel{\text{def}}{=} \left(\sqrt{\frac{2}{r}} \sin \frac{\pi}{r}\right)^n \left(e\left(\frac{-3(r-2)}{8r}\right)\right)^\sigma \tag{5}$$

and

$$[\mathbf{k}] = \prod_{i=1}^{n} [k_i]. \tag{6}$$

The sum is over all colorings $\mathbf{k}$ of L.

Remark: The invariant in [**RT2**] also contains the multiplicative factor c^ν where ν is the rank of $H_1(M;Z)$ (equivalently, the nullity of the linking matrix). If this factor is included, then (2) above does not hold, so for this reason and simplicity we prefer the definition in (4).

Recall that every closed, oriented, connected 3-manifold M can be described by surgery on a framed link L in S^3, denoted by M_L [**L1**] [**Wa**]. Adding 2-handles to the 4-ball along L produces an oriented 4-manifold W_L for which $\partial W_L = M_L$, and the intersection form (denoted by $x \cdot y$) on $H_2(W_L;Z)$ is the same as the linking matrix for L so that σ is the index of W_L. Also recall that if $M_L = M_{L'}$, then one can pass from L to L' by a sequence of K-moves [**K1**] [**F-R**] of the form

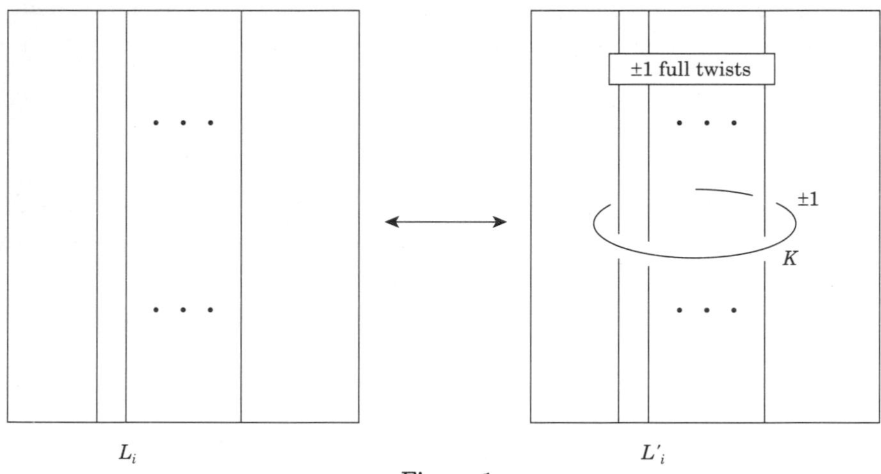

Figure 1

where $L'_i \cdot L'_i = L_i \cdot L_i + \left(L'_i \cdot K)\right)^2 K \cdot K$.

The constants α_L and $[\mathbf{k}]$ in (4) are chosen so that $\tau_r(M)$ does not depend on the choice of L, i.e. $\tau_r(M)$ does not change under K-moves. In fact, one defines $J_{L,\mathbf{k}}$ (below), postulates an invariant of the form of (4), and then uses the K-move for one strand only to solve uniquely for α_L and $[\mathbf{k}]$. It is then a theorem [**RT2**] that $\tau_r(M)$ is invariant under many stranded K-moves.

To describe $J_{L,\mathbf{k}}$, begin by orienting L and projecting L onto the plane so that for each component L_i, the sum of the self-crossings is equal to the framing $L_i \cdot L_i$.

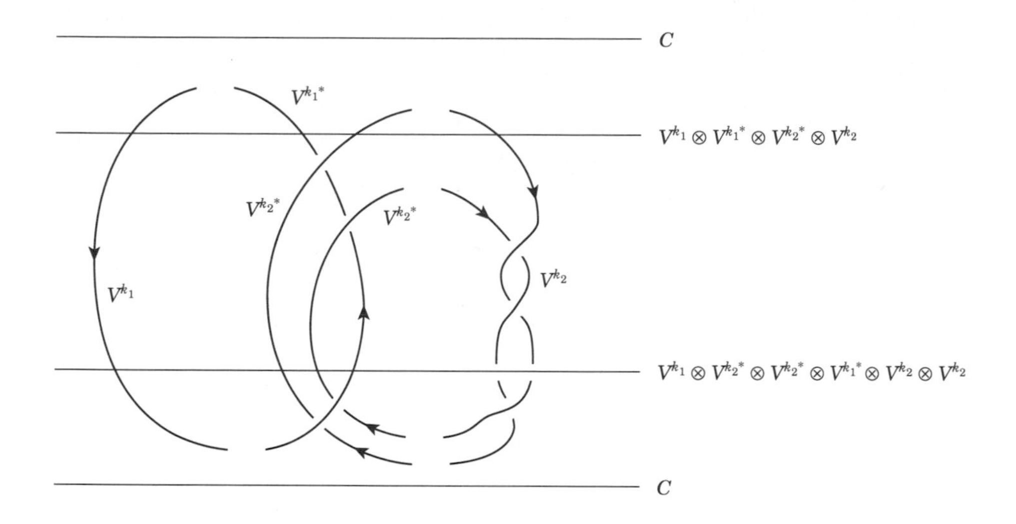

Figure 2

Removing the maxima and minima, assign a vector space V^{k_i} to each downward oriented arc of L_i, and its dual $V^{k_i^*}$ to each upward oriented arc as in Figure 2.

Each horizontal line λ which misses crossings and extrema hits L in a collection of points labeled by the V^{k_i} and their duals, so we associate to λ the tensor products of the vector spaces in order. To each extreme point and to each crossing, we assign an operator from the vector space just below to the vector space just above. The composition is a (scalar) operator from $\mathbf{C}$ to $\mathbf{C}$, and the scalar is $J_{L,\mathbf{k}}$. The vector spaces and operators are provided by representations of $\mathcal{A}$.

To motivate $\mathcal{A}$, recall that the universal enveloping algebra U of $sl(2,\mathbf{C})$ is a 3-dimensional complex vector space with preferred basis X, Y, H and a multiplication with relations $HX - XH = 2X$, $HY - YH = -2Y$ and $XY - YX = H$. To quantize, U, consider the algebra U_h of formal power series in a variable h with coefficients in U, with the same relations as above except that $XY - YX = \frac{\sinh \frac{hH}{2}}{\sinh \frac{h}{2}} = H + \frac{H^3 - H}{24}h^2 + \dots$. Setting $q = e^h$, and then by analogy with the above notation, $s = e^{h/2}$, $t = e^{h/4}$, $\bar{s} = e^{-h/2}$, and $[H] = \frac{s^H - \bar{s}^H}{s - \bar{s}}$, the relations can be written $HX = X(H+2)$, $HY = Y(H-2)$, $XY - YX = [H]$. It is convenient to introduce the element $K = t^H = e^{\frac{hH}{4}}$ and $\bar{K} = \bar{t}^h$. Note that $\bar{K} = K^{-1}$, $KX = sXK$,

$KY = \bar{s}YK$ and $XY - YX = \frac{K^2 - \bar{K}^2}{s - \bar{s}} = [H]$.

We want to specialize U_h at $h = \frac{2\pi i}{r}$ (so $q = e^{2\pi i/r}$) and look for complex representations, but there are difficulties with divergent power series. It seems easiest to truncate, and define $\mathcal{A}$ to be the finite dimensional algebra over $\mathbf{C}$ generated by X, Y, K, $\bar{K}$ with the above relations

$$
\begin{aligned}
K\bar{K} &= 1 = \bar{K}K \\
KX &= sXK \\
KY &= \bar{s}YK \\
XY - YX &= [H] = \frac{K^2 - \bar{K}^2}{s - \bar{s}}
\end{aligned}
\tag{7}
$$

as well as

$$
\begin{aligned}
X^r &= Y^r = 0 \\
K^{4r} &= 1.
\end{aligned}
$$

$\mathcal{A}$ is a complex Hopf algebra with comultiplication Δ, antipode S and counit ε given by

$$
\begin{aligned}
\Delta X &= X \otimes K + \bar{K} \otimes X \\
\Delta Y &= Y \otimes K + \bar{K} \otimes Y \\
\Delta K &= K \otimes K \qquad (\Delta H = H \otimes 1 + 1 \otimes H) \\
SX &= -sX \\
SY &= -\bar{s}Y \\
SK &= \bar{K} \qquad (SH = -H) \\
\varepsilon(X) &= \varepsilon(Y) = 0 \\
\varepsilon(K) &= 1.
\end{aligned}
\tag{8}
$$

There are representations V^k of $\mathcal{A}$ in each dimension $k > 0$ given by

$$
\begin{aligned}
Xe_j &= [m + j + 1]e_{j+1} \\
Ye_j &= [m - j + 1]e_{j-1} \\
Ke_j &= s^j e_j
\end{aligned}
\tag{9}
$$

where V^k has basis $e_m, e_{m-1}, \ldots, e_{-m}$ for $m = \frac{k-1}{2}$. The relations in $\mathcal{A}$ are easily verified using the identity $[a][b] - [a+1][b-1] = [a - b + 1]$. For example, the 2 and

3 dimensional representations are

$$X = \begin{pmatrix} 0 & 1 \\ 0 & 0 \end{pmatrix}, \quad Y = \begin{pmatrix} 0 & 0 \\ 1 & 0 \end{pmatrix} \quad \text{and} \quad K = \begin{pmatrix} t & 0 \\ 0 & \bar{t} \end{pmatrix}$$

$$X = \begin{pmatrix} 0 & [2] & 0 \\ 0 & 0 & [1] \\ 0 & 0 & 0 \end{pmatrix}, \quad Y = \begin{pmatrix} 0 & 0 & 0 \\ [1] & 0 & 0 \\ 0 & [2] & 0 \end{pmatrix} \quad \text{and} \quad K = \begin{pmatrix} s & 0 & 0 \\ 0 & 0 & 0 \\ 0 & 0 & \bar{s} \end{pmatrix}$$

respectively.

It is useful to represent V^k by a graph in the plane with one vertex at height j for each basis vector e_j, and with oriented edges from e_j to $e_{j\pm 1}$ labeled by $[m \pm j + 1]$ if $[m \pm j + 1] \neq [r] = 0$, indicating the actions of X and Y on V^k. Figure 3 gives some examples, using the identities $[j] = [r - j] = -[r + j]$.

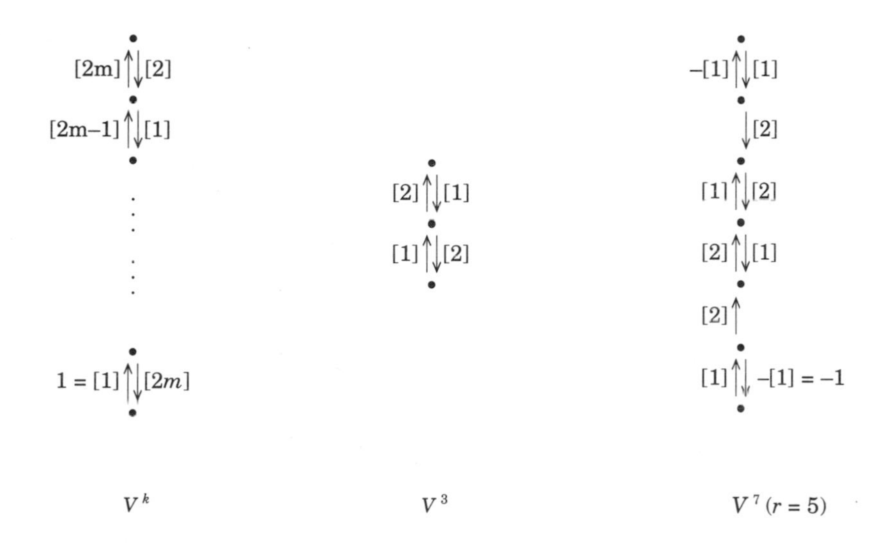

Figure 3

The Hopf algebra structure on $\mathcal{A}$ allows one to define $\mathcal{A}$-module structures on the duals $V^* = \mathrm{Hom}_{\mathbf{C}}(V, \mathbf{C})$ and tensor products $V \otimes W = V \otimes_{\mathbf{C}} W$ of $\mathcal{A}$-modules V and W. In particular, $(Af)(v) = f(S(A)v)$ and $A(v \otimes w) = \Delta A \cdot (v \otimes w)$ for $A \in \mathcal{A}$, $f \in V^*$, $v \in V$, $w \in W$. Thus the vector spaces in Figure 2 will be $\mathcal{A}$-modules and the operators will be $\mathcal{A}$-linear.

The structure of $\mathcal{A}$-modules for $k \leq r$ and their tensor products $V^i \otimes V^j$ for $i+j-1 \leq r$ is parallel to the classical case and is well known:

(10) THEOREM [**RT2**]. *If $k \leq r$, then the representations V^k are irreducible and self dual. If $i+j-1 \leq r$, then $V^i \otimes V^j = \bigoplus_k V^k$ where k ranges by twos over $\{i+j-1,\ i+j-3,\ i+j-5, \ldots, |i-j|+1\}$.*

(11) COROLLARY. *If $k < r$, then*

$$V^k = \sum_j (-1)^j \binom{k-1-j}{j} (V^2)^{\otimes k-1-2j},$$

where the sum is over all $0 \leq j < \frac{k}{2}$.

Here we have written $U = V - W$ to mean $U \oplus W = V$, $jV = \underbrace{V \oplus \cdots \oplus V}_{j \text{ times}}$ and $V^{\otimes j} = \overbrace{V \otimes \cdots \otimes V}^{j \text{ times}}$. This corollary is the key to our later reduction from arbitrary colorings to 2-dimensional ones.

The Hopf algebra $\mathcal{A}$ has the additional structure of a quasi-triangular Hopf algebra [**D**], that is, there exists an invertible element R in $\mathcal{A} \otimes \mathcal{A}$ satisfying

$$\begin{aligned} R\Delta(A)R^{-1} &= \check{\Delta}(A) \quad \text{for all } A \text{ in } \mathcal{A} \\ (\Delta \otimes \mathrm{id})(R) &= R_{13}R_{23} \\ (\mathrm{id} \otimes \Delta)(R) &= R_{13}R_{12} \end{aligned} \tag{12}$$

where $\check{\Delta}(A) = P(\Delta(A))$ and $P(A \otimes B) = B \otimes A$, $R_{12} = R \otimes 1$, $R_{23} = 1 \otimes R$ and $R_{13} = (P \otimes \mathrm{id})(R_{23})$. R is called a *universal R-matrix*. Historically, R-matrices have been found for U_h, $\mathcal{A}$ and other Hopf algebras by Drinfeld [**D**], Jimbo [**J**], Reshetikhin and Turaev [**RT2**] and others. We look for an R of the form $R = \sum c_{nab} X^n K^a \otimes Y^n K^b$, and recursively derive the constants c_{nab} from the defining relation $R\Delta(A)R^{-1} = \check{\Delta}(A)$. This approach was suggested to us by A. Wasserman who had carried out a similar calculation.

(13) THEOREM. *A universal R-matrix for $\mathcal{A}$ is given by*

$$R = \frac{1}{4r} \sum_{n,a,b} \frac{(s-\bar{s})^n}{[n]!} \bar{t}^{\,ab+(b-a)n+n} X^n K^a \otimes Y^n K^b$$

where the sum is over all $0 \leq n < r$ and $0 \leq a,\, b < 4r$ and $[n]! = [n][n-1]\ldots[2][1]$.

(14) COROLLARY. *R acts in the module $V^k \otimes V^{k'}$ by*

$$Re_i \otimes e_j = \sum_n \frac{(s-\bar{s})^n}{[n]!} \cdot \frac{[m+i+n]!}{[m+i]!} \frac{[m'-j+n]!}{[m'-j]!} t^{4ij-2n(i-j)-n(n+1)} e_{i+n} \otimes e_{j-n}$$

where $k = 2m+1$, $k' = 2m'+1$, *and* $\frac{[p]!}{[n]!} = [p][p-1]\dots[n+1]$.

EXAMPLES: In $V^2 \otimes V^2$, the R-matrix is

$$(t) \oplus \begin{pmatrix} \bar{t} & \bar{t}(s-\bar{s}) \\ 0 & \bar{t} \end{pmatrix} \oplus (t)$$

with respect to the basis $e_{1/2} \otimes e_{1/2}$, $e_{1/2} \otimes e_{-1/2}$, $e_{-1/2} \otimes e_{1/2}$, and $e_{-1/2} \otimes e_{-1/2}$, and

$$\check{R} = PR = (t) \oplus \begin{pmatrix} 0 & \bar{t} \\ \bar{t} & \bar{t}(s-\bar{s}) \end{pmatrix} \oplus (t). \tag{15}$$

In $V^3 \otimes V^3$,

$$\check{R} = (q) \oplus \begin{pmatrix} 0 & 1 \\ 1 & q-\bar{q} \end{pmatrix} \oplus \begin{pmatrix} 0 & 0 & \bar{q} \\ 0 & 1 & 1-\bar{q} \\ \bar{q} & (q-\bar{q})(1+\bar{q}) & (q-\bar{q})(1-\bar{q}) \end{pmatrix} \oplus \begin{pmatrix} 0 & 1 \\ 1 & q-\bar{q} \end{pmatrix} \oplus (q).$$

It is now possible to assign operators to the following elementary colored tangles [**RT1**]:

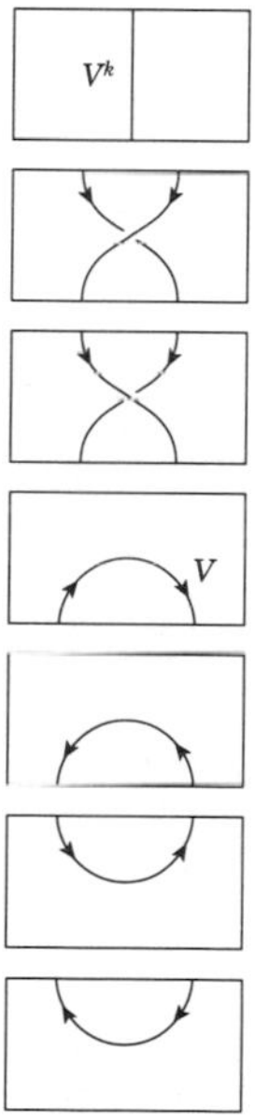

V^k $\rightarrow$ id

$\rightarrow \check{R} = PR$

$\rightarrow \check{R}^{-1} = R^{-1}P$

V $\rightarrow E$ where $E(f \otimes x) = f(x)$, $f \in V^*$, $x \in V$

$\rightarrow E_{K^2}$ where $E_{K^2}(x \otimes f) = f(K^2 x)$

$\rightarrow N$ where $N(1) = \Sigma_i\, e_i \otimes e^i$

$\rightarrow N_{\bar{K}^2}$ where $N_{\bar{K}^2}(1) = \Sigma_i\, e^i \otimes \bar{K}^2 e_i$

Figure 4

From these elementary tangle operators, define [**RT1**] $\mathcal{A}$-linear operators $J_{T,\mathbf{k}}$ for arbitrary oriented, colored, framed tangles $T, \mathbf{k}$. If T is a link L, then we obtain the scalar $J_{L,\mathbf{k}}$. The invariance of $J_{L,\mathbf{k}}$ under Reidemeister moves on L is well known; the Yang–Baxter equation $(\mathrm{id}\otimes\check{R})(\check{R}\otimes\mathrm{id})(\mathrm{id}\otimes\check{R}) = (\check{R}\otimes\mathrm{id})(\mathrm{id}\otimes\check{R})(\check{R}\otimes\mathrm{id})$ is the key ingredient, and it follows easily from the defining properties (12) for R. Note that $J_{L,\mathbf{k}}$ is independent of choice of orientation of L.

EXAMPLES: The following examples are easily derived from the R-matrix and the irreducibility of V^k:

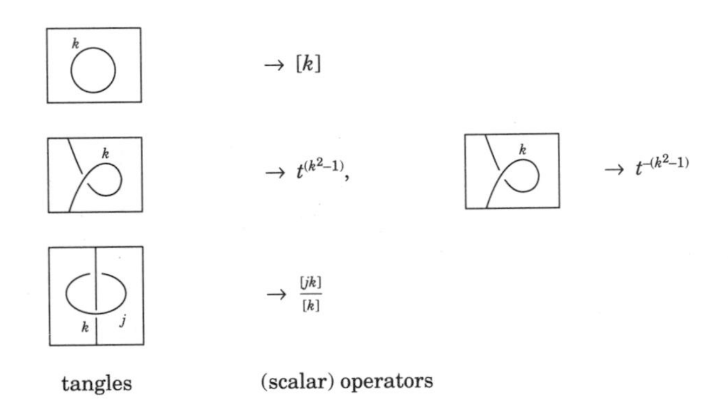

Figure 5

With this definition of $J_{L,\mathbf{k}}$, we have completed the definition of $\tau_r(M_L)$. The examples in Figure 5 can be used to check the one-strand K-move, or conversely, they may be used to solve for the coefficients of $J_{L,\mathbf{k}}$ in the definition of $\tau_r(M_L)$.

When all components of L are colored by the 2-dimensional representation, then $J_{L,\mathbf{2}} \overset{\text{def}}{=} J_L$ is just a variant of the Jones polynomial. First note that from (15) $\check{R}$ on $V^2 \otimes V^2$ satisfies the characteristic polynomial

$$t\check{R} - \bar{t}\check{R}^{-1} = (s - \bar{s})I.$$

Then, adjusting for framings, J_L satisfies the skein relation

$$qJ_{L_+} - \bar{q}J_{L_-} = (s - \bar{s})J_{L_0} \tag{16}$$

(see [**L2**] for background on skein theory). If $\tilde{V}_L = \tilde{V}(q)$ is the version of the Jones polynomial (for oriented links) satisfying this skein relation and $\tilde{V}_{\text{unknot}} = 1$, then it follows that

$$J_L = J_{L,\mathbf{2}} = [2]t^{3L\cdot L}\tilde{V}_L. \tag{17}$$

Remark: the relation between $J_{L,2}$ and $\tilde{V}_L$ is important because the values of $\tilde{V}_L$ at certain roots of unity have a topological description, as they do for the usual Jones polynomial V_L **[LM1]**, **[Lip]**, **[Mur]**. In particular, the values at $q = e(1/r)$, for $r = 1$, 2, 3, 4 and 6, are as follows:

(18)

r	$\tilde{V}_L$	V_L
1	2^{n-1}	$(-2)^{n-1}$
2	$\det L$	$\det L$
3	1	$(-1)^{n-1}$
4	$a\sqrt{2}^{n-1}$	$a(-\sqrt{2})^{n-1}$
6	$\sqrt{3}^d(-i)^\omega$	$(-\sqrt{3})^d(-i)^\omega$

where n is the number of components of L, $\det L$ is the value at -1 of the (normalized) Alexander polynomial of L, a is $(-1)^{\mathrm{Arf}(L)}$ when L is proper (so the Arf invariant is defined) and 0 otherwise, d is the nullity of $Q(\bmod 3)$ where Q is the quadratic form of L (represented by $S + S^t$ for any Seifert matrix S of L), and ω is the Witt class of $Q(\bmod 3)$ in $W(\mathbf{Z}/3\mathbf{Z}) = \mathbf{Z}/4\mathbf{Z}$. It is well known that $|\det L| = |H_1(M)|$, where M is the 2-fold branched cover of S^3 along L, and $d = \dim H_1(M;\mathbf{Z}/3\mathbf{Z})$ (since any matrix representing Q is a presentation matrix for $H_1(M)$).

PROPOSITION. *If S is a sublink of L obtained by removing some 1-colored components, then*

$$J_{L,\mathbf{k}} = J_{S,\mathbf{k}|S}. \tag{19}$$

Using (11) and (17) we obtain a formula for the general colored framed link invariant $J_{L,\mathbf{k}}$ in terms of J or $\tilde{V}$ for certain cables of L. In particular, a cabling $\mathbf{c}$ of a framed link L is the assignment of non-negative integers c_i to the L_i, and the associated cable of L, denoted $L^{\mathbf{c}}$, is obtained by replacing each L_i with c_i parallel pushoffs (using the framing).

(20) THEOREM. *Using multi-index notation,*

$$\begin{aligned} J_{L,\mathbf{k}} &= \sum_{\mathbf{j}} (-1)^{\mathbf{j}} \binom{\mathbf{k}-\mathbf{1}-\mathbf{j}}{\mathbf{j}} J_{L^{\mathbf{k}-\mathbf{1}-2\mathbf{j}}} \\ &= [2] \sum_{\mathbf{j}} (-1)^{\mathbf{j}} \binom{\mathbf{k}-\mathbf{1}-\mathbf{j}}{\mathbf{j}} t^{3L^{\mathbf{k}-\mathbf{1}-2\mathbf{j}} \cdot L^{\mathbf{k}-\mathbf{1}-2\mathbf{j}}} \tilde{V}_{L^{\mathbf{k}-\mathbf{1}-2\mathbf{j}}} \end{aligned}$$

for any orientation on $L^{\mathbf{k}-\mathbf{1}-2\mathbf{j}}$, where the sum is over all $\mathbf{0} \le \mathbf{j} < \frac{\mathbf{k}}{2}$.

EXAMPLES: If $L = K =$ framed knot, then

$$\begin{aligned} J_{K,3} &= J_{K^2} - 1 \\ J_{K,4} &= J_{K^3} - 2J_K \\ J_{K,5} &= J_{K^4} - 3J_{K^2} + 1. \end{aligned} \tag{21}$$

(22) THEOREM. *$\tau_r(M_L) = \alpha_L \sum_{\mathbf{c}} \langle \mathbf{c} \rangle J_{L^{\mathbf{c}}}$ where the sum is over all cables $\mathbf{c} = (c_1, \ldots, c_n)$, $0 \le c_i \le r-2$, α_L is as in (5), and $\langle \mathbf{c} \rangle = \sum_{\mathbf{j}} [\mathbf{c} + 2\mathbf{j} + 1](-1)^{\mathbf{j}} \binom{\mathbf{c}+\mathbf{j}}{\mathbf{j}}$ where the sum is over all $\mathbf{j} \ge 0$ with $\mathbf{c} + 2\mathbf{j} + 1 < r$.*

Remark: a formula like this motivated Lickorish [**L3**] to give an elementary and purely combinatorial derivation of essentially the same 3-manifold invariant as τ_r. The proof reduced to a combinatorial conjecture whose proof has been claimed by Koh and Smolinsky [KS]. This elegant approach is much shorter and simpler. However it may be less useful because the above algebra involving $\mathcal{A}$ organizes a great deal of combinatorial information.

For example, using (20) one can give a recursive formula for $J_{H_n,2} = J_{H_n}$ for the unoriented, n-component, 1-framed, right-handed Hopf link H_n:

$$\begin{aligned} J_{H_0} &= 1 \\ J_{H_1} &= t^3[2] \\ J_{H_n} &= t^{n^2+1}[2n] + \sum_{k=1}^{n/2} (-1)^{k-1} \binom{n-k-1}{k} J_{H_{n-2k}}. \end{aligned} \tag{23}$$

Using deeper properties of $\mathcal{A}$ [**RT2**], one obtains a closed formula:

$$J_{H_n} = t^2 \sum_{k=0}^{n-1} \binom{n-2}{k} [2(n-2k)] t^{(n-2k)^2}. \tag{24}$$

It is not clear how to derive such formulae in a combinatorial way from skein theory.

(25) SYMMETRY PRINCIPLE: Suppose we are given a framed link of $n+1$ components, $L \cup K$, $L = L_1 \cup \cdots \cup L_n$, with colors $\mathbf{l} = l_1 \cup \cdots \cup l_n$ on L and k on K. If we switch the color k to $r - k$, then

$$J_{L \cup K, \mathbf{l} \cup r-k} = \gamma J_{L \cup K, \mathbf{l} \cup k}$$

where $\gamma = i^{ra}(-1)^{\lambda + ka}$, a is the framing on K and $\lambda = \sum_{\text{even } l_i} K \cdot L_i \underset{\text{mod } 2}{\equiv} K \cdot (\mathbf{l} + \mathbf{1})L$.

Use of the Symmetry Principle enables one to cut the number of terms in $\tau_r(M_L)$ from the order of $(r-1)^n$ to $(\frac{r}{2})^n$. It also has interesting topological implications.

EXAMPLE: For $r = 5$ and $L = K$ with framing $a > 0$, then

$$\begin{aligned}
\tau_5(M_K) &= \sqrt{\frac{2}{5}} \sin\frac{\pi}{5} e\left(-\frac{9}{40}\right)\sum_{k=1}^{4}[k]J_{K,\mathbf{k}} \\
&= \alpha_K(1 + [2]J_k + [3]i^{5a}(-1)^{2a}J_K + i^{5a}(-1)^a) \\
&= \alpha_K(1 + i^a + ([2] + [3]i^a)J_K) \quad \text{for } a \text{ even} \\
&= \alpha_K(1 + i^a)(1 + [2]^2 t^{3a}\tilde{V}_K) \quad \text{since } [3] = [2] \\
&= 0 \quad \text{for } a \equiv 2 \mod 4.
\end{aligned} \tag{26}$$

For $a \not\equiv 2 \mod 4$, this shows that $\tilde{V}_k$ is an invariant of M_K.

Next we discuss the evaluations of $\tau_r(M)$ when $r = 3$, 4 and 6. Note that $\tau_2(M) = 1$.

For $r = 3$,

$$\tau_3(M) = \frac{1}{\sqrt{2}^n} c^\sigma \sum_{S<L} i^{S \cdot S} \tag{27}$$

where $M = M_L$, $c = e\left(-\frac{1}{8}\right) = \frac{1-i}{\sqrt{2}}$ and $<$ denotes sublink and we sum over all sublinks including the empty link ($\phi \cdot \phi = 0$). It is not hard to see how the formula follows from (4) since components with color 1 are dropped (19); it also follows from the cabling formula (22).

Evidently, Formula 27 depends only on the linking matrix A of L. It is not hard to give an independent proof of the well definedness of (27) by checking its invariance under blow ups and handle slides as in the calculus of framed links [**K1**]. This means that $\tau_3(M)$ is an invariant of the stable equivalence class of A (where stabilization means $A \oplus (\pm 1)$). It follows that $\tau_3(M)$ is a homotopy invariant determined by rank $H_1(M; Z)$ and the linking pairing on $\operatorname{Tor} H_1(M; Z)$, for these determine the stable equivalence class of A.

The cumbersome sum in (27) can be eliminated by using Brown's $Z/8Z$ invariant β associated with A. View A as giving a $Z/4Z$-valued quadratic form on a $Z/2Z$-vector space by reducing mod 4 along the diagonal (to get the form) and reducing mod 2 (to get the inner product on the vector space). A is stably equivalent to a diagonal matrix and then $\beta = n_1 - n_3 \mod 8$ where n_i is the number of diagonal entries congruent to i (mod 4). Observe that

$$\rho(M) = \sigma - \beta \ (\mathrm{mod}\, 8) \tag{28}$$

is an invariant of $M = M_L$.

(29) THEOREM. *If all μ-invariants of spin structures on M are congruent* (mod 4), *then*

$$\tau_3(M) = \sqrt{2}^{b_1(M)} c^{\rho(M)}$$

where $b_1(M) = \operatorname{rank} H_1(M; Z/2Z)$, $c = e\left(-\frac{1}{8}\right)$ *and* $\rho(M)$ *is as above. Otherwise,* $\tau_3(M) = 0$.

(30) COROLLARY. *If M is a $Z/2Z$-homology sphere, then*

$$\tau_3(M) = \pm c^{\mu(M)}$$

where $c = e\left(-\frac{1}{8}\right)$, $\mu(M) = \mu$*-invariant of M and the $\pm$ sign is chosen according to whether* $|H_1(M; Z)| \equiv \pm 1$ *or* ± 3 (mod 8).

(31) REMARK: $\tau_3(M)$ is not always determined by $H_1(M; Z)$ and the μ-invariants of M (although τ_4 is, see below). For example, if $M = L(4,1) \# L(8,1)$, then $\rho(\pm M) = \pm 2$ so $\tau_3(\pm M) = \pm 2i$, yet M and $-M$ have the same homology and μ-invariants.

(32) THEOREM. $\tau_4(M^3) = \sum_\Theta c^{\mu(M,\Theta)}$ *where* $c = e\left(-\frac{3}{16}\right)$, $\mu(M, \Theta)$ *is the* μ*-invariant of the spin structure Θ on M and the sum is over all spin structures on M.*

The keys to the proof are these: use the cabling formula (20) to drop 1-colored components, keep 2-colored components and double 3-colored components; the undoubled components turn out to be a characteristic sublink and hence to correspond to a spin structure; at $r = 4$, the Arf invariant (18) comes into play; finally,

$$\mu(M, \Theta) = \sigma - C \cdot C + 8 \operatorname{Arf}(C) \mod 16 \tag{33}$$

is a crucial congruence where C is a characteristic sublink corresponding to Θ.

The congruence (33) is well known in 4-manifold theory [**K2**], being a generalization of Rohlin's Theorem. It turns out, motivated by (32) above that we can give a purely combinatorial proof of (33) without reference to 4-manifolds.

At present, we have no general formula for $\tau_6(M)$ in terms of "classical" invariants of M, although it is plausible that one exists. Indeed, it is immediate from the Symmetry Principle (25) and the cabling formula (20) that $\tau_6(M_L)$ can be expressed in terms of Jones polynomials of cables of L with each component at most doubled. (If the linking number of each component L_i of L with $L - L_i$ is odd, then doubled components may also be eliminated.) Now, since the Jones polynomial of a link at $e\left(\frac{1}{6}\right)$ is determined by the quadratic form of the link (see 18) it would suffice to show that the quadratic forms of these cables are invariants of M.

In particular if M is obtained by surgery on a knot K with framing a, then it can be shown that

$$\tau_5(M) = \frac{(-i)^\sigma}{\sqrt{3}}(1 + 2t^{8a} + \frac{3}{2}(1 + (-1)^a)t^{3a}\tilde{V}_K)$$

where σ is 0 if $a = 0$, 1 if $a > 0$ and -1 if $a < 0$. It follows that $\tau_6(M)$ is determined by a and the Witt class of the quadratic form Q of K. Thus, for odd a, or $a = 0$, $\tau_6(M)$ is determined by $H_1(M; Z)$ (with its torsion linking form, needed to determine the sign of a when a is divisible by 3). For even a, one also needs to know $H_1(\tilde{M}; Z)$ with its torsion linking form (which determines the Witt class of Q) where $\tilde{M}$ is the canonical 2-fold cover of M.

We are especially grateful to N. Yu. Reshetikhin for his lectures and conversations on [**RT1**] and [**RT2**],. and to Vaughn Jones, Greg Kuperberg and Antony Wasserman for valuable insights into quantum groups.

References

[**D**] V. G. Drinfel'd, *Quantum groups*, Proc. Int. Cong. Math. 1986 (Amer. Math. Soc. 1987), 798–820.

[**FR**] R. Fenn and C. Rourke, *On Kirby's calculus of links*, Topology **18** (1979), 1–15.

[**J**] M. Jimbo, *A q-difference analogue of $U(\mathcal{G})$ and the Yang–Baxter equation*, Letters in Math. Phys. **10** (1985), 63–69.

[**K1**] R. C. Kirby, *A calculus for framed links in S^3*, Invent. Math. **45** (1978), 35–56.

[**K2**] ————, "The Topology of 4-Manifolds," Lect. Notes in Math., v. 1374, Springer, New York, 1989.

[**KM**] R. C. Kirby and P. M. Melvin, *Evaluations of new 3-manifold invariants*, Not. Amer. Math. Soc., **10** (1989), p. 491, Abstract 89T-57-254.

[**KS**] H. K. Ko and L. Smolinsky, *A combinatorial matrix in 3-manifold theory*, to appear Pacific J. Math..

[**L1**] W. B. R. Lickorish, *A representation of orientable, combinatorial 3-manifolds*, Ann. Math. **76** (1962), 531–540.

[**L2**] ————, *Polynomials for links*, Bull. London Math. Soc. **20** (1988), 558–588.

[**L3**] ————, *Invariants for 3-manifolds from the combinatorics of the Jones polynomial*, to appear Pacific J. Math.

[LM1] W. B. R. Lickorish and K. C. Millett, *Some evaluations of link polynomials*, Comment. Math. Helv. **61** (1986), 349–359.

[Lip] A. S. Lipson, *An evaluation of a link polynomial*, Math. Proc. Camb. Phil. Soc. **100** (1986), 361–364.

[Mur] H. Murakami, *A recursive calculation of the Arf invariant of a link*, J. Math. Soc. Japan **38** (1986), 335–338.

[RT1] N. Yu. Reshetikhin and V. G. Turaev, *Ribbon graphs and their invariants derived from quantum groups*, MSRI preprint (1989).

[RT2] ———————, *Invariants of 3-manifolds via link polynomials and quantum groups*, to appear Invent. Math.

[Wa] A. H. Wallace, *Modifications and cobounding manifolds*, Can. J. Math. **12** (1960), 503–528.

[W] Ed Witten, *Quantum field theory and the Jones polynomial*, Comm. Math. Phys. **121** (1989), 351–399.

Department of Mathematics
University of California
Berkeley, CA 94720

Department of Mathematics
Bryn Mawr College
Bryn Mawr, PA 19010

Representations of Braid Groups

Lecture by **M.F.ATIYAH**

The Mathematical Institute, Oxford

Notes by **S.K.DONALDSON**

The Mathematical Institute, Oxford

In this lecture we will review the theory of Hecke algebra representations of braid groups and invariants of links in 3-space, and then describe some of the results obtained recently by R.J. Lawrence in her Oxford D.Phil. Thesis [**3**].

(a)Braid group representations, Hecke algebras and link invariants. We begin by recalling the definition of the braid groups, and their significance for the theory of links. The braid group on n strands, B_n, can be defined as the fundamental group of the *configuration space* C_n of n distinct points in the plane. Thus C_n is the quotient:

$$C_n = \tilde{C}_n / S_n, \tag{1}$$

where $\tilde{C}_n = \{(x_1, \ldots, x_n) \in (\mathbf{R}^2)^n | x_i \neq x_j \text{ for } i \neq j\}$, and the symmetric group S_n acts on $\tilde{C}_n$ in the obvious way. Elements of the braid group ("braids ") can be described by their graphs in $\mathbf{R}^2 \times [0,1] \subset \mathbf{R}^3$, as in the diagram.
The action of the braid on its' endpoints defines a homomorphism from B_n to S_n, and this is just the homomorphism corresponding to the Galois covering (1).
For any braid β we can construct a link β' in the 3−manifold $S^1 \times \mathbf{R}^2$ by identifying the top and bottom slices in the graph. The link has an obvious monotonicity property: the projection from the link to S^1 has no critical points. It is easy to see that if α is another braid then $(\alpha\beta\alpha^{-1})'$ is isotopic to β', and then to show that this construction sets up a 1-1 correspondence

$$\bigcup_n \text{Conjugacy classes in } B_n \equiv \text{Isotopy classes of monotone links in } S^1 \times \mathbf{R}^2.$$

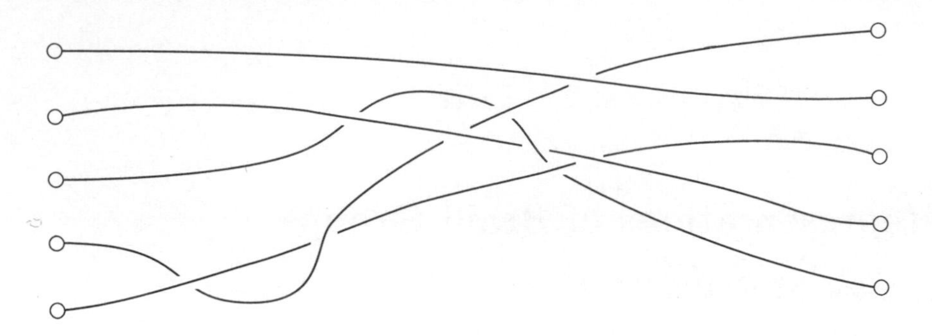

The graph of a braid

Thus an invariant of monotone links in $S^1 \times \mathbf{R}^2$ is just the same thing as a set of *class functions* on the braid groups. In particular, for any finite dimensional linear representation ρ of B_n we obtain an $S^1 \times \mathbf{R}^2$ link-invariant ρ' through the character

$$\rho'(\beta') = \mathrm{Tr}\ \rho(\beta).$$

It is also possible to associate a link $\hat{\beta}$ in S^3 to a braid β, using the standard embedding $S^1 \times \mathbf{R}^2 \subset \mathbf{R}^3$. All links in the 3- sphere are obtained in this way and the isotopy classes of links in S^3 can be regarded as obtained from the braids by imposing an equivalence relation generated by certain "Markov moves ". We get invariants of links in the 3-sphere from representations of the braid groups which are "consistent" with these moves.

The representations of the braid groups which we will discuss have the feature that they depend on a continuous parameter $q \in \mathbf{C}$. When $q = 1$ the representations are just those coming from the symmetric group and in general they factor through an intermediate object ; the *Hecke algebra* $H_n(q)$. The idea of using Hecke algebra representations to obtain link invariants is due to Vaughan Jones and we refer to the extremely readable Annals of Mathematics paper by Jones [**2**] for a beautiful account of the more algebraic approach to his theory. Here we will just summarise the basic facts and definitions. To define the algebra $H_n(q)$ we recall that there is a standard system of generators σ_i $i = 1, \dots, n-1$ for the braid group B_n which are lifts of transpositions in S_n, as pictured in the diagram.
The Hecke algebra $H_n(q)$ is the quotient of the group algebra $\mathbf{C}[B_n]$:

$$H_n(q) = \mathbf{C}[B_n] / < (\sigma_i - 1)(\sigma_i + q) = 0 \ ; \ i = 1, \dots, n-1 > .$$

In fact the braid group can be described concretely as the group generated by the σ_i subject to the relations

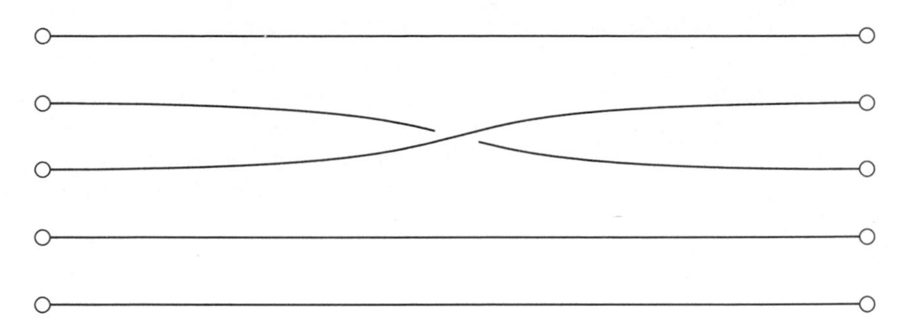

The braid σ_2

$$\sigma_{i+1}\sigma_i\sigma_{i+1} = \sigma_i\sigma_{i+1}\sigma_i$$
$$\sigma_i\sigma_j = \sigma_j\sigma_i \quad \text{for } |i-j| > 1.$$

So $H_n(q)$ is the algebra generated by the σ_i subject to these relations and the further conditions $(\sigma_i - 1)(\sigma_i + q) = 0$. If $q = 1$ we obtain the relations $\sigma_i^2 = 1$ satisfied by the transpositions in S_n, and it is easy to check that the algebra $H_n(1)$ is canonically isomorphic to the group algebra $\mathbf{C}[S_n]$. For any q, a representation of $H_n(q)$ gives a representation of the group algebra $\mathbf{C}[B_n]$ and hence a representation of B_n and from the discussion above we see that when $q = 1$ these are indeed just the representations which factor through the symmetric group.

These Hecke algebras arise in many different areas in mathematics, and the parameter q can play quite different roles. In algebra we take q to be a prime power and let $\mathbf{F}$ be the field with q elements ; then we obtain $H_n(q)$ as the double coset algebra of the group $G = SL(n, \mathbf{F})$ with respect to the subgroup B of upper-triangular matrices. This is the sub-algebra of $\mathbf{C}[G]$ generated by the elements

$$\tau_D = \sum_{x \in D} x$$

for $D \in B\backslash G/B$. On the other hand in physical applications one should think of q as e^{ih} where h is Planck's constant ; the limit $q \to 1$ then appears as the classical limit of a quantum mechanical situation.

There is an intimate relationship between the representations of the Hecke algebra for general q and the representations of the symmetric group, i.e. of $H_n(1)$. This relationship can be obtained abstractly using the fact that $H_n(1)$ – the group algebra of the symmetric group – is semi-simple. We then appeal to a general "rigidity" property: a small deformation of a semi-simple algebra does not change the isomorphism class of the algebra. Hence the $H_n(q)$ are isomorphic, as abstract algebras, to $\mathbf{C}[S_n]$ for all q sufficiently close to 1 – in fact this is true for all values of q except a finite set E of roots of unity. So for these generic values of q the representations of $H_n(q)$ can be identified with those of S_n and we obtain a family of representations $\rho_{q,\lambda}$ of B_n, with characters $\chi_{q,\lambda}$, indexed by the irreducible representations λ of S_n and a complex number $q \in \mathbf{C} \setminus E$. These are the representations whose characters are used to obtain the new link invariants. More precisely, the two-variable "HOMFLY" polynomial invariant of links can be obtained from a weighted sum of characters of the form:

$$X(q, z) = \sum_{\lambda} a_{\lambda}(q, z)\chi_{q,\lambda},$$

for certain rational functions a_λ of the variables q and z. The earlier 1-variable Jones polynomial $V(q)$ is obtained from $X(q, z)$ by setting $q = z$.

We will now recall some of the rudiments of the representation theory of symmetric groups, and the connection with the representations of unitary groups. The irreducible representations of S_n are labelled by "Young Diagrams " , or equivalently by partitions $n = p_1 + \cdots + p_r$, with $p_1 \geq \cdots \geq p_r > 0$. For example the trivial representation corresponds to the "1-row " diagram, or partition $n = n$, the 1– dimensional parity representation to the 1-column diagram, or partition $n = 1 + \cdots + 1$. Now let V be the standard l dimensional representation space of the unitary group $U(l)$. There are natural commuting actions of $U(l)$ and S_n on the tensor product $V^{\otimes n} = V \otimes \cdots \otimes V$, so there is a joint decomposition :

$$V^{\otimes n} = \bigoplus_{\lambda} A_{\lambda} \otimes B_{\lambda},$$

where A_λ is a representation space of $U(l)$ and B_λ is a represenation space of S_n. The index λ runs over the irreducible representation spaces of S_n, i.e. over the Young diagrams. So these Young diagrams also label certain representations A_λ of the unitary group. It is a fundamental result that the A_λ are zero except when the diagram has l rows or fewer (that is, for partitions with at most l terms), in particular the irreducible representations of $U(2)$ are labelled precisely by the 1 and 2-row diagrams.

The co-efficients $a_\lambda(q, z)$ have the property that they vanish when $q = z$ for all diagrams λ with more than 2 rows. Thus the 1-variable polynomial $V(q)$ uses only the representations of the Hecke algebra associated to the 1- and 2- row diagrams. These are the diagrams which label the representations of $U(2)$ and this ties in with the quantum-field theory approach of Witten, in which the V-polynomial is obtained in the framework of a gauge theory with structure group $U(2)$(or rather $SU(2)$). The more general X polynomial is obtained from gauge theory using structure groups $SU(l)$, for all different values of l.

(b) Geometric constructions of representations.
It is natural to ask for direct geometric constructions of these representations of the braid group. We now change our point of view slightly : the braid group B_n is the fundamental group of the configuration space C_n, so linear representations of B_n are equivalent to *flat vector bundles* over C_n. Thus we seek flat vector bundles whose monodromy yields the representations $\rho_{q,\lambda}$, and we pay particular attention to the 2- row diagrams which appear in the V-polynomial.
Constructions of these flat bundles are already known in the context of conformal field theory [**4**], using complex analysis , and these tie in well with Witten's quantum field theory interpretation of the Jones' invariants. In her thesis, Ruth Lawrence developed more elementary constructions which used only standard topological notions, specifically homology theory with twisted co-efficients. To describe her construction we begin with a fundamental example which yields the "Burau" representation and the Alexander polynomial of a link.

Let $X = \{x_1, \ldots, x_n\}$ be a point in the configuration space C_n. The complement $\mathbf{R}^2 \setminus X$ retracts on to a wedge of circles, so its' first homology is $\mathbf{Z}^n$ and there is an n-dimensional family of flat complex line bundles over the complement , i.e representations $\pi_1(\mathbf{R}^2 \setminus X) \to \mathbf{C}^*$. There is a preferred 1- dimensional family of representations ν_q, which send each of the standard generators of π_1 to the same complex number q. These are preserved by the action of the diffeomeorphism group of $\mathbf{R}^2 \setminus X$, and thus extend to families, as X varies in C_n. More precisely, let W_n be the space

$$W_n = \{(X, y) \in C_n \times \mathbf{R}^2 | \ y \in \mathbf{R}^2 \setminus X\}.$$

It is not hard to see that $H_1(W_n)$ is $\mathbf{Z}^2$, generated by a loop in which the point y encircles one of the points of X and a loop in which one of the points of X encircles another. We consider the 1-parameter family of representations $\tilde{\nu}_q : \pi_1(W_n) \to \mathbf{C}^*$ which map the first generator to q and the second generator to 1. These restrict to the representations ν_q over the punctured planes $\mathbf{R}^2 \setminus X$, regarded as the fibres of the natural map

$$p : W_n \to C_n.$$

We let $\mathcal{L}_q$ be the flat complex line bundle over W_n associated to the representation $\tilde{\nu}_q$.
Recall now the following general construction. If $f : E \to B$ is a fibration and $\mathcal{M}$ is a local-coefficient system over E then for fixed r and for each $b \in B$ we can obtain a vector space $V_b = H^r(f^{-1}(b); \mathcal{M})$. The spaces V_b fit together to define a vector bundle over B, and this bundle has a natural flat connection, since homology is a homotopy invariant. The monodromy of this connection then gives a representation of $\pi_1(B)$, the action on the cohomology of the fibres. We apply this in the situation above with the map p and the co-efficient system $\mathcal{L}_q$ (or, more

precisely, the sheaf of locally constant sections of the flat bundle $\mathcal{L}_q$) taking $r = 1$. This gives us a flat bundle over C_n, with fibres $H^1(\mathbf{R}^2 \setminus X; \mathcal{L}_q)$. (Note that we could consider a 2-parameter family of representations of $\pi_1(W_n)$, using the extra generator for $H^1(W_n)$, but this would give no great gain in generality, since it would just correspond to taking the tensor product with the 1-dimensional representations of the braid group.)

To identify the representation which is obtained in this way we begin by looking at the twisted cohomology of $\mathbf{R}^2 \setminus X$. We can replace this punctured plane by a wedge of n circles, and use the corresponding cellular cochains :

$$C^0 = \mathbf{Z} \quad , \quad C^1 = \mathbf{Z}^n,$$

with twisted co-boundary map $\delta : C^0 \to C^1$ given by $\delta = ((1-q), (1-q), \ldots, (1-q))$. For $q \neq 1$ the 1-dimensional twisted cohomology has dimension $n-1$. It is not hard to identify the $n-1$ dimensional representation of the braid group which this leads to. It is obtained from a representation on $\mathbf{C}^n$ by restricting to the subspace of vectors whose entries sum to 0. The standard generator σ_i of B_n acts on $\mathbf{C}^n$ by fixing all the basis vectors $e_1, \ldots e_n$ except for e_i, e_{i+1} and acting by the matrix

$$\begin{pmatrix} 1-q & q \\ 1 & 0 \end{pmatrix}$$

on the subspace spanned by e_i, e_{i+1}. The representation on the vectors whose entries sum to zero is the *reduced Burau representation*, and this is in fact the representation $\rho_{\lambda,q}$ obtained from the partition $n = (n-1)+1$. (There is some choice in sign convention in here : the automorphism $\sigma_i \mapsto -\sigma_i$ of $H_n(q)$ switches rows and columns in our labelling of representations by Young diagrams.) The representation is clearly a deformation of the reduced permutation representation of S_n, which is obtained by taking $q = 1$. The Alexander polynomial appears in the following way : if β is a braid and ψ_β is the matrix given by the Burau representation then

$$\det(1 - \psi_\beta(q)) = (1 + q + \cdots + q^{n-1})\Delta_{\hat{\beta}}(q),$$

where $\Delta_{\hat{\beta}}$ is the Alexander polynomial of the knot $\hat{\beta}$ in the 3-sphere.

Lawrence extends this idea to obtain other representations of the braid group. The extension involves iteration of the configuration space construction. Let $C_{n,m}$ be the space :

$$C_{n,m} = \{ \ (\{x_1, \ldots, x_n\}, \{y_1, \ldots, y_m\}) \in C_n \times C_m | x_i \neq y_j \text{ for any } \ i, j \ \}.$$

There is an obvious fibration $p_{n,m} : C_{n,m} \to C_n$. Notice that $C_{n,1}$ is just the space W_n which we considered before, and $p_{n,1} = p$. In general we will obtain representations of the braid group from the twisted cohomology of the fibres of $p_{n,m}$.

For $m > 1$ the group $H_1(C_{n,m})$ is $\mathbf{Z}^3$, with generators represented by loops in which

(1) one of the points y_j encircles one of the points x_i,
(2) one of the points y_j encircles another,
(3) one of the points of x_i encircles another.

For the same reason as before we may restrict attention to representations which are trivial on the third generator. Thus we consider a 2-parameter family of representations $\tilde{\nu}_{q,\alpha}$ of $\pi_1(C_{n,m})$, which map the first generator to α and the second to q. We let $\mathcal{L}_{q,\alpha}$ be the corresponding flat line bundle over $C_{n,m}$, then for each α and q we have a representation $\phi_{m,q,\alpha}$ of the braid group B_n on the middle cohomology of the fibre:

$$H^m(p_{n,m}^{-1}(X)\ ;\mathcal{L}_{q,\alpha}).$$

Lawrence proves that these yield all the representations corresponding to 2-row Young diagrams. More precisely, we have:

THEOREM,[3].

Suppose $2m \leq n$ and let λ be the representation of the symmetric group corresponding to the partition $n = (n - m) + m$ (a 2-row Young diagram). If $\alpha = q^{-2}$ the representation $\phi_{m,q,\alpha}$ of B_n contains with multiplicity 1 the irreducible Hecke algebra representation $\rho_{q,\lambda}$.

Remarks.

(1) This irreducible piece makes up "most " of the space. For example, when $m = 2$ the representation $\rho_{q,\lambda}$ has dimension $(1/2)n(n-3)$ and the remaining piece has dimension n.
(2) Lawrence also studies the representations $\phi_{m,q,\alpha}$ when α is not equal to q^{-2}. For generic α and q they are irreducible and she analyses the degeneration of the representation when $\alpha \to q^{-2}$.
(3) We have seen that the 2-row Young diagrams covered by the theorem arise naturally in the Jones theory as the representations associated to the group $SU(2)$ which appear in the one variable V-polynomial. It seems very probable that the other representations, corresponding to Young diagrams with more rows and to the groups $SU(l)$, will be obtained in the same fashion by considering configuration spaces of l disjoint sets of points.
(4) There are striking general similarities between this approach and an idea proposed by Hitchin for studying the moduli space of rank 2 holomorphic bundles over a Riemann surface, which comes to the fore in Witten's interpretation of the Jones invariants. (See the reports on the contributions of Witten and Hitchin in these Proceedings.) Hitchin's idea is to represent rank 2 bundles on a Riemann surface Σ as the direct images of line bundles on a fixed branched cover $\tilde{\Sigma} \to \Sigma$ (cf.[1]). The branched cover is specified by a configuration of points in Σ, and the general effect is to reduce

the non-Abelian theory on Σ to the Abelian theory (of line bundles) on the cover.

References

1. N.J.Hitchin, *Stable bundles and integrable systems*, Duke Mathematical Jour. **54** (1987), 91-114.
2. V.R.F.Jones, *Hecke algebra representations of braid groups and link polynomials*, Annals of Mathematics **126** (1987), 335-388.
3. R.J.Lawrence, "Homology Representations of Braid Groups," D.Phil. Thesis, Oxford, 1989.
4. A.Tsuchiya and Y.Kanie, *Vertex operators in conforma field theory on* $\mathbf{P}^1$ *and monodromy representations of braid groups*, Advanced Studies in Pure Maths. **16** (1988), 297-372.

PART 3

THREE-DIMENSIONAL MANIFOLDS

At present the study of 3-dimensional manifolds is dominated by Thurston's conjecture that a compact 3-maniold can be decomposed into a finite family of summands, which are either bounded by spheres (and can be closed), or bounded by tori, each summand admitting one of eight geometric structures. And although this conjecture is far from being proved —for example if the universal covering manifold $\tilde{M}^3$ is diffeomeorphic to S^3, is M^3 elliptic ? —it does seem to provide the framework in which to pose interesting questions. In his lecture H. Rubinstein asks to what extent the rigidity proved by Mostow for hyperbolic manifolds extends to other geometries. Starting with an irreducible manifold M^3 with infinite fundamental group, his method is to construct a non-positively curved metric with singularities, using a restricted type of polyhedral decomposition. This exists, for example, if M^3 is a covering of S^3 branched along the figure of eight knot with all degrees greater than or equal to 3. In this situation it is possible to prove that a homotopy equivalence $f : M \to M'$ (irreducible) can be replaced by a homeomorphism. Note that we only need to assume that one manifold M admits a suitable polyhedral metric, and that we do not restrict attention to Haken manifolds. This result shows that in some sense hyperbolic 3-manifolds are generic. Their automorphisms are the subject of the article by Thomas, in which two proofs are given of the result that an arbitrary finite group G can be realised as a subgroup of the (finite) isometry group $I(M)$ of some hyperbolic 3-manifold M of finite volume. The motivation of this paper is the famous result of A. Hurwitz that the order of the group of conformal automorphisms of a closed Riemann surface of genus $g \geq 2$ is bounded above by $84(g-1)$.

The remaining paper in this section, by Kirby and Taylor, studies *Pin* structures on low-dimensional manifolds. For example they show that, if Pin^+ denotes the usual central extension of $\mathbf{Z}/2$ by $O(n)$ classified by w_2, then $\Omega_4^{Pin^+}$ is cyclic of order 16 and generated by the class of $\mathbf{RP}^4$. Furthermore the fake projective space of Cappell and Shaneson represents ± 9, the Kummer surface ± 8, and these relations survive in the topological category. Since $\Omega_3^{Spin} = 0$, any oriented 3-manifold M^3 with a chosen spin structure σ bounds a spin manifold W^4 the signature of which, reduced modulo 16, gives the classical μ-invariant of the pair (M^3, σ). Kirby and Taylor now introduce a function $\beta : H_2(M; \mathbf{Z}/2) \to \mathbf{Z}/8$, related to the symmetric, trilinear, intersection map τ, which up to a factor of 2 detects the difference between the μ-invariants associated to different spin structures. This enables them to give an extension and geometric interpretation of Turaev's work on τ. They also consider the bordism of characteristic pairs (M^4, F^2), where F^2 is an embedded surface dual to $w_2 + w_1^2$, showing that the group $\Omega_4^!$ is isomorphic to $\mathbf{Z}/8 \oplus \mathbf{Z}/4 \oplus \mathbf{Z}/2$. And in the same way that the μ-invariant can be defined for pairs (M^3, σ), rather than just for $\mathbf{Z}/2$-homology spheres, they extend Robertello's description of the Arf invariant of a knot to a suitable class of links L embedded in M^3 with fixed structure σ. It is clear that this paper is once again the starting point for new developments — for example, does the definition of the Casson invariant extend in the same way as that of μ, and what classes in $\Omega_4^{Pin^+}$ are represented by the exotic projective spaces constructed by Fintushel and Stern ?

An Introduction to Polyhedral Metrics of Non-Positive Curvature on 3-Manifolds

I. R. AITCHISON

University of Melbourne

J. H. RUBINSTEIN

Institute for Advanced Study and
University of Melbourne

§0 INTRODUCTION

Polyhedral differential geometry has been an active area of resarch for a long time. In general relativity it is often called Regge calculus. In the 1960's, work was done by T. Banchoff and D. Stone. More recently, beautiful results have been obtained by J. Cheeger [9], M. Gromov [13] and M. Gromov and W. Thurston [14].

The Geometrization Programme of Thurston (see [33], [34] and the survey article of P. Scott [29]) seeks to classify all closed 3-manifolds by dividing them canonically into pieces, which admit locally symmetric Riemannian metrics called *geometries*. There are eight geometries S^3, $S^2 \times \mathbb{R}$, $\mathbb{R}^3$, Nil, Solv, $\widetilde{\mathrm{PSL}(2,\mathbb{R})}$, $\mathbb{H}^2 \times \mathbb{R}$ and $\mathbb{H}^3$.

Our aim is to introduce polyhedral metrics which are applicable to the geometries $\mathbb{R}^3$, $\mathbb{H}^2 \times \mathbb{R}$ and $\mathbb{H}^3$. In particular, these metrics have non-positive curvature, in the sense of polyhedral differential geometry. It is easy to show that only the three geometries indicated admit such metrics. However this is sufficient to describe generic 3-manifolds. We quickly review the basic strategy for classifying 3-manifolds.

By Kneser [25] and Milnor [28], there is essentially a canonical way of decomposing any closed 3-manifold into a finite connected sum, i.e. $M = M_1 \# M_2 \# \ldots \# M_k$. Each M_i is *prime*, i.e. if M is expressed as a connected sum, $M = P \# Q$, then either P or Q is a 3-sphere. It is elementary to show that if N is a prime 3-manifold, then either N is a 2-sphere bundle over S^1 or N is *irreducible*, meaning that every embedded 2-sphere in N bounds a 3-cell. From now on we will always assume that 3-manifolds under consideration are irreducible and orientable, to simplify the discussion.

Suppose V is a closed, embedded surface in M and V is not a 2-sphere or projective plane. Then V is called *incompressible* if the map $\pi_1(V) \to \pi_1(M)$ of fundamental groups induced by the embedding is one-to-one. By Dehn's lemma and the loop theorem [31], an orientable V is incompressible if and only if whenever D is a 2-disk embedded in M with $D \cap V = \partial D$ (the boundary of D), then ∂D is contractible in V. We will call a 3-manifold *Haken* if it contains such an imcompressible surface.

The *characteristic variety* of a Haken 3-manifold M comes from the work of W. Jaco and P. Shalen [23] and also K. Johannson [24]. A *Seifert fibered* 3-manifold N has a foliation by circles. Each leaf has a foliated neighborhood which is a solid torus $D \times S^1$. The foliation can be described by gluing the ends together of $D \times [0,1]$, where $(z,0)$ is identified with $(\exp(2\pi iq/p)z, 1)$. Note that here p, q are relatively prime positive integers and D is viewed as the unit disk in the complex plane. The leaves are the unions of finitely many arcs of the form $\{z\} \times [0,1]$. Suppose that N is a compact orientable Seifert fibered 3-manifold and ∂N is a non-empty collection of tori. If N is embedded in M, we say that N is incompressible if every torus of ∂N is incompressible in M.

The *characteristic variety Theorem* ([23], [24]) can now be summarized. If M is Haken, then either M is a Seifert fiber space or M has a maximal incompressible Seifert fibered submanifold N. N is unique up to isotopy and every map $f\colon N' \to M$ where N' is Seifert fibered and $f_*\colon \pi_1(N') \to \pi_1(M)$ is one-to-one, can be homotoped to have image in N. N is called the characteristic variety of M.

If M is a Seifert fiber space, then we can call M itself the characteristic variety. Thurston's uniformization theorem [33], [34], [35] then shows that if M is a Haken 3-manifold with empty characteristic variety then M has a metric of constant negative curvature, i.e. is hyperbolic. Also if M is Haken, not Seifert fibered and has a non-empty characteristic variety N, then $M - N$ has a complete metric which is hyperbolic. Also N admits a metric of type $\mathbf{H}^2 \times \mathbf{R}$, or $\mathbf{R}^3$, i.e. of non-positive curvature. R. Schoen and P. Shalen (unpublished) have shown that if N is of type $\mathbf{H}^2 \times \mathbf{R}$ and $M - N$ has finite volume then M admits a Riemannian metric of non-positive curvature.

Finally the Geometrization Programme conjectures that if M is irreducible, has infinite fundamental group and is neither Haken nor Seifert fibered, then M always should admit a hyperbolic metric. So metrics of non-positive curvature should occur on most 3-manifolds which are irreducible and have infinite fundamental group.

Our approach is to describe polyhedral metrics with non-positive curvatures on several large classes of 3-manifolds. Our constructions include various branched coverings over knots and links, Heegaard splittings, groups generated by reflections (Coxeter groups), surgery on knots and links and singular incompressible surfaces. The last method seems to have special significance. We are able to derive a strong *topological rigidity* result. Assume a 3-manifold M admits a polyhedral metric of non-positive curvature coming from decomposing M into regular Euclidean cubes. If M' is irreducible and there is a homotopy equivalence between M and M' then M is homeomorphic to M'.

Note that Mostow rigidity shows that if M and M' are *both* hyperbolic and homotopy equivalent then M and M' are isometric. Here we only need to assume one of the manifolds has a special polyhedral metric of non-positive curvature. Notice also that from a purely topological point of view, F. Waldhausen [39] proved that if M is Haken, M' is irreducible and M is homotopy equivalent to M' then M and M' are homeomorphic. In our case we do not suppose M or M' is Haken. Since A. Hatcher [19] has shown most surgeries on links yield non-Haken 3-manifolds, it is most useful to drop the assumption of existence of embedded incompressible surfaces.

To deal with surgery on knots and links, we discuss also polyhedral metrics of non-positive curvature on knot and link complements. One example is given in detail – a two component link in $\mathbf{R}P^3$ (real projective 3-space) which is obtained by identifying the faces of a *single* regular ideal cube in $\mathbf{H}^3$. We show how this link lifts to a simple four component link in S^3. Also by branched coverings we get an infinite collection of examples formed from gluing cubes with some ideal vertices. We sketch the interesting result that nearly all surgeries on all components of such links give 3-manifolds which satisfy the conclusions of the topological rigidity theorem.

In this paper, our aim is to give an overview of this subject. Arguments are only summarized. For more details, the reader is referred to [1], [2], [3]. In the final section, we give some extremely optimistic conjectures. It would certainly be possible to study polyhedral metrics adapted to the other five geometries. Especially in view of the work already done by R. Hamilton [15], [16] and P. Scott [30], this does not appear to be quite as significant as the case of non-positive curvature.

We would like to thank J. Hass, C. Hodgson, P. Scott and G. Swarup for many helpful conversations.

§1 POLYHEDRAL METRICS OF NON-POSITIVE CURVATURE

We begin by discussing examples of such metrics in dimension two, i.e. on surfaces.

The boundary of the regular Euclidean cube defines a polyhedral metric on the 2-sphere. At each vertex, the three squares give total dihedral angle of $3\pi/2$. If we attribute a positive curvature of $\pi/2$ at each vertex, then the total curvature is 4π, as given by Gauss-Bonnet. Similarly if a flat torus is formed as usual by identifying opposite edges of a regular square, then the dihedral angle at the vertex is 2π and so the vertex has zero curvature. Finally if an octagon has edges identified by the word $aba^{-1}b^{-1}cdc^{-1}d^{-1}$, in cyclic order, the result is a closed orientable surface of genus two. We find it very useful to ascribe dihedral angles of $\pi/2$ at each vertex of the octagon. There is a natural way to do this, so that the metric is Euclidean except at the center of the octagon. Join the midpoint of each edge to the center of the octagon. This divides the polygon into eight quadrilaterals. We view each of the latter as a regular Euclidean square. The result is the dihedral angle at the center is 4π, so there is curvature of -2π there. Similarly after identification, the single vertex has dihedral angle 4π and curvature -2π as well. Again the total curvature agrees with Gauss-Bonnet.

Note that a key feature of such metrics is that geodesics diverge *at least linearly.* If a line segment meets a point p where the dihedral angle is $2\pi + k$, for $k > 0$, then it is easy to see that all possible geodesic extensions form a cone subtending an angle k at p. So if two points move out along geodesic rays emanating from a single point, then the points travel away from each other at least at linear speed. In fact, working in the universal cover of the octagon surface described above, we see that such geodesic rays actually diverge exponentially. (See also M. Gromov [13].)

Definition. A polyhedral metric of non-positive curvature on a closed orientable surface is a metric which is locally Euclidean except at a finite number of points, where the dihedral angle is greater than 2π.

Suppose a closed orientable n-dimensional manifold M is formed by gluing together the codimension one faces of a finite collection of compact Euclidean polyhedra $\Sigma_1, \Sigma_2, \dots, \Sigma_t$. Assume that the face identifications are achieved by Euclidean isometries. If Q is an r-dimensional face of Σ_i, let B^{n-r} be a small ball of dimension $n-r$ which is orthogonal to Q and is centered at a point x in int Q. We call ∂B^{n-r} the *link* of Q and denote it by $\mathrm{lk}(Q)$. Clearly the metric on the $(n-r-1)$-sphere $\mathrm{lk}(Q)$ changes only by a homothety if we vary x and the size of B^{n-r}. At any point of $\mathrm{lk}(Q) \cap \mathrm{int}\, \Sigma_i$ or $\mathrm{lk}(Q) \cap \mathrm{int}\, F$, where F is a codimension one face of Σ_i, it is easy to see that $\mathrm{lk}(Q)$ is locally spherical. So $\mathrm{lk}(Q)$ is locally spherical away from a codimension two complex. We again require geodesic rays which meet Q

orthogonally to diverge as in the surface case.

Definition. M has a polyhedral metric with non-positive curvature if every closed embedded geodesic loop in $\mathrm{lk}(Q)$ has length at least 2π, for every face Q, assuming that the metric on $\mathrm{lk}(Q)$ is scaled to have curvature one at the locally spherical points.

Theorem (Cartan-Hadamard). *If M has a polyhedral metric with non-positive curvature, then the universal cover of M is diffeomorphic to Euclidean space.*

In [7], manifolds with such metrics are called Cartan-Hadamard spaces. Note that if a 3-manifold has a non-positively curved polyhedral metric then it is irreducible, as can be seen by lifting an embedded 2-sphere to the universal cover.

To finish this section we give two important methods for presenting 3-manifolds with these metrics. This viewpoint works well in all dimensions but is particularly useful in dimension three.

I. Cubings with non-positive curvature. Suppose M is obtained by gluing together faces of a finite collection of regular Euclidean cubes, all with the same edge length. There are two conditions to achieve a polyhedral metric of non-positive curvature on M from such a cubing:

(a) Each edge must belong to at least four cubes. (This is equivalent to the link of the edge has length at least 2π, as each cube contributes $\pi/2$.)

(b) Let F, F',F'' be three faces of cubes at a vertex v. Assume the faces have edges e_i, e_i', e_i'' respectively at v, for $i = 1, 2$. Finally suppose all edges are oriented with v as initial point and for (i)–(iii) below, identifications are orientation-preserving. Then we exclude the following three types of gluings of edges.

   (i) e_1 is identified to e_2.

   (ii) e_1 (resp. e_2) is identified to e_1' (resp. e_2').

   (iii) e_1 (resp. e_1', e_1'') is identified to e_2' (resp. e_2'', e_2).

Remarks. 1) Condition (b)(iii) above does *not* apply if F, F', F'' are all faces of a single cube and $e_1 = e_2'$, $e_1' = e_2''$, $e_1'' = e_2$.

2) It turns out that an embedded closed geodesic of length less than 2π can only arise on $\mathrm{lk}(v)$ if one of the gluings (i), (ii), (iii) occurs.

II. Generalized cubings with non-positive curvature. There is an obvious rotation of order three about the diagonal of a cube. The quotient space (orbifold) is homeomoprhic to a ball and is locally Euclidean, except along the diagonal, where the dihedral angle is clearly $2\pi/3$. Suppose we now take a cyclic branched cover of this orbifold, with branch set the diagonal and having degree $d \geq 3$. The resulting space is called a *flying saucer*. It can be viewed as having top and bottom hemispheres, dividing up the 2-sphere boundary into two disks. Each disk consists of d Euclidean squares with a common vertex (the two ends of the diagonal). For example, if $d = 4$ then each hemisphere has the same induced metric as for the octagon at the beginning of this section. If $d = 3$ we get a cube back. Note that all the dihedral angles at the edges of a flying saucer are still $\pi/2$ and all faces are squares.

We can now define a generalized cubing with non-positive curvature exactly as for cubings. The conditions which are needed for the gluing of faces are identical. Here M is constructed by attaching together faces of a finite number of flying saucers, where all edges have the same length. Note again that some of these flying saucers could be cubes.

§2 CONSTRUCTING 3-MANIFOLDS WITH NICE METRICS

Our first result is a general criterion for constructing polyhedral metrics of non-positive curvature by gluing together a finite set of compact 3-dimensional Euclidean polyhedra $\Sigma_1, \Sigma_2, \ldots, \Sigma_t$. Assume that for each Σ_i, no loop C on $\partial\Sigma_i$ crosses at most three edges at one point each, unless C bounds a disk on $\partial\Sigma_i$ containing a vertex of degree three and C meets the three edges ending at v.

Theorem 1. *Suppose a closed orientable 3-manifold M is obtained by identifying faces of $\Sigma_1, \Sigma_2, \ldots, \Sigma_t$. Assume:*

(i) *Each face of Σ_i has at least four edges.*

(ii) *Each edge in M belongs to at least four of the Σ_i.*

(iii) *Let v be a vertex of M and F, F', F'' any three faces at v. Suppose the same gluings of edges of these faces, as in I(b) of the previous section, are excluded. Note that of course F, F', F'' could be all the faces of some Σ_i at v. Then M has a polyhedral metric of non-positive curvature.*

Remarks. This result is proved by dividing each Σ_i into cubes (respectively flying saucers) if every vertex of Σ_i has degree three (resp. degree three or greater). We illustrate this with the example of the Weber-Seifert hyperbolic dodecahedral space [40].

Example. Construct a regular hyperbolic dodecahedron with all dihedral angles $2\pi/5$. Identify opposite pairs of faces by rotation of $3\pi/5$. Then all edges have degree five in the resulting manifold M which thus has a hyperbolic metric.

Now the dodecahedron can be decomposed into twenty cubes as follows. Join the center of each face to the midpoints of all edges of the face. This divides each pentagonal face into five quadrilaterals (squares – c.f. the octagon example in §1). Now join the center of each face to the center of the dodecahedron. The neighborhood of every original vertex on the boundary of the dodecahedron is three squares. when these three squares are "coned" to the center of the dodecahedron, the result is a cube. It is easy to check that every edge of these cubes has degree four or five in M, checking I(a) in §1. On the other hand, there is a unique vertex in M coming from the vertices of the dodecahedron. This vertex has a link in M which has the structure of an icosahedron (the dual tesselation to the dodecahedral tesselation of H^3). It is easy to check that the other vertices have links which are either octahedral or are formed from ten triangles arranged like two pentagons joined along their boundaries. Therefore I(b) follows immediately and M has a polyhedral metric of non-positive curvature.

The next construction of nice metrics can be viewed as dual to Theorem 1.

As is well-known, every closed orientable 3-manifold M can be constructed by identifying the boundaries of two 3-dimensional handlebodies, by an orientation-reversing homeomorphism. The *genus* is the number of handles and such a decomposition is called a Heegaard splitting of M. There is no procedure known for constructing hyperbolic metrics from such splittings.

Before stating the result, we need to define Heegaard diagrams. Assume $M = Y \cup Y'$, where Y, Y' are handlebodies of genus g with $\partial Y = \partial Y' = L$, a closed orientable surface. let $D_1, D_2, \ldots, D_m$ (resp. $D'_1, \ldots, D'_n$) be a collection of disjoint *meridian disks* for Y (resp. Y'). This means that D_i (resp. D'_j) is properly embedded in Y (resp. Y') with ∂D_i (resp. $\partial D'_j$) a non-contractible loop in ∂Y (resp. $\partial Y' = \partial Y = L$). We say that the collection $D_1, D_2, \ldots, D_m$ is *full* if $\operatorname{int} Y - D_1 \ldots D_m$ is a set of open 3-cells. The meridian disks will *never* be chosen so that D_i and D_k are parallel in Y, for $i \neq k$. Finally denote ∂D_i by C_i, $\partial D'_j$ by C'_j and let $\mathcal{C} = \{C_1, C_2, \dots, C_m\}$, $\mathcal{C}' = \{C'_1, C'_2, \dots, C'_n\}$, $\mathcal{D} = \{D_1, D_2, \dots, D_m\}$, $\mathcal{D}' = \{D'_1, D'_2, \dots, D'_n\}$.

Definition. The triple $(L, \mathcal{C}, \mathcal{C}')$ is called a *Heegaard diagram* for M if both collections of meridian disks $\mathcal{D}$ and $\mathcal{D}'$ are full.

Obviously M can be assembled from the Heegaard diagram by attaching 2-handles to L along $\mathcal{C}$ and $\mathcal{C}'$, then filling in by 3-handles. Without loss of generality, assume $\mathcal{C}$ and $\mathcal{C}'$ are isotoped to be transverse and to have minimal intersection, i.e. no pair of arcs in $\mathcal{C}$ and $\mathcal{C}'$ have common endpoints and bound a 2-gon in L.

Suppose R is the closure of a component of $L - \mathcal{C} - \mathcal{C}'$. We call R a *region* of the Heegaard diagram. A boundary component of R can be viewed as a polygon, with an equal number of arcs coming from $\mathcal{C}$ and $\mathcal{C}'$.

Theorem 2. *Suppose a Heegaard diagram $(L, \mathcal{C}, \mathcal{C}')$ for M satisfies the following conditions:*

(a) *Every curve C_i (resp. C'_j) meets $\mathcal{C}'$ (resp. $\mathcal{C}$) at least four times. Moreover, every non-contractible loop on L must intersect $\mathcal{C} \cup \mathcal{C}'$ in at least four points.*

(b) *Every region of the Heegaard diagram is a disk and has at least six boundary edges.*

Then M has a polyhedral metric of non-positive curvature. Moreover M has a cubing (resp. generalized cubing) if all regions are hexagons and every curve C'_j meets $\mathcal{C}$ exactly in four points (resp. regions have six or more edges and $C'_j \cap \mathcal{C}$ has at least four points).

Remarks. Note that we are *not* assuming that any region R is an *embedded* disk. So an edge may occur twice in the boundary of R. Also splittings as in the theorem are often *reducible*, i.e. have trivial handles. So there can be curves C_i and C'_j in the diagram which cross exactly once.

Examples. 1) Heegaard diagrams corresponding to cubings come from torsion-free subgroups of finite index in the Coxeter group constructed from the tesselation of the hyperbolic plane by the regular right-angled hyperbolic hexagon. Note that such subgroups yield similar tilings of Riemann surfaces by right-angled hexagons, which is condition (b) in the theorem, in the case of a cubing. For more information, see also [5].

2) Generally, suppose as in Theorem 1 that M is formed by identifying faces of $\Sigma_1, \Sigma_2, \ldots, \Sigma_t$. Let us truncate each Σ_i by chopping off a piece at each vertex v by a plane chosen to pass through each edge e at v at $3/4$ of the distance along e away from v. Then we obtain a truncated polyhedron $\widehat{\Sigma}_i$ as illustrated in Figure 1 for the cube.

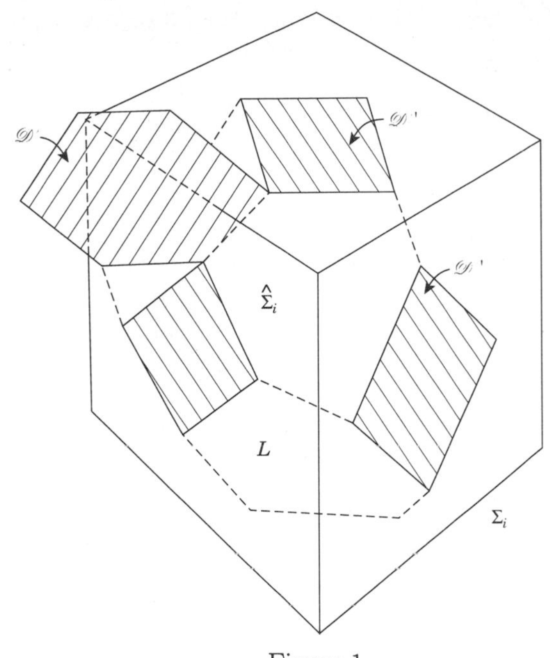

Figure 1

This shows how to obtain the surface L (the hexagonal faces in Figure 1). Y' is the union of the truncated polyhedra $\widehat{\Sigma}_i$ and Y is the closure of $M - Y'$. The disks $\mathcal{D}'$ are the quadrilateral faces in Figure 1 and the disks $\mathcal{D}$ are dual to the edges of the polyhedra Σ_i. Moreover, to go from the Heegaard diagram to the cubing or generalized cubing, we merely reverse the process of truncation, i.e. expand $\widehat{\Sigma}_i$ back to Σ_i.

Another standard method of constructing 3-manifolds is via *branched coverings*. W. Thurston [37] has introduced the concept of a *universal knot or link* $\mathcal{L}$ in S^3, which is disjoint union of embedded circles so that *every* closed orientable 3-manifold is some branched cover of S^3 over $\mathcal{L}$. H. Hilden, M. Lozano and J. Montesinos [20], [21] have shown that various knots and links are universal. Among these are the Whitehead link, Borromean rings, Figure 8 knot and the 5_2 knot.

Theorem 3. *Suppose $\mathcal{L}$ is a link in S^3 and a 3-manifold M is constructed which is a branched coverings of S^3 over $\mathcal{L}$, with all components of the branch set having degree at least d. Then M has a polyhedral metric of non-positive curvature if $\mathcal{L}$ is the Whitehead link or 5_2 knot and $d = 4$, $\mathcal{L}$ is the Figure 8 knot and $d = 3$ or $\mathcal{L}$ is the Borromean rings and $d = 2$. Moreover if $\mathcal{L}$ is the Whitehead link, 5_2 knot or Borromean rings (resp. Figure 8 knot) then all such M have cubings (resp. generalized cubings) as in §1.*

Remarks. We sketch the argument for the Whitehead link $\mathcal{L}$. The hyperbolic dodecahedral space M has a cyclic group action of order five given by rotation of the dodecahedron about an axis through the centers of opposite faces. This is shown in [40] to exhibit M as a 5-fold cyclic branched cover of S^3 over $\mathcal{L}$. Clearly this action permutes the twenty cubes of M described previously. So we see that S^3 can be built from four cubes. Also $\mathcal{L}$ is contained in the edges of these cubes and each edge in $\mathcal{L}$ has dihedral angle $\pi/2$. All other edges in S^3 still have degree four or five. We conclude that any branched cover of S^3 over $\mathcal{L}$, where all components have degree at least four, yields a cubing with non-positive curvature.

It is also sometimes useful to construct metrics by gluing together *hyperbolic* polyhedra and also to look at branched coverings for polyhedral hyperbolic metrics.

Suppose $\mathcal{L}$ is a simple link which is not a 2-bridge link, torus link or Montesinos link. (Here we include knots as examples of links.) Also assume that if an embedded 2-sphere S meets $\mathcal{L}$ in four points, then S bounds a 3-cell B with $\mathcal{L} \cap B$ consisting of two unknotted arcs. Then by the orbifold theorem of W. Thurston [36] (see also [22]), the 2-fold branched cover M of S^3 over $\mathcal{L}$ has a hyperbolic metric and the covering transformation is an isometry. Projecting to S^3, we conclude that S^3 has a polyhedral hyperbolic metric with singular set $\mathcal{L}$ and dihedral angle π along $\mathcal{L}$. Using the smoothing technique of [14], we easily deduce the following result.

Theorem 4. *Any branched cover of S^3 over $\mathcal{L}$, where all components of the branch set have degree at least two, has a Riemannian metric with strictly negative curvature.*

Corollary. *Any such a branched cover has universal cover $\mathbb{R}^3$, by the Cartan-Hadamard theorem.*

An important class of hyperbolic 3-manifolds are surface bundles over the circle with pseudo-Anosov monodromy (see [38]). We describe two closely related constructions of such bundles with polyhedral metrics of non-positive curvature.

The first is given in the article of D. Sullivan [32] on Thurston's work. The 2-fold cyclic branched cover of S^3 over the Borromean rings $\mathcal{L}$ is a flat 3-manifold M. In Figure 2, S^3 is formed by folding a cube along six edges bisecting the faces. This gives a polyhedral metric which is flat except along $\mathcal{L}$, where the dihedral angle is π. The flat 3-manifold M has a fibering as a 2-torus bundle over a circle. The fibers are easily seen to be transverse to the pre-image $\widehat{\mathcal{L}}$ of $\mathcal{L}$.

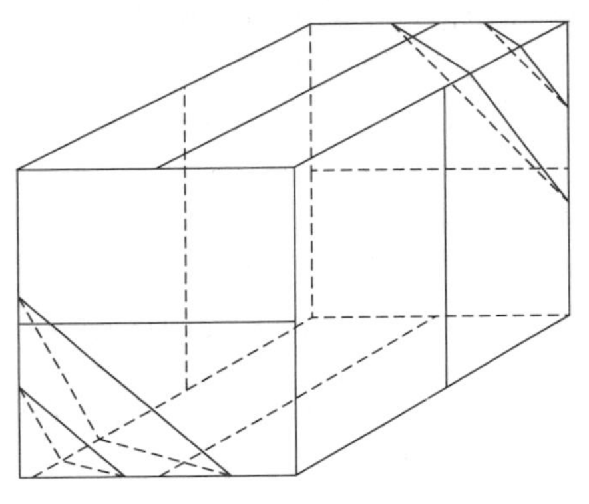

Figure 2

Then this fibering lifts to *any* branched covering $\bar{M}$ of M over $\widehat{\mathcal{L}}$. Such an $\bar{M}$ can be shown to be a hyperbolic bundle [38] and M obviously has a cubing satisfying the conditions of §1. This completes the first construction.

For the second method, start with a product $L \times S^1$, where L is a closed orientable surface of genus at least two. As in §1, we put a polyhedral metric of non-positive curvature on L by decomposing it into Euclidean squares, with all vertices having degree at least four. Then $L \times S^1$ has a cubing in an obvious way, by dividing S^1 into intervals. Let us form a link $\mathcal{L}$ in $L \times S^1$ as a union of embedded geodesic loops in this metric. Start with a diagonal of a cube and note that at an end which lies over a vertex with degree greater than four, there is a cone of choices of continuing the geodesic. In this way, we can build complicated loops which are *braids*, in the sense that the loops are transverse to the fibering by copies of L. So any branched cover M of $L \times S^1$ over $\mathcal{L}$ is a fiber bundle over S^1 with a polyhedral metric of non-positive curvature. Negative curvature occurs along the pre-image $\widehat{\mathcal{L}}$ of $\mathcal{L}$ and along the vertical loops projecting to vertices with degree greater than four in L.

To finish the discussion, we need to decide when the bundle M is hyperbolic, i.e. has pseudo-Anosov monodromy. If the monodromy is *not* pseudo-Anosov, then a finite power takes some non-contractible loop C in the fiber to itself, homotopically. Hence there is a map $f\colon T^2 \to M$ so that $f_*\colon \pi_1(T^2) \to \pi_1(M)$ is one-to-one. By standard methods, we can homotop f to a map g realizing the smallest area in the homotopy class. Then g is a minimal immersion, so by Gauss-Bonnet $g(T^2)$ is flat, due to the non-positivity of the curvature. In particular $g(T^2)$ cannot cross the graph, described above, along which negative curvature is concentrated. However it is straightforward to choose $\mathcal{L}$ so that the complement of the graph is a handlebody which contains no (singular) incompressible tori. So we can arrange that M is of hyperbolic type.

There is a nice connection between hyperbolic Coxeter groups and polyhedral metrics of non-positive curvature. Suppose that G is a group generated by a finite number of hyperbolic reflections and that G is cocompact, i.e. H^3/G is compact. (We will briefly discuss the finite volume case in §5.) Assume that a fundamental domain for G is a hyperbolic polyhedron Σ with all dihedral angles of the form π/m, for some integer $m \geq 2$. Andreev's theorem ([6], [33]) gives a characterization of such Σ. There are three conditions to be satisfied:

(1) At each vertex of Σ, the sum of the dihedral angles is greater than π.

(2) If a loop C on $\partial\Sigma$ meets three edges exactly once each and C does not bound a disk with a single vertex belonging to these edges, then the sum of the dihedral angles at the edges is less than π.

(3) If a loop C on $\partial\Sigma$ meets four edges in one point each, then either the sum of the dihedral angles of the edges is less than 2π or C bounds a disk containing two vertices of degree three joined by an edge and the dihedral angles add to 2π.

The following result is easy to check.

Lemma. *If three planes of faces of Σ meet in a triple point in H^3 or S^3_∞, then the triple point is a vertex of Σ. If the planes do not intersect then there is a common perpendicular plane which crosses Σ in a triangular region.*

Corollary. *If Σ has a face of degree three, then either Σ is one of nine possible hyperbolic simplices (see [8]) or Σ contains a triangular hyperbolic disk which is properly embedded and orthogonal to $\partial\Sigma$.*

Assume Σ has a disk D as in the corollary. It is clear that any 3-manifold M whose fundamental group is a torsion free subgroup of finite index in G will have an embedded totally geodesic incompressible surface built from copies of D. So we leave this case aside, since it is difficult to handle and it is not so interesting to construct nice polyhedral metrices, as Waldhausen's theorem applies.

On the other hand, we have:

Theorem 5. *Suppose G is a hyperbolic Coxeter group with fundamental domain Σ having dihedral angles of the form π/m, for $m \geq 2$. Assume that Σ is not a simplex and Σ contains no properly embedded hyperbolic triangular disk perpendicular to $\partial\Sigma$. Then if M is a 3-manifold with $\pi_1(M)$ isomorphic to a torsion free subgroup of finite index in G then M has a polyhedral metric of non-positive curvature and a cubing.*

Remarks. 1) The theorem follows immediately from Theorem 1. Compare also with the example of the Weber-Seifert hyperbolic dodecahedral space.

2) Six of the nine simplices can be dealt with very simply. A fundamental domain for the action of the dihedral group of order six on the cube is a Euclidean simplex Δ with dihedral angles (2, 2, 4, 2, 3, 4), using the notation of [8]. The rotation of order three is about a diagonal and the reflection is in a plane containing this diagonal of the cube. The integers m in the bracket are dihedral angles of π/m for the six edges of Δ.

Now it is easy to verify that for the simplices $T3$, $T4$, $T6$, $T8$, $T9$ on the list of [8], tahe dihedral angles can be *increased* to give Δ. In other words, the simplices can be given the metric of Δ and negative curvature will be concentrated along edges which are on the cube. This latter observation turns out to be important in the next section. Finally $T1$ is constructed by gluing two copies of Δ by reflection in the face with dihedral angles (2, 2, 4).

The simplices $T2$, $T5$ and $T7$ are more complex. Note that 120 copies of $T2$ give the regular hyperbolic icosahedron with dihedral angle $2\pi/3$ (see [8]). This is the fundamental domain for the fivefold cyclic branched cover M of S^3 (see [41]), over the Figure 8 knot. Hence there is an induced polyhedral metric of non-positive curvature on M, which is more difficult to relate back to the simplex $T2$. Note also that the Weber Seifert hyperbolic dodecahedral manifold M has $\pi_1(M)$ a torsion free subgroup of index 120 in the Coxeter group for $T4$ (see [8]). The polyhedral metric we have just described using Δ is the same as previously.

§3 SINGULAR INCOMPRESSIBLE SURFACES

Definition. Suppose V is a closed surface. If $f\colon V \to M$ is continuous and $f_*\colon \pi_1(V) \to \pi_1(M)$ is one-to-one then we call f or $f(V)$ a (singular) incompressible surface.

Our aim is to analyze the relationship between polyhedral metrics of non-positive curvature arising from cubings or generalized cubings and singular incompressible surfaces. Also some remarks will be made about the characteristic variety in 3-manifolds with (generalized) cubings.

Let $p\colon \widetilde{M} \to M$ be the universal covering. By M. Freedman, J. Hass and P. Scott [12], if an incompressible surface f has least area in its homotopy class, then the preimage $p^{-1}(f(V))$ consists of embedded planes. The following important properties were introduced by P. Scott [30].

Definition. Suppose $f\colon V \to M$ is a (singular) incompressible surface. f has the 4-*plane* property if for any least area map g homotopic to f and four planes P_1, P_2, P_3, P_4 in $p^{-1}(g(V))$, at least one pair of planes are disjoint. f has the 1-*line* property if $P_1 \cap P_2$ is either empty or a single line, for all pairs of planes in $p^{-1}(g(V))$.

Remark. These properties depend only on the homotopy class of f, i.e. the subgroup $f_*\pi_1(V)$ in $\pi_1(M)$.

Definition. Suppose $f\colon V \to M$ is a (singular) incompressible surface. f has the *triple point property* if for any map g homotopic to f so that $p^{-1}(g(V))$ consists of planes, any three such planes meeting pairwise must intersect in at least one triple point.

Remark. This property can also be stated in terms of $f_*\pi_1(V)$. Suppose H_1, H_2, H_3 are subgroups of $\pi_1(M)$ which are conjugate to $f_*(V)$ in $\pi_1(M)$ and have non-trivial intersections in pairs. Then f has the triple point property is equivalent to $H_1 \cap H_2 \cap H_3 = \{1\}$.

Theorem 6. *Suppose M has a cubing (respectively generalized cubing) giving a polyhedral metric of non-positive curvature. Then M has a (singular) incompressible surface $f\colon V \to M$ satisfying the 4-plane, 1-line and triple point (resp. 4-plane and triple point) properties. In addition, f can be realized as a totally geodesic surface in the polyhedral metric.*

Remarks. 1) In the case of a cubing, to build $f(V)$ start with a square parallel to and midway between opposite faces of a cube. Follow this square to similar squares in adjacent cubes, via the exponential map. The result is eventually a totally geodesic closed orientable surface immersed in M, since there are only finitely many squares of this type.

For a generalized cubing, the method is similar except we begin with a square D which is a face of a flying saucer. Choose a preferred normal direction to this square which we call "up" for convenience. Let e be an edge of D. Then e belongs to d squares, where $d \geq 4$. If $d = 4$ then there is a unique choice of adjacent square to D at e so as to obtain a totally geodesic surface. If $d > 4$ we proceed as follows. Let F_1 be the flying saucer on the "up" side of D and let D_1 be the second face of F_1 containing e. Let F_2 be the adjacent flying saucer containing D_1 and let D_2 be the other face of F_2 including e. We continue on from D to D_2.

Roughly speaking, make a "turn" of π on the "up" side at each edge to go from a square to the next. Again this constructs a totally geodesic surface which is immersed but is in general *not* self transverse as in the case of a cubing. The surface can use a square twice, since the "up" direction may reverse as a loop is traversed in the manifold.

2) We discuss the 4-plane and triple point properties first, in the context of a cubing. Suppose four planes P_1, P_2, P_3, P_4 all intersect in pairs. It is easy to see that one of the planes, say P_4, can be chosen so that the other three planes meet P_4 in lines which cross in a triangle. However the planes meet at right angles (the squares are parallel to cubical faces). So a totally geodesic right angled triangle is obtained which contradicts Gauss-Bonnet, since the curvature is non-positive.

Similarly if P_1, P_2, P_3 are planes so that each pair meets in lines but there is no triple point, then there are infinite triangular prisms. Choose a plane orthogonal to these three planes. This yields a similar contradiction.

For a generalized cubing, it can be verified that if two planes meet, then they intersect in a 2-complex which has a spine which is a tree. We could say that f has the 1-*tree* property rather than the 1-line property. However by essentially the same arguments as above, the 4-plane and triple point properties follow. Finally, for a cubing, if two planes meet in at least two lines, then choosing a plane orthogonal to both planes gives a right angled

2-gon, a contradiction. Note that, since the planes are totally geodesic, they meet along geodesics which are lines, not loops in the universal cover, as the curvature is non-positive.

Theorem 7. *Suppose M has a cubing with non-positive curvature and M' is irreducible with M homotopy equivalent to M'. Then M and M' are homeomorphic.*

Remarks. 1) This follows immediately from the result of J. Hass and P. Scott [18]. In fact they only need to assume M has an incompressible surface satisfying the 4-plane, 1-line properties and must work hard to deal with triangular prisms.

2) The same result should be true for generalized cubings, but dropping the 1-line property causes considerable problems in the technique in [18].

We now turn to the converse of Theorem 7.

Definition. An immersed surface $f\colon V \to M$ is called *filling* if $M - f(V)$ consists of open cells. Note that we do *not* assume that V is connected.

Theorem 8. *Suppose M is closed, orientable, irreducible and contains a singular incompressible surface $f\colon V \to M$ so that f is filling and satisfies the 4-plane, 1-line and triple point properties. Then M admits a cubing which induces a polyhedral metric with non-positive curvature.*

Remarks. The proof proceeds by checking that the closures of the components of $M - f(V)$ are polyhedra satisfying the conditions of Theorem 1. Note that the union of all the surfaces constructed in Theorem 6 is filling.

The analogous result for generalized cubings should be that a filling incompressible surface satisfying the 4-plane, "1-tree" and triple point properties, suffices. However there are difficulties in moving such a surface to a canonical position as in [18]. To precisely define the 1-tree property, note we need to give a characterization depending only on the homotopy class of f.

To complete this section, we comment on characteristic varieties in a 3-manifold M with a (generalized) cubing. Let T be an embedded incompressible torus in M. Then T can be isotoped to be minimal relative to the polyhedral metric. By Gauss-Bonnet, as the curvature is non-positive it follows that this minimal surface, which we again denote by T, is totally geodesic and flat. This implies the following:

- T is disjoint from the interior of any edge at which negative curvature is concentrated, unless T contains such an edge and locally has dihedral angle π at the edge.
- If T passes through a vertex v, then $T \cap \mathrm{lk}(v)$ is a geodesic loop of length 2π.

Using these two facts, we can work out where the characteristic variety is located. To be precise, remove the interiors of all edges with negative curvature from M. Also delete all vertices v for which there is no geodesic loop in $\mathrm{lk}(v)$ with length 2π. Finally suppose a number of negatively curved edges end at some vertex v and $\mathrm{lk}(v)$ has a finite number of geodesic loops of length 2π, denoted by $C_1, C_2, \ldots, C_k$. (It is easy to show infinitely many such curves is impossible.) There are two possibilities for a pair of negatively curved edges e_1 and e_2 which meet $\mathrm{lk}(v)$ at x_1 and x_2 respectively.

If x_1 and x_2 are separated by some C_i, then we "split apart" e_1 and e_2 at v. Conversely, if x_1 and x_2 are on the same side of every C_j, then we join e_1 and e_2 at v. In this way a graph Γ is constructed in M by joining or pulling slightly apart the edges of negative curvature at such vertices. The closure of the complement $M - \Gamma$ is the region in which the characteristic variety is located. For example, in many cases it is very easy to observe that the characteristic variety must be empty, since $M - \Gamma$ is an open handlebody or is a product of a surface and an open interval. Note also that the characteristic variety can "touch" Γ, but not cross it. So we must take the closure of $M - \Gamma$.

§4 STRUCTURE OF THE SPHERE AT INFINITY

We begin by describing the ideal boundary or sphere at infinity, then will discuss some important properties of it. Let M be a closed orientable 3-manifold with a cubing or generalized cubing of non-positive curvature. Let $\widetilde{M}$ denote the universal cover of M. Following M. Gromov [13], let $C(\widetilde{M})$ be the continuous functions on $\widetilde{M}$ with the compact open topology. The map $x \to d_x$ defines an embedding of $\widetilde{M}$ in $C(\widetilde{M})$, where $d_x(y) = d(x, y)$ is the distance between points in $\widetilde{M}$. Finally let $C_*(\widetilde{M})$ be the quotient of $C(\widetilde{M})$ by dividing out by the subspace of constant functions. Then the ideal boundary of $\widetilde{M}$ is $bd(\widetilde{M}) = \mathrm{cl}(\widetilde{M}) - \widetilde{M}$ in $C_*(\widetilde{M})$, where cl denotes closure. A function h in $C(\widetilde{M})$ which projects to $\bar{h}$ in $bd(\widetilde{M})$ is called a *horofunction* centered at $\bar{h}$.

In P.Eberlein, B. O'Neill [11], $bd\,\widetilde{M}$ is defined more geometrically as follows. Suppose geodesics $c_1, c_2\colon \mathbb{R} \to \widetilde{M}$ are parametrized by arc length. Then c_1, c_2 are *asymptotic* if $d(c_1(t), c_2(t))$ is a bounded function of t. The equivalence classes of geodesics using this relation are the *points at infinity* and are denoted $\widetilde{M}(\infty)$. It is easy to verify that these points are in one-to-one correspondence with all the geodesic rays $c(t)$ starting at any fixed point x in $\widetilde{M}$. To put a topology on $\widetilde{M}(\infty)$, we need to build the tangent space $T(\widetilde{M})_x$ as a topological space (which is homeomorphic to $\mathbf{R}^3$).

Suppose for convenience that x is in the interior of some cube. Locally, the geodesics starting at x are straight lines. As these lines are extended, they hit vertices and edges with negative curvature. As described in §1, for surfaces, the geodesic segments which continue a line of this type form a cone. As viewed from x, an edge (resp. a vertex) pulls back to an arc of a great circle (resp. a point) in the unit sphere S^2 of the usual tangent space at x. We cut S^2 open along this arc (resp. point) and insert a suitable disk, representing the cone of extensions of lines to the edge (resp. line to the vertex). The size of the cone is bounded by the maximal degree of edges (resp. vertices) in $\widetilde{M}$.

Since $\widetilde{M}$ covers M, which is compact, such degrees are bounded. Also the inserted disk must be scaled down by dividing by the distance of the edge (resp. vertex) from x. In this way, S^2 is modified by adding infinitely many disks with diameters converging to zero. The result of all the insertions is a new 2-sphere $\widehat{S}^2$, which is called the unit sphere of the tangent space of $\widetilde{M}$ at x.

The topology on $\widetilde{M}(\infty)$ is induced by the one-to-one correspondence with $\widehat{S}^2$. Note that $T(\widetilde{M})_x$ can be viewed as the infinite cone on $\widehat{S}^2$. It follows that $\widetilde{M}(\infty)$ and $bd\,\widetilde{M}$ are homeomorphic, as in [7]. This completes the description of the sphere at infinity for M. Finally, note that a horofunction at $c(\infty)$, where c is a geodesic ray from x, can also be described as a Busemann function, $h_c(y) = \lim_{t \to \infty} (d(y, c(t)) - t)$.

Now the Geometrization Programme conjectures that if M is irreducible, has infinite fundamental group and is *atoroidal* (i.e. has empty characteristic variety) then M should admit a hyperbolic metric. In [11] it is shown there is a close connection between metrics of strictly negative curvature and visibility manifolds.

Definition. If M has a (generalized) cubing of non-positive curvature, then M is a *visibility* manifold if given any two points z_1 and z_2 on $\widetilde{M}(\infty)$, there is a geodesic c in $\widetilde{M}$ with $z_1 = c(\infty)$ and $z_2 = c(-\infty)$.

We note the following simple result.

Proposition. *Suppose M has a (generalized) cubing of non-positive curvature. Then M is a visibility manifold if and only if M is atoroidal.*

Proof: Suppose M has a non-trivial characteristic variety. Then M has an immersed incompressible torus. (If M is Seifert fibered, there may not be any embedded such tori.) As previously, using least area maps, we may assume $f\colon T^2 \to M$ is totally geodesic and flat. If $\tilde{f}\colon \mathbf{R}^2 \to \widetilde{M}$ is a lift of f to $\widetilde{M}$, then $\tilde{f}$ is a totally geodesic, flat embedding. It is simple to check that if c_1, c_2 are non-parallel geodesics in $\tilde{f}(\mathbf{R}^2)$, then $z_1 = c_1(\infty)$ and $z_2 = c_2(\infty)$ cannot be joined by a geodesic. Consequently M is not a visibility manifold.

Conversely, if M does not have the visibility property, as in [7] or [11] two geodesic rays c_1 and c_2 can be found so that $c_1(0) = c_2(0) = x$ and the horofunction h_{c_1} is bounded on c_2. We can choose x to be an interior point of a cube. Let v_1 and v_2 be unit vectors in the directions of c_1 and c_2 respectively at x. Also let $\mathcal{C}$ denote the union of all geodesic rays from x in the direction of $\lambda v_1 + (1-\lambda)v_2$, for $0 \leq \lambda \leq 1$. If $\mathcal{C}$ meets edges (and vertices) of negative curvature transversely at a sequence of points with distance to x converging to infinity, then it is easy to see that $c_1(t)$ and $c_2(t)$ diverge faster than any linear function in t. Consequently $\lim\limits_{t\to\infty} h_{c_1}(c_2(t)) = \infty$, a contradiction.

We conclude that $\mathcal{C}_t = \{y\colon y \text{ is in } \mathcal{C} \text{ and } d(y,x) \geq t\}$ is flat and totally geodesic, for t sufficiently large. Suppose $\mathcal{C}_t$ is projected to M. Then it can be shown that either the image is compact or by taking limit points, a compact totally geodesic flat surface is obtained. In either case, this shows that the characteristic variety of M is non-empty and the proposition is established.

The final topic for this section is the limit circles of the singular incompressible surfaces described in §3. The first result shows these surfaces are quasi-Fuchsian with regard to a hyperbolic metric. Next, the 4-plane, 1-line, triple point and filling properties are translated to the intersection pattern for the limit circles. We would like to thank Peter Scott for a very helpful conversation which led to this viewpoint. The final theorem gives an interesting converse; the picture of the circles on the 2-sphere at infinity completely determines a 3-manifold with a cubing of non-positive curvature. So this class of 3-manifolds can be derived from appropriate group actions on the 2-sphere, with invariant graphs which are the union of all the limit circles.

Suppose $f\colon V \to M$ is a surface in a cubing with non-positive curvature, obtained from squares which are parallel to faces and bisect cubes. Then we know that a component P of the pre-image of $f(V)$ in the universal cover $\widetilde{M}$ is an embedded plane. The geodesics c lying in P define a limit circle.

Definition. $bd(P) = P(\infty) = \{c(\infty)\colon c \text{ is in } P\}$ is a *limit circle* of $f\colon V \to M$.

Similarly, if M has a generalized cubing, then $f\colon V \to M$ can be chosen as a union of squares which are faces of flying saucers, as discussed in §3. Then the same definition of limit circle applies.

Lemma. *If M has a hyperbolic metric and a (generalized) cubing of non-positive curvature and $f\colon V \to M$ is a totally geodesic surface in the polyhedral metric as above, then f is quasi-Fuchsian in the hyperbolic metric.*

Proof: By a result of W. Thurston [33], either f is quasi-Fuchsian or f lifts to $\bar{f}\colon \bar{V} \to \bar{M}$, where $\bar{M}$ is a fiber bundle finitely covering M and $\bar{f}$ is an embedding giving a fiber. In the latter case, $\bar{M}$ has a (generalized) cubing given by lifting the structure in M. Let C be the union of all the line segments in the squares of $\bar{f}(\bar{V})$, which bisect the squares and are parallel to the edges. Then C is a collection of immersed geodesics.

Cut $\bar{M}$ along $\bar{f}(\bar{V})$ to obtain $\bar{V} \times [0,1]$. We can form a surface in $\bar{V} \times [0,1]$ by starting with squares intersecting $\bar{f}(\bar{V})$ orthogonally along C, then following via the exponential map to construct a properly immersed totally geodesic surface in $\bar{V} \times [0,1]$. Then it can be shown that this surface is homeomorphic to $C \times [0,1]$ and meets both $\bar{V} \times \{0\}$ and $\bar{V} \times \{1\}$ in C. (Suppose some arc in the surface has endpoints in, say, $\bar{V} \times \{1\}$, but is not homotopic along the surface into $\bar{V} \times \{1\}$, keeping its ends fixed. Then a right angled 2-gon can be formed by projecting the arc to $\bar{V} \times \{1\}$ in $\bar{V} \times [0,1]$. This is a contradiction, by Gauss Bonnet.) Hence $\bar{V} \times [0,1]$ has the product polyhedral metric and the monodromy for $\bar{M}$ is periodic. This implies that neither $\bar{M}$ nor M has a hyperbolic metric.

Remark. There is a simple proof that each circle $bd(P)$ in $\widetilde{M}(\infty)$ satisfies $\widetilde{M}(\infty) - bd(P)$ is a pair of open disks. In fact, the Gauss or normal map, when suitably defined over P, gives a homeomorphism between P and each of these two regions.

Suppose that $f\colon V \to M$ is a totally geodesic surface constructed as above from a cubing with non-positive curvature. Let $\{P_i\colon i \in I\}$ denote the plane components of the pre-image $p^{-1}(f(V))$ in $\widetilde{M}$.

Definition. f satisfies the 4 *circle property*, if given limit circles $P_1(\infty)$, $P_2(\infty)$, $P_3(\infty)$, $P_4(\infty)$, at least one pair has no transverse intersections. f satisfies the 2 *point property* if given limit circles $P_1(\infty)$ and $P_2(\infty)$, then $P_1(\infty)$ and $P_2(\infty)$have either zero or two transverse intersection points. f satisfies the *triple region property* if whenever limit circles $P_1(\infty)$, $P_2(\infty)$ and $P_3(\infty)$ meet in pairs with two transverse crossing points, then their intersection pattern is as in Figure 3(a) and never as in Figure 3(b). The limit circles of f are called *filling* if for any z in $\widetilde{M}(\infty)$, $z = \bigcap_j \{D_j \colon D_j$ is a disk bounded by $P_j(\infty)$ in $\widetilde{M}(\infty)$ and z is in $\operatorname{int} D_j\}$. Finally the limit circles of f are said to be ***discrete*** if they form a discrete subset of the space of embedded circles in $\widetilde{M}(\infty)$.

Remark. It is easy to show the limit circles have no transverse triple points, by the filling and 4-plane properties.

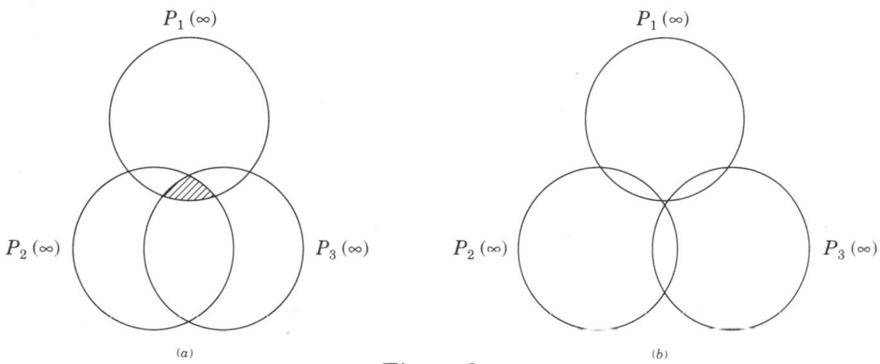

Figure 3

Theorem 9. *Assume M has a cubing of non-positive curvature and $f \colon V \to M$ is a standard filling totally geodesic surface. Then f has the 4 circle, 2 point and triple region properties. Moreover, the limit circles of f are filling and discrete.*

Proof: The first three properties of f follow immediately from the corresponding properties of the plane components of $p^{-1}(f(V))$ (see Theorem 6). To see that the limit circles are filling, assume first that M is atoroidal. Then by the proposition above, M is a visibility manifold. Assume there are points z, z' in the intersection of all disks D_j satisfying $\partial D_j = P_j(\infty)$ and z is in $\operatorname{int} D_j$. Let c be the geodesic in $\widetilde{M}$ with $c(\infty) = z$ and $c(-\infty) = z'$. It is easy to see that c cannot meet any component of $p^{-1}(f(V))$ transversely, since no circle $P_j(\infty)$ can separate z and z'. Consequently the closure of the component of $\widetilde{M} - p^{-1}(f(V))$ containing c is non-compact. This contradicts the assumption that f is filling, since then the complementary regions of $f(V)$ and $p^{-1}(f(V))$ are all (compact) cells.

Assume M has a non-empty characteristic variety N. Then $p^{-1}(N)$ consists of disjoint copies of the universal cover $\widetilde{N}$ of N embedded in $\widetilde{M}$. Again we suppose that there are distinct points z, z' in the intersection of all disks bounded by limit circles and containing a fixed point in their interiors. The limit circles corresponding to the tori boundary components of N divide $\widetilde{M}(\infty)$ into two regions, one of which is a union of copies of $\widetilde{N}(\infty)$. (Note we can have $M = N$, so N has no boundary.) If z, z' lie outside these copies of $\widetilde{N}(\infty)$, the same argument as before applies, as $M - N$ has the visibility property. On the other hand, if z, z' are in a copy of $\widetilde{N}(\infty)$, we can use the fact that N has an $\mathbb{R}^3$ or $\mathbb{H}^2 \times \mathbb{R}$ geometric structure. (By H. Lawson, S. T. Yau [26], the latter polyhedral metric on N splits as a product metric.) So we can explicitly separate z and z' by a limit circle.

Finally, suppose there is a sequence of limit circles $P_n(\infty)$ converging to a limit circle $P_0(\infty)$. Let γ_n be covering transformations in $\pi_1(M)$ such that $P_n = \gamma_n P_0$. γ_n is defined modulo the subgroup $f_*\pi_1(V)$ in $\pi_1(M)$, i.e. lies in a coset of this subgroup. Coset representatives can be chosen so that for some x in P_0, $\gamma_n x \to x$ as $n \to \infty$. This contradicts the fact that γ_n is a covering transformation, for all n.

Theorem 10. *Suppose a 2-sphere S^2 has a family of embedded circles $\{C_i : i \in I\}$ satisfying the 4 circle, 2 point, triple region, filling and discrete properties. Then there is a canonical way of constructing a 3-cell B^3 with $\partial B^3 = S^2$ so that each C_i bounds a properly embedded disk $\bar{P}_i$ in B^3, the disks intersect minimally and all the complementary regions of $\bigcup_i \bar{P}_i$ are 3-cells. In particular, if the C_i are limit circles for a standard, filling totally geodesic surface $f: V \to M$ in a cubed manifold with non-positive curvature, then $\pi_1(M)$ acts as a covering transformation group on int B^3. Moreover, $\bigcup_i \bar{P}_i$ is invariant and there is a homeomorphism from int $B^3/\pi_1(M)$ to M mapping $\bigcup_i P_i/\pi_1(M)$ to $f(V)$, where $P_i =$ int $\bar{P}_i$.*

Proof: We sketch the construction of the 3-cell B^3 and the disks $\bar{P}_i$. The homeomorphism between int $B^3/\pi_1(M)$ and M follows from [18], as in Theorem 7. The model to have in mind is a collection of totally geodesic planes in hyperbolic 3-space. The strategy is to show $\bigcup_i \bar{P}_i$ can be realized as a 2-complex so that each face lies in a pair of 2-spheres, one on each side of the face. Capping off these 2-spheres with polyhedral 3-cells (the complementary regions of $\bigcup_i \bar{P}_i$) gives B^3.

(i) A vertex v of a polyhedral 3-cell corresponds to a triple region of three limit circles, as in Figure 3a. We view a disk D_j bounded by a circle C_j in S^2 as representing a half-space of B^3 bounded by $\bar{P}_j$. Similarly two circles with transverse intersections, C_i and C_j, define a line c where $\bar{P}_i$ and $\bar{P}_j$ cross. c has endpoints at the two transverse points of $C_i \cap C_j$. So a triple region corresponds to three such lines meeting at a triple point v. We can schematically project onto the 2-sphere S^2 "at infinity" as in Figure 4.

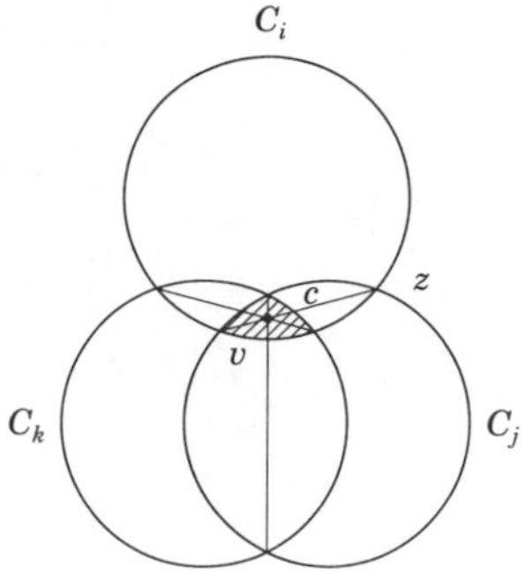

Figure 4

(ii) An edge e of a polyhedral 3-cell is an arc of a line c defined by intersecting circles C_i and C_j. e ends at vertices v and v' defined by three circles C_i, C_j, C_k and C_i, C_j, C_m respectively. The picture is drawn in Figure 5. Note by the 4-circle property that C_k and C_m have no transverse crossing points.

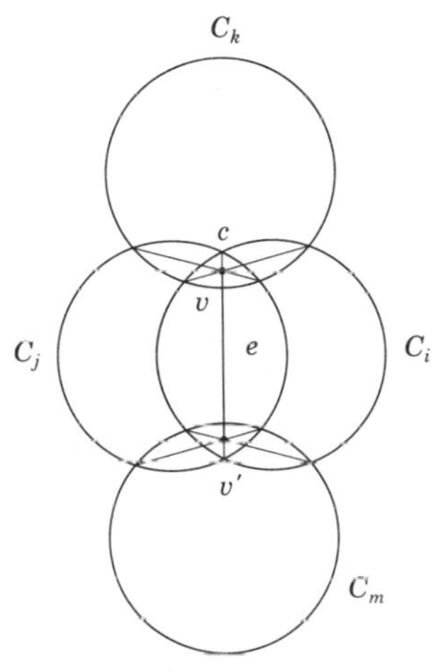

Figure 5

(iii) A face f of a polyhedral 3-cell is a finite-sided polygon bounded by edges $e_1, e_2, \ldots, e_p$ running around a circle C_i. (See Figure 6.)

f is uniquely specified by a choice of a single vertex v and a triple region as in Figure 4 containing v. Consider the point z in Figure 4. By the filling property, there is a circle C_m with z in $\operatorname{int} D_m$, where D_m is a disk bounded by C_m in S^2. By the 4 circle property, C_m cannot transversely cross C_k and if C'_m is another such circle, then C_m and C'_m have no transverse intersection points. So these circles are nested and by discreteness, we can choose C_m so

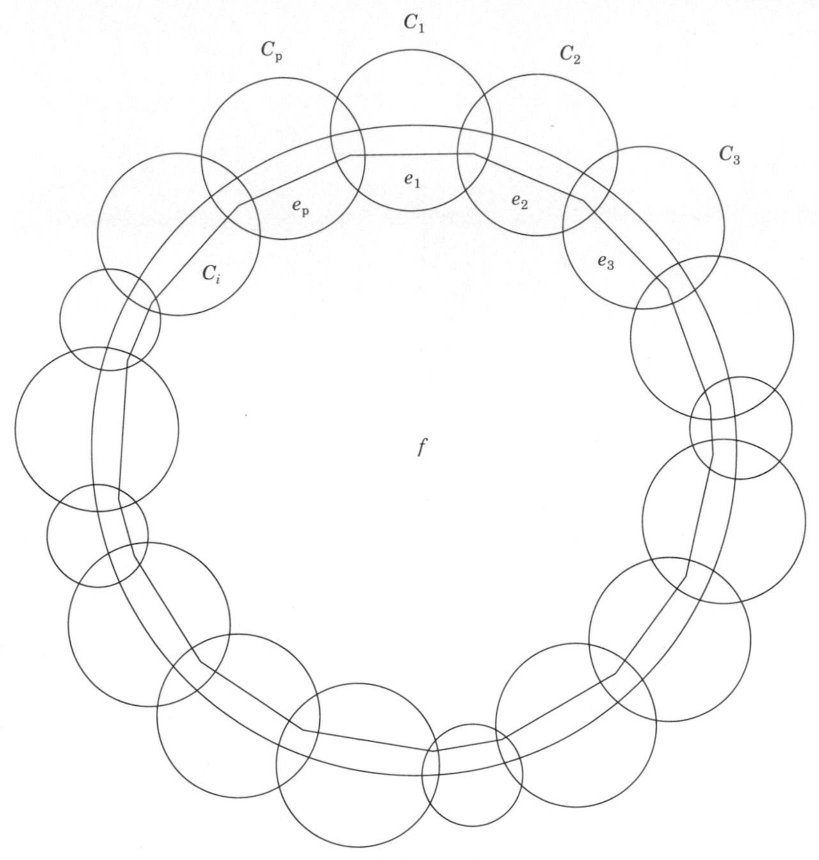

Figure 6

that D_m is maximal. This gives a unique prescription for forming the edge e from v in the direction of the specified line c. (See Figure 5.) e runs from v (corresponding to C_i, C_j, C_k) to v' (corresponding to C_i, C_j, C_m). By this process, we can generate the picture of f as in Figure 6. We need only check that there is a *finite* chain of circles $C_1, C_2, \ldots, C_p$ produced, so that f is a finite polygon.

Assume, on the contrary, as in Figure 7, that an infinite pattern of circles is constructed. By the 4-circle property, the circles are linearly ordered along C_i as in Figure 7.

Let $C_1, C_2, C_3, \ldots$ denote the circles and let $D_1, D_2, D_3, \ldots,$ be the disks bounded by the respective circles, as in Figure 7. Assume that a point y_j is chosen in D_j so that $y_j \to y$ as $j \to \infty$. By the filling property, there is a circle C' bounding a disk D' with y in $\operatorname{int} D'$. Now C' must transversely meet some circle C_k. Since C' intersects C_i, by the 4 circle property, C' does

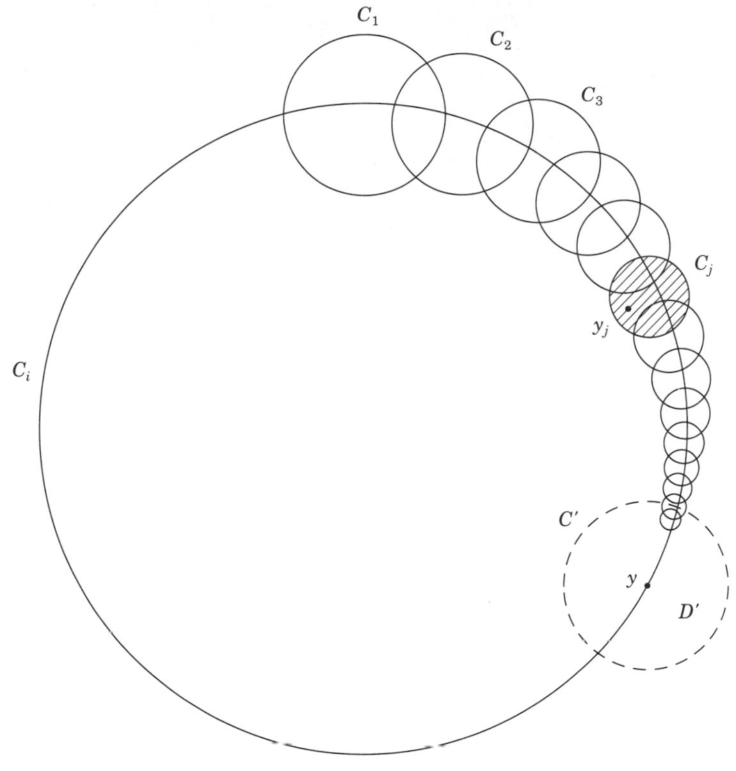

Figure 7

not transversely cross C_{k-1} and C_{k+1}. It can be checked that C_{k+1} must be inside D'. This contradicts our choice of C_{k+1} as bounding a maximal disk. So f is finite-sided.

(iv) To complete the construction, all the adjoining faces to f are combined to give a finite-sided polygonal 2-sphere. See Figure 8.

We need only check finitely many adjacent faces are obtained which cover S^2. As before, suppose faces $f_1, f_2, f_3, \ldots$ are produced containing a sequence $y_1, y_2, y_3, \ldots$, of points with $y_n \to y$ as $n \to \infty$. By the filling property, y is in the interior of a disk D' bounded by C'. This gives a contradiction to the maximality of the chosen circles, by the 4-circle property.

To show the result of attaching all the complementary 3-cells to $\dot{\cup} \bar{P}_i$ is B^3 we employ the argument in [17].

Remarks. We would like to thank B. Maskit for a helpful comment about non-transverse intersections of the limit circles. Also there is an analogous construction of n-manifolds which are cubed with non-positive curvature, for $n \geq 4$, using

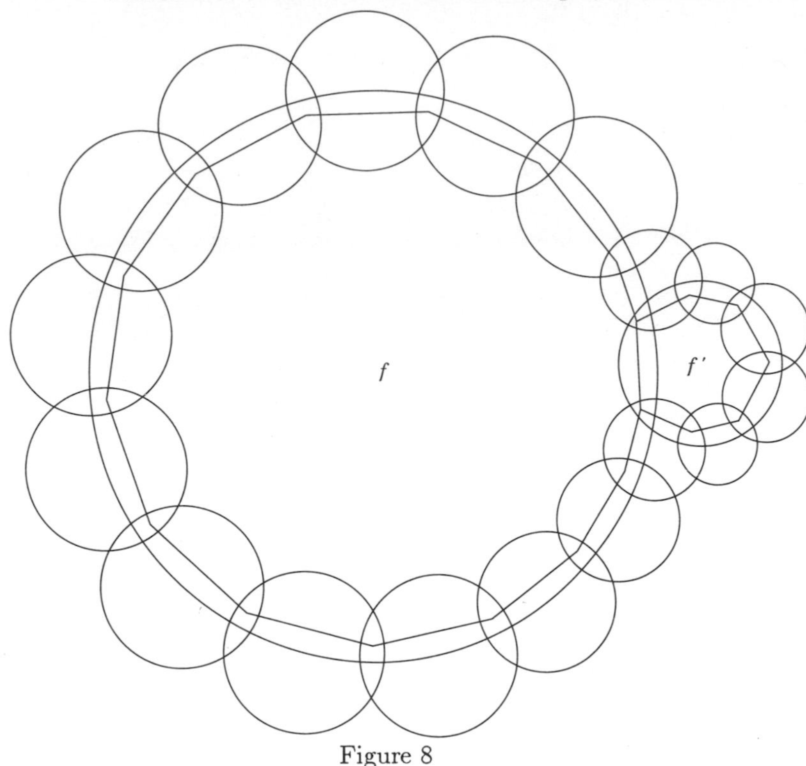

Figure 8

codimension one spheres in S^{n-1} which are equivariant under a torsion free group action.

§5 COMPLETE POLYHEDRAL METRICS OF FINITE VOLUME AND DEHN SURGERY ON CUSPS

In this section, we describe the polyhedral analogue of complete non-compact hyperbolic 3-manifolds with finite volume. A brief discussion is given of the translation of results in the previous sections to this setting. Finally a detailed description is given of a specific example, a two component link in $\mathbb{R}P^3$ formed by identifying faces of a *single cube*. This example is double covered by a simple four component link in the 3-sphere and also has a hyperbolic metric given by the regular ideal cube in hyperbolic 3-space. Negatively curved Dehn surgery of M. Gromov is described and applied to this example. In particular, we show the remarkable result that for all but a very small number of surgeries on each component of the link, the result is a closed 3-manifold with a Riemannian metric of strictly negative curvature and the surgered manifold satisfies the topological rigidity result of Theorem 7. In [3], a large class of alternating links in S^3 are discussed which have similar properties to this example.

Assume M is a compact orientable 3-manifold with finitely many singularities with the link type of a torus. This means that there is a finite number of points $v_1, v_2, v_3, \ldots, v_k$ in M, where a neighborhood of v_i is a cone on a torus rather than an open 3-cell. We say that M has a polyhedral metric with non-positive curvature if M is formed by gluing together finitely many Euclidean polyhedra $\Sigma_1, \Sigma_2, \ldots, \Sigma_t$ as in §1, so that each v_i is a vertex and the link of each edge and vertex, different from all v_i, has no closed geodesic loops of length strictly less than 2π. If M is neither a union of a solid torus and a cone on a torus nor the suspension of a torus, nor the union of a twisted line bundle over a Klein bottle and a cone on a torus, then we say M has finite volume (by analogy with the hyperbolic case). Note that this includes also the possibility that M has a characteristic variety. (For such manifolds, there are incompressible tori which are embedded and not parallel to links of the v_i.)

We will be interested in the case that M has a (generalized) cubing of non-positive curvature and finite volume. Denote $M - \{v_1, v_2, \ldots, v_k\}$ by M_0, the manifold part of M. Then M_0 has universal cover $\mathbf{R}^3$ and is a Cartan-Hadamard space. (We can rescale near v_i so that the v_i are at "infinity", i.e. the metric on M_0 is complete.) The conditions that the (generalized) cubing must satisfy are as in Theorem 1, where vertices refer to points in M_0. Notice that at a singular vertex v_i, in the case of a cubing the average degree of edges at v_i must be six, so that the Euler characteristic of $\mathrm{lk}(v_i)$ is zero. For a generalized cubing, $\mathrm{lk}(v_i)$ may contain many-sided polygons, so the average degree can be much less.

There is a nice counterpart to Theorem 2. Suppose L is a closed orientable surface of positive genus. A *compression body* is the result of attaching disjoint 2-handles to $L \times \{0\}$ in $L \times [0, 1]$ and capping off any boundary 2-spheres by adding 3-cells. Suppose M has a finite number of toral singularities. It is easy to show that M can be formed by a *Heegaard splitting*, i.e. a compression body with boundary consisting of k tori and a copy of L, can be glued to a handlebody along L. Then cones on the tori can be attached to the remaining boundary components, resulting in M. A *Heegaard diagram* $(L, \mathcal{C}, \mathcal{C}')$ for M then has the property that $\mathcal{C}$ is full (i.e. every component of $L - \mathcal{C}$ is planar) but $L - \mathcal{C}'$ has k regions which are punctured tori and the rest are planar. With this modification, the remaining conditions on the diagram are as in Theorem 2 to ensure a polyhedral metric of non-positive curvature on M, *except* that the only non-contractible loops on L which can cross $\mathcal{C} \cup \mathcal{C}'$ in three or fewer points, are non-separating curves in $L - \mathcal{C}'$.

An illustration of the technique in Theorem 3 will be given in the following example of the two component link in $\mathbb{R}P^3$. Theorem 5 also has an analogue; consider hyperbolic Coxeter groups where the fundamental domain has some ideal vertices. Again a cubing can be constructed with the toral cone singularities at the ideal vertices and non-positive curvature, so long as the domain is not an ideal simplex and there are no orthogonal totally geodesic triangular disks in the domain. The case of ideal simplices requires a special argument, as for the compact case.

A very important result is that if M has a (generalized) cubing with finite volume and non-positive curvature, then M has a singular incompressible totally geodesic surface $f\colon V \to M$ missing the singular vertices and satisfying the same properties as in Theorem 6. f is called *filling* if the closures of the components of $M - f(V)$ are cells and cones on tori. Then, exactly as in Theorem 8, a finite volume M with a cubing of non-positive curvature arises from a filling singular incompressible surface $f\colon V \to M$ satisfying the 4-plane, 1-line and triple point properties. The characteristic variety of M can be located as in §3.

The results in §4 also apply in the finite volume case. Here M is *atoroidal* means that any embedded incompressible torus in M_0 is parallel to the link of some singular vertex v_i. As in Theorems 9 and 10, we can characterize the limit circles of the totally geodesic, filling surface $f\colon V \to M$ and can reconstruct M from these circles.

Example. Suppose a cube has faces identified in pairs as in Figure 9. The resulting 3-manifold M has two vertices v_1 and v_2 with toral links and two edges a and b, both having degree six. It is immediate that M_0 is a complete hyperbolic 3-manifold with finite volume and two cusps, by using the regular hyperbolic cube metric with all dihedral angles $\pi/3$.

M can be identified by drawing a neighborhood of the dual 1-skeleton as a genus three handlebody and attaching a pair of 2-handles dual to a and b. This is a Heegaard splitting as discussed above. By a sequence of handle slides, it can be shown that M is the complement of a two component link in $\mathbb{R}P^3$. We draw its double cover in Figure 10, as a four component link in S^3. (See [2] for more details.)

We content ourselves here with algebraicaly decribing $\pi_1(M)$ in terms of generators and relations. Also we identify the cusps as subgroups of $\pi_1(M)$ and give the (meridianal) surgery yielding $Z_2 = \pi_1(\mathbb{R}P^3)$. The generators X, Y, Z of $\pi_1(M)$ are the face identifications shown in Figure 9. They can also be viewed as loops built by joining the centers of a pair of matched faces to the center of the cube. So X,

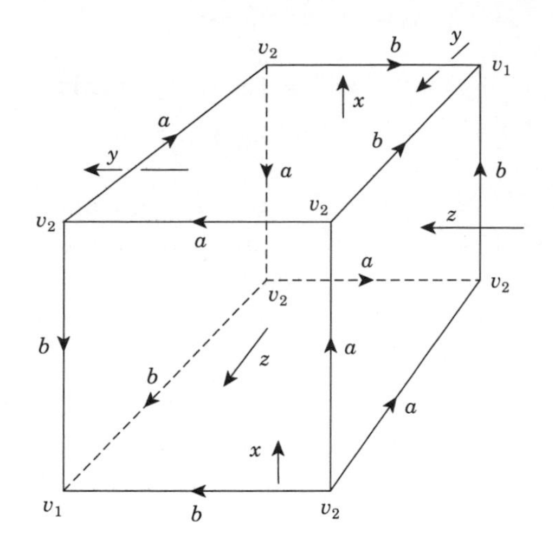

Figure 9

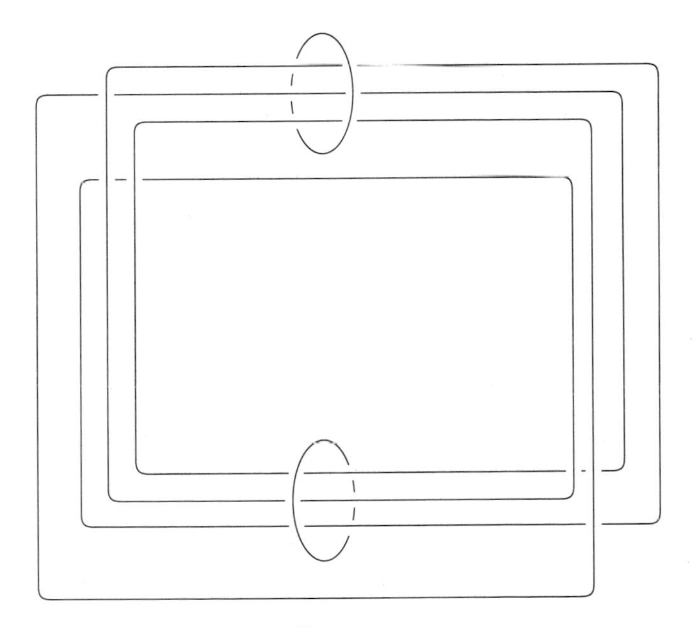

Figure 10

Y, Z generate the fundamental group of the dual 1-skeleton, which is a bouquet of three circles. Next, the attaching circles of the 2-handles dual to a and b can be pushed into the dual 1-skeleton to give relations:

$$XZ^{-1}Z^{-1}XY^{-1}Y^{-1} = 1, \quad \text{for the dual to } a$$
$$XZY^{-1}X^{-1}Z^{-1}Y = 1, \quad \text{for the dual to } b$$

It is immediate that $H_1(M) = Z \oplus Z \oplus Z_2$. The peripheral or cusp subgroups are the free abelian subgroups of rank two in $\pi_1(M)$ given by loops on the links of v_1 and v_2. For $\mathrm{lk}(v_1)$ we compute generators $\{Z^{-1}X^{-1}, Z^{-1}Y\}$, by choosing a suitable base point. Similarly for $\mathrm{lk}(v_2)$, a generating set is $\{Y^2X^{-1}Y, YZX^{-1}Y\}$. Finally if solid tori are attached to M_0 at v_1 and v_2 so that meridian disks have boundaries $Z^{-1}Y$ and $Y^2X^{-1}Y$, the result is a closed orientable 3-manifold M^* with $\pi_1(M^*) = Z_2$. By handle slides, M^* is actually homeomorphic to $\mathbf{R}P^3$.

Consider the hyperbolic, totally geodesic surface $f\colon V \to M$ with image the three squares bisecting the cube and equidistant from opposite faces. V has three faces, six edges and two vertices, hence Euler characteristic -1. We conclude that V is non-orientable with three cross caps.

To complete the discussion of this example, we briefly indicate the idea of negatively curved Dehn surgery attributed to M. Gromov. Suppose a horospherical neighborhood of one of the cusps of M is removed. Choose a geodesic loop on the flat horotorus boundary with length strictly larger than 2π. Then a suitable negatively curved solid torus can be attached with appropriate smoothing along a collar of the torus boundary, so that the geodesic curve bounds a disk in the solid torus and the resulting metric has strictly negative curvature everywhere. The key point is that by Gauss-Bonnet, the boundary of a negatively curved disk on a horotorus which has constant mean curvature one must be longer than 2π. (The boundary loop will have geodesic curvature one in M.) Now since the cusp neighorhood can be chosen disjoint from $f(V)$ and all sufficiently long geodesics are permissible, we deduce the main result of this section.

Theorem 11. *Let M_s be the closed orientable 3-manifold obtained by Dehn surgery on both cusps of M. Then for all but a finite number of choices for each cusp, M_s has a Riemannian metric of strictly negative curvature. Moreover, if M' is an irreducible 3-manifold which is homotopy equivalent to such an M_s, then M' is homeomorphic to M_s.*

Proof: The idea is that since the metric is negatively curved and has not changed near $f(V)$, we see that f is still totally geodesic and hence incompressible. In addition, the angles of intersection between the plane components of the pre-image of $f(V)$ in the universal cover of M_s are all still $\pi/2$. We conclude that f still satisfies the 4-plane, 1-line and triple point properties. Therefore, the topological rigidity Theorem 7 applies (i.e. the main result of [18]). Note that f is not filling in M_s, but this is unnecessary here.

Remarks. 1) We can construct infinitely many other examples from this one as follows. Let C be the vertical line through the center of the cube in Figure 9, i.e. C joins the centers of the pair of faces identified by X. Then C is a geodesic loop. Take any branched cover of M over C. The result is a 3-manifold with a cubing of non-positive curvature and finite volume again. Note that here we *cannot* directly use a regular hyperbolic cube metric. However, we can divide such a cube into eight congruent cubes, using the usual three bisecting squares. Put this metric on each cube in the branched covering manifold. This is exactly like a polyhedral metric in that sums of dihedral angles around edges are at least 2π, etc. To find $f(V)$, we must use the flying saucer technique, since the cubes are no longer regular. However, the 1-line property *is* valid here, since we can *also* use Euclidean cubes and push $f(V)$ into the middle of the Euclidean cubes! So Theorem 7 applies to all these examples.

2) Using a computer, we have generated all single cube manifolds with toral links and metrics of non-positive curvature (see [2]).

§6 PROBLEMS

1) Two basic pieces have been used to construct polyhedral metrics of non-positive curvature; the cube and the flying saucer. Are there any other suitable Euclidean polyhedra? It is clearly an advantage for a polyhedron to be regular, i.e. have congruent faces, and to have dihedral angles associated with a tesselation of Euclidean space.

2) Describe appropriate theories of polyhedral metrics for the other five geometries of Thurston.

3) Suppose M has a polyhedral metric of non-positive curvature and M is atoroidal. There are three ways one might tackle the problem of constructing a hyperbolic metric on M.

   a) The Ricci flow of R. Hamilton [15] may deform the polyhedral metric to the hyperbolic one. Note that the polyhedral metric is instantaneously smoothed by the Ricci flow to a Riemannian metric. If M is atoroidal, is this metric strictly negatively curved? There is also an interesting "cross-curvature" flow discussed in notes of B. Chow and R. Hamilton. This parabolic flow has a quadratic expression for the time derivative of the metric and only makes sense in dimension three! This appears to be a desirable feature, given the results of M Gromov and W. Thurston [14] showing that no flows can deform even nearby metrics to hyperbolic ones in higher dimensions.

b) The action of $\pi_1(M)$ on $\widetilde{M}$ extends to an action on the 2-sphere at infinity $\widetilde{M}(\infty)$. If an invariant conformal structure on $\widetilde{M}(\infty)$ could be found, then we would obtain an embedding of $\pi_1(M)$ in PSL$(2, \mathbb{C})$. So there would be a hyperbolic 3-manifold M' with the same fundamental group as M. By the rigidity Theorem 7, at least if M is cubed then we obtain that M and M' are homeomorphic. Consequently M is hyperbolic.

c) D. Long [27] has recently shown that if a closed orientable 3-manifold M is hyperbolic and has an immersed totally geodesic surface V, then M has a finite sheeted cover $\widehat{M}$ in which V lifts to an embedding. Suppose an analogous result could be established for cubed M with a polyhedral metric of non-positive curvature. By Thurston's uniformization theorem [34], $\widehat{M}$ has a hyperbolic metric since it is atoroidal. By a result of M. Culler, P. Shalen [10], $\pi_1(M)$ embeds in PSL$(2, \mathbb{C})$ as it is a finite extension of $\pi_1(\widehat{M})$. Again as in b), we conclude that M is hyperbolic.

4) Does every hyperbolic 3-manifold admit a (generalized) cubing of non-positive curvature? We do not know any obvious obstruction.

5) Extend the result of J. Hass, P.Scott [18] to cover generalized cubings. So the 1-line property is replaced by the 1-tree property.

6) Does every Haken 3-manifold have a (generalized) cubing? Does every surface bundle over a circle with pseudo Anosov monodromy have such a structure?

7) We have several constructions of extensive classes of 4-manifolds with cubings of non-positive curvature [4]. Such a 4-manifold M^4 has an immersed cubed 3-manifold V^3 and the map $\pi_1(V) \to \pi_1(M)$ is one-to-one. By Theorem 7

we know that V is essentially determined by $\pi_1(V)$. Also V has a variety of special properties, such as the 5-plane property. Is there a topological rigidity result for such 4-manifolds?

References

[1] Aitchison, I. R. and Rubinstein, J. H., *Polyhedral metrics of non-positive curvature on 3-manifolds,* in preparation.

[2] Aitchison, I. R. and Rubinstein, J. H., *Polyhedral metrics of non-positive curvature on 3-manifolds with cusps,* in preparation.

[3] Aitchison, I. R., Lumsden, E. and Rubinstein, J. H., *Polyhedral metrics on alternating link complements,* in preparation.

[4] Aitchison, I. R. and Rubinstein, J. H., *Polyhedral metrics of non-positive curvature on 4-manifolds,* in preparation.

[5] Aitchison, I. R. and Rubinstein, J. H., *Heaven and Hell,* Cursos y Congresos (Santiago de Compostela, Spain) 10 (1989), 5–25.

[6] Andreev, E., *On convex polyhedra in Lobacevskii spaces,* Math. USSR Sbornik 10 (1970), 413–446 (English translation).

[7] Ballman, W., Gromov, M. and Schroeder, V., *Manifolds of nonpositive curvature,* Birkhäuser, Boston 1985.

[8] Best, L. A., *On torsion-free discrete subgroups of $PSL(2,\mathbb{C})$ with compact orbit space,* Can. J. Math. 23 (1971), 451–460.

[9] Cheeger, J., Müller, W. and Schrader, R., *On the curvature of piecewise flat spaces,* Comm. Math. Phys. 92 (1984), 405–454.

[10] Culler, M. and Shalen, P., *Varieties of group representations and splittings of 3-manifolds,* Annals of Math. 117 (1983), 109–146.

[11] Eberlein, P. and O'Neill, B., *Visibility manifolds,* Pac. J. Math. 46 (1973), 45–103.

[12] Freedman, M., Hass, J. and Scott, P., *Least area incompressible surfaces in 3-manifolds,* Invent. Math. 71 (1983), 609–642.

[13] Gromov, M., *Hyperbolic manifolds, groups and actions, Riemann surfaces and related topics,* Stonybrook Conference Proceedings, ed. I. Kra and B. Maskit, Annals of Math. Studies 97, Princeton University Press, NJ 1981, 183–214.

[14] Gromov, M. and Thurston, W., *Pinching constants for hyperbolic manifolds,* Invent. Math. 89 (1987), 1–12.

[15] Hamilton, R. S., *Three manifolds with positive Ricci curvature,* J. Diff. Geom. 17 (1982), 255–306.

[16] Hamilton, R. S., *Four-manifolds with positive curvature operator,* J. Diff. Geom. 24 (1986), 153–179.

[17] Hass, J., Rubinstein, J. H. and Scott, P., *Compactifying converings of closed 3-manifolds,* J. Diff. Geom. 30 (1989), 817–832.

[18] Hass, J. and Scott, P., *Homotopy equivalence and homeomorphism of 3-manifolds,* preprint MSRI July 1989.

[19] Hatcher, A., *On the boundary curves of incompressible surfaces,* Pac. J. Math. 99 (1982), 373–377.

[20] Hilden, H., Lozano, M. T. and Montesinos, J. M., *The Whitehead link, the Borromean rings and the Knot* 9_{46} *are universal,* Collect. math. 34 (1983), 19–28.

[21] Hilden, H., Lozano, M. T. and Montesinos, J. M., *On knots that are universal,* Topology 24 (1985), 499–504.

[22] Hodgson, C., *Notes on the orbifold theorem,* in preparation.

[23] Jaco, W. and Shalen, P., *Seifert fibered spaces in 3-manifolds,* Memoirs of A.M.S. 220, A.M.S., Providence, RI 1980.

[24] Johannson, K., *Homotopy equivalences of 3-manifolds with boundary,* Springer Lecture Notes in Math 761, Springer-Verlag, Berlin 1979.

[25] Kneser, H., *Geschlossen Flächen in dreidimensionalen Mannigfaltigkeiten,* Jahr. der Deutsher math. Ver. 38 (1929), 248–260.

[26] Lawson, H. B. and Yau, S. T., *Compact manifolds of nonpositive curvature,* J. Diff. Geom. 7 (1972), 211–228.

[27] Long, D., *Immersions and embeddings of totally geodesic surfaces,* Bull. London Math. Soc. 19 (1987), 481–484.

[28] Milnor, J. W., *A unique factorization theorem for 3-manifolds,* Amer. J. Math. 84 (1962), 1–7.

[29] Scott, P., *The geometries of 3-manifolds,* Bull. London Math. Soc. 15 (1983), 401–487.

[30] Scott, P., *There are no fake Seifert fibre spaces,* Annals of Math. 117 (1983), 35–70.

[31] Stallings, J., *On the loop theorem,* Annals of Math. (2) 72 (1960), 12–19.

[32] Sullivan, D., *Travaux de Thurston sur les groupes quasi-Fuchsiens et les variétés hyperboliques de dimension 3 fibrées sur S^1,* Seminaire Bourbaki 1979/80 No. 554, Springer Lecture Notes in math. 842, Springer-Verlag, Berlin 1980.

[33] Thurston, W., *The geometry and topology of 3-manifolds,* Princeton University Lecture Notes 1978.

[34] Thurston, W., *Three dimensional manifolds, Kleinian groups and hyperbolic geometry,* Bull, Amer. Math. Soc. 6 (1982), 357–381.

[35] Thurston, W., *Hyperbolic structures on 3-manifolds I: Deformation of acylindrical manifolds,* Annals of Math. 124 (1986), 203–246.

[36] Thurston,W., *An orbifold theorem for 3-manifolds,* preprint.

[37] Thurston, W., *Universal links,* preprint.

[38] Thurston, W., *Hyperbolic structures on 3-manifolds II: Surface groups and 3-manifolds which fiber over the circle,* preprint.

[39] Waldhausen, F., *On irreducible 3-manifolds which are sufficiently large,* Annals of Math. 87 (1968), 56–88.

[40] Weber, C. and Seifert, H., *Die Beiden Dodekaederräume,* Math. Z. 37 (1933), 237–253.

[41] Richardson, J. S. and Rubinstein, J. H., *Hyperbolic manifolds from regular polyhedra,* preprint 1982.

FINITE GROUPS OF HYPERBOLIC ISOMETRIES

C.B. Thomas*

1. INTRODUCTION

Our starting point is the theorem of Hurwitz, which gives the bound $84(g-1)$ for the group of orientation preserving hyperbolic isometries of a closed surface of genus ≥ 2. In section 2 we will sketch a proof of this result in terms of hyperbolic tesselations, since this is the method which seems most apt for generalising to higher dimensions. It is also clear that one should define a Hurwitz group to be a finite group G for which there exists a closed Riemann surface S_g with the property that $|I^+(S_g)| = |G| = 84(g-1)$. And since any Hurwitz group G maps onto a non-abelian *simple* Hurwitz group G/K, it is an interesting question to ask which of the now-classified simple groups satisfies the Hurwitz condition. For example, among the 26 sporadic groups, 11 are definitely Hurwitz, 13 are not, and for the remaining 2 groups the question has yet to be answered. The two exceptions are G equal to the Baby Monster B or the Monster M, and for the latter the problem reduces to finding the smallest value of c such that M is a homomorphic image of the triangle group $T(2,3,c)$. The best value known to me at the time of writing is $c = 29$.

It is natural to ask if there are analogous results in dimension 3, given that, if M^3 is hyperbolic, then $I(M^3)$ is finite, and has order closely related to the volume, which in turn belongs to a well-ordered subset of the real numbers $\mathbb{R}$. It is obvious that if a 3-dimensional Hurwitz group G is defined to be a finite homomorphic image of the figure of eight knot group, and if G is non-abelian and simple, then G has order divisible by 12. To some extent 12 plays a similar role in dimension 3 to 84 in dimension 2; at this point it is instructive to go back to Hurwitz' original argument, in which he considers the orbifold obtained by allowing the finite group $I^+(S_g)$ to act on S_g. The factor 84 emerges from a case by case comparison of $\chi(S_g)$ and $\chi(S_g/I^+(M))$, depending on the number of branch points, and the genus of the base orbifold. The factor 84 is only needed for the case when the genus equals 0 and the number of branch points 3, otherwise it can be replaced by 12.

A second 3-dimensional problem is the realisability of an arbitrary finite group G as a (full?) group of hyperbolic isometries. We give two different proofs of

* This survey is based on a lecture given at the Max-Planck-Institut in Bonn, in April 1989, rather than during the Durham Symposium in July.

the fact that a cofinal family of alternating groups A_n, and hence any finite group, is realisable in the weaker sense that $A_n \subseteq I(M^3)$ for a suitable manifold M^3 of finite volume. We also consider the same problem for *closed* manifolds, but the result (Theorem 5) needs to be sharpened along the following lines. C. Adams has recently announced that the smallest limit volume for a hyperbolic 3-orbifold is attained by $H^3/PSL_2(\mathbf{Z}(i))$, hence the best realisability results will be obtained for finite coverings of this space. Now there is a link group of index 48 in $PSL_2(\mathbf{Z}(i))$, with a presentation which seems well adapted for finding homomorphic images among the simple groups, for example A_n with n not too small. As a hyperbolic manifold the link complement has six cusps, each of which can be closed by a Dehn surgery, introducing new relations into the fundamental group. Modulo interesting numerical details, which we hope to publish at length elsewhere, it seems clear that there is a wide class of simple groups G acting as hyperbolic isometries on a closed manifold of volume close to $|G|(14.655\ldots)$. Furthermore the index of G in the full group is isometries will be small.

Because it is concerned with algebraic surfaces we wish to mention another direction, in which our methods, and in particular those of section 2 can be applied. For surfaces of general type there is a very crude bound on the order of the group of birational self-maps, first obtained by Andreotti, and briefly discussed by Kobayashi in [9]. By taking products and using Theorem 2 below we see that any finite group is realisable in the weaker sense, but given the unmapped nature of the terrain, it is hard to see how to be more precise. It may be more profitable to consider the following problem: in [20] Wall shows that there is a class of elliptic surfaces, with Kodaira dimension equal to 1 and odd first Betti number, which have a geometric structure modelled on $\widetilde{SL}_2(\mathbf{R}) \times \mathbf{R}$ rather than on $H^2 \times H^2$ or $H^2(\mathbf{C})$. The fundamental group Γ is an extension of a free abelian group of rank 2 by a Fuchsian group F. Question: to what extent can one realise finite quotients of F as groups of "horizontal" isometries in the full, infinite, isometry group of the surface?

Apart from the ideas sketched in this introduction, this lecture is mainly a survey of known results in dimensions 2 and 3. The most interesting open question remains the determination of the genus of an arbitrary finite simple group, and in particular the completion of the list of Hurwitz groups. In dimensions 3 and 4 (for surfaces of general type) it would also be interesting to know if an arbitrary finite group can always be realised as the full structural automorphism group. What, for this author at least, gives the subject its special flavour is the mixture of algebra and geometry, which is so characteristic of the beautiful book of W. Magnus on hyperbolic tesselations [12].

2. CONFORMAL AUTOMORPHISMS OF SURFACES

Let a, b, c be integers ≥ 2, such that $\frac{1}{a} + \frac{1}{b} + \frac{1}{c} < 1$, and let Δ be the hyperbolic triangle with angles π/a, π/b, π/c. If A_0, B_0, C_0 denote the sides opposite the appropriate angles, let $T^*(a, b, c)$ denote the group generated by the reflections A, B, C of the hyperbolic plane H^2 in these sides. Then $T^*(a, b, c)$ contains a subgroup $T(a, b, c)$ of index 2 consisting of all words of even length in A, B, C, which is well-known to have a presentation of the form

$$T = < x, y, z : x^a = y^b = z^c = xyz = 1 > .$$

The two following results are well-known. We include sketch proofs of both of them in order to illustrate both the similarities, and the problems which arise, in connection with their extension to dimension three. Where necessary we say that the group G is an (a, b, c)-group if G is a homomorphic image of $T(a, b, c)$.

THEOREM 1. (A. Hurwitz) *Let S_g be a closed Riemann surface of genus $g \geq 2$, $\pi_1(S_g) = \Phi_g$. The group $I(S_g)$ of conformal automorphisms is finite and has order $\leq 84(g-1)$. $I(S_g)$ is isomorphic to the quotient of a Fuchsian group Γ by a normal subgroup isomorphic to Φ_g. The maximal value for the order of $I(S_g)$ occurs if F is isomorphic to $T(2, 3, 7)$.*

Idea of the proof: The finiteness of $I(S_g)$ is a special case of a theorem on the isometries of negatively curved manifolds, see [9, III.2.1], although Hurwitz' original proof is still interesting to read, see [8, I.1]. Now map the surface S_g (suitably dissected) conformally onto a fundamental region for Φ_g in such a way that the *finite* extension $< \Phi_g, I(S_g) >$ can be described as a Fuchsian group F. The fundamental region for F defines a subtesselation of that of Φ_g and $|I(S_g)|$ equals the quotient of the areas. The value of the bound and the condition for its attainment now follow from the facts that the hyperbolic area of the fundamental region of $T(2, 3, 7)$ equals twice the area of Δ, i.e. $\pi/21$, and that this is minimal.

We also have the following realisation theorem for an arbitrary finite group G.

THEOREM 2. (L. Greenberg) *Let S_g be a closed Riemann surface and G an arbitrary finite group. Then there is a regular covering map $f : R \to S$, such that the group of covering transformations is both isomorphic to G, and equal to the full automorphism group $I(R)$.*

Idea of the proof: Use a Fuchsian group Γ with signature $(g; \nu_1 \ldots \nu_k)$, where $x_1, \ldots, x_k$ generate G and have orders $\nu_1, \ldots, \nu_k$. (For technical reasons suppose that $k > 2g-3$.) It is almost immediate that allowing Γ to act on the unit disc gives a Riemann surface with $G \subseteq I(R)$. The hard part is to show that G exhausts the automorphisms; for this we need Γ to satisfy a maximal condition, which is only available in a weak form for the larger class of Kleinian groups.

Theorems 1 and 2 suggest the following definitions:

If G is an arbitrary finite group, the *genus* of G equals the smallest value of g, such that G acts effectively and conformally on the closed Riemann surface S_g.

Note that we do not require G to be maximal for a genus action. In point of fact for many interesting groups the index of a genus action in the full group of automorphisms is small. Thus A. Woldar [22] has shown that if G is a finite simple $(2, b, c)$-group with genus action on S, then G is normal in $I(S)$ and $I(S)$ is a subgroup of $Aut\,(G)$.

The finite group G is *Hurwitz* if $G = I(S_g)$ for some closed Riemann surface with $g = \frac{|G|}{84} + 1$.

It follows from Theorem 1 and the properties of triangle groups that G is Hurwitz if and only if G is a homomorphic image of $T(2,3,7)$, that is G is generated by elements x and y of orders 2 and 3, with xy or order 7.

Elementary properties of Hurwitz groups:

(i) If G is Hurwitz, then so is G/K for all proper normal subgroups K of G.

(ii) If G is Hurwitz of order n, there is a simple Hurwitz group G, of order dividing n.

(iii) No solvable group G can be Hurwitz.

(iv) The order of a Hurwitz group is divisible by 84.

A purely algebraic proof of (iv) follows from the fact that 4 must divide the order of a finite non-abelian simple group. It is clearly an interesting problem to list all simple Hurwitz groups.

EXAMPLES

(i) M. Conder [2]: For all $n \geq 168$ the alternating groups A_n are Hurwitz. The same is true for all but 62 values of n in the range $5 \leq n \leq 167$.

Remark: A_{10} and A_{13} are $(2,3,8)$-groups.

(ii) A. Macbeath [10]: The projective special linear groups $PSL_2(\mathbf{R}_q)$ are Hurwitz in the following cases,

$$q = 7,$$
$$q = p \equiv \pm 1 (mod\ 7) \text{ and } q = p^3,\ p \equiv \pm 2 \text{ or } \pm 3 (mod\ 7).$$

More generally H. Glover and D. Sjerve [5] have calculated the genus of all members of this family. For example, they show that generically $PSL_2(\mathbf{F}_p)$ is a $(2,3,d)$-group, where $d = min\{e : e \geq 7$ and e divides $\frac{p-1}{2}$ or $\frac{p+1}{2}\}$.

(iii) The group of Lie type $G_2(\mathbf{F}_q)$ is Hurwitz for all $q \geq 5$. I am indebted to G. Malle, see [13], for supplying me with this information.

(iv) Among the 26 sporadic simple groups it is easy to read off from the Atlas [3] that J_1, J_2, Co_3 and Ly are Hurwitz groups. However none of the five Mathieu groups is a (2,3,7)-group, and as stated in the introduction the best value which I know for the Monster M is (2,3,29). It would however be most interesting to know the genus of this group, and to investigate a genus action on the appropriate Riemann surface.

Question: What is the genus of the projective symplectic group $PSp_4(\mathbf{F}_p)$, when p is an odd prime? The group has order $p^4(p^4-1)(p^2-1)/2 = 2p^2(p^2+1)|PSL_2(\mathbf{F}_p)|^2$, and so 84 divides the order under the same conditions as in (ii) above. However it seems most unlikely that the group is of type (2,3,7) in many of these cases.

3. HYPERBOLIC AUTOMORPHISMS IN DIMENSION 3

In the upper half-space model $H^3 = \{(z,t) \in \mathbf{C} \times \mathbf{R} : t > 0\}$ of hyperbolic 3-space the group of orientable isometries can be identified with $PSL_2(\mathbf{C})$. If Γ is a discrete torsion-free (Kleinian) subgroup of $PSL_2(\mathbf{C})$, then H^3/Γ is a complete orientable manifold. Allowing elements of finite order leads to hyperbolic orbifolds, but in what follows we will be mainly interested in the case when H^3/Γ has no singularities, and is non-compact of finite volume. Discussion of the closed case is postponed until the next section. As a topological invariant the volume $v(M^3)$ replaces the Euler characteristic $\chi(S_g) = 2 - 2g$; the set of volumes forms a well-ordered subset of the real numbers, and there are only finitely many manifolds with a given volume v, see [19]. Furthermore as a consequence of the general result on negatively curved manifolds already referred to, the isometry group $I(M^3)$ is finite.

Assuming that M^3 is non-compact the group Γ contains parabolic elements γ, with the property of having a fixed point or cusp with respect to the action of Γ extended from H^3 to $H^3 \cup \mathbf{C} \cup \{\infty\}$. Such an element γ has trace equal to ± 2, and if Γ is torsion free the isotropy subgroup of the cusp is parabolic and free abelian of rank 2. This provides the link with knot spaces; we write πK for $\pi_1(S^3 \setminus K)$ for a knot K in S^3. The fundamental group πK is said to have an *excellent representation* if and only if $S^3 \setminus K$ is isometric with some H^3/Γ, and a maximal peripheral subgroup of πK corresponds to the isotropy subgroup of the single cusp. Thurston proves that πK is excellent provided that K is neither a torus nor a satellite knot, and manifolds of this kind admit triangulation by ideal hyperbolic tetrahedra. The following data are taken from a computer print-out of J. Weeks, dated 12.1.89, kindly provided for me by D.B.A. Epstein. The print-out lists all cusped hyperbolic manifolds obtainable from at most five ideal tetrahedra; the entries in the first column below refer to the position in Weeks' ordering, and where necessary we will write Γ_i for $\pi_1(M_i)$.

Label	Orientable?	Volume	No. of cusps O	n	Tetrahedra	Symmetries	H_1
0	n	1.0149...	—	1	1	2	$\mathbb{Z}$
1	n	1.8319...	—	1	2	2	$\mathbb{Z}/2\times\mathbb{Z}$
4	O	2.0298...	1	—	2	8	$\mathbb{Z}$
125	O	3.6638...	2	—	4	8	$\mathbb{Z}\times\mathbb{Z}$

The manifold M_4 is particularly significant, and can be identified with the complement of the figure of eight knot, see below. This particular excellent representation is due to R. Riley [17], and can be summarised as follows:

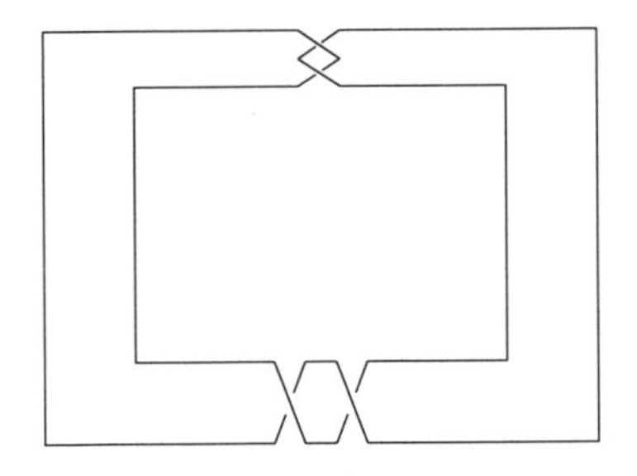

$\pi K = \{x_1, x_2 : wx_1w^{-1} = x_2,\ w = x_1x_2^{-1}x_1^{-1}x_2\}$, with peripheral subgroup $< \ell, m >$, $x_1 = m$ and $\ell = [x_2^{-1}, x_1][x_1^{-1}, x_2]$.

The excellent representation θ is then given by $\theta(x_1) = \begin{pmatrix} 1 & -1 \\ 0 & 1 \end{pmatrix}$ and $\theta(x_2) = \begin{pmatrix} 1 & 0 \\ \omega & 1 \end{pmatrix}$, where ω is a primitive cube root of unity. Indeed the image of θ is a subgroup of index 12 in $PSL_2(\mathbb{Z}(\omega))$.

Remark: M_4 is the only hyperbolic knot complement, which is arithmetic, i.e. such that its defining Kleinian group is an arithmetic subgroup of $PSL_2(\mathbb{C})$, see [16].

It is implicit in Weeks' calculations, see also [1], that M_4 realises the smallest possible volume for an orientable cusp, and that M_4 double covers the non-orientable manifold M_0. As such that pair (M_4, M_0) can be said to play a corresponding role in dimension 3 to the pair of triangle groups ($T(2,3,7)$, $T^*(2,3,7)$) for surfaces. We also have the following elementary result.

PROPOSITION 3. *Let $p : M^3 \to N^3$ be a finite covering map of hyperbolic manifolds, such that $\pi_1 M^3 \triangleleft \pi_1 N^3$. Then the order of the covering transformation group equals $v(M^3)/v(N^3)$.*

Given the non-closed form of the hyperbolic volume function it is hard to be more precise. However for a cusped orientable manifold analogy with Theorem 1 suggests the existence of some constant A, related to the minimal volume for a suitable class of *orb*ifolds, such that

$$|I(M^3)| \leq Av(M^3).$$

(Here in contrast to the 2-dimensional case we include the orientation reversing isometries.) Given that $v(M_4) = 2.0298\ldots$, the following calculation shows that $A > 3.9\ldots$.

EXAMPLE: The full isometry group $I(M_4)$ is a dihedral group of order 8.

Proof: An isometry induces an automorphism of π_1. Conversely such an automorphism can be realised as a homotopy equivalence, well-defined up to inner automorphisms. By Mostow's rigidity theorem this homotopy equivalence can be deformed to an isometry, showing that $I(M_4) \cong Out(\pi_1 M)$. In [12] W. Magnus proved directly that the outer automorphism group is dihedral.

This rather informal discussion suggests the following tentative definition: the finite group G is a *3-dimensional Hurwitz group* if and only if there is an epimorphism $\phi : \Gamma_4 \twoheadrightarrow G$.

EXAMPLE: If p is an odd prime, $p \geq 5$, then reduction modulo p defines homomorphisms

$$\phi : \Gamma_4 \twoheadrightarrow \begin{cases} PSL_2(\mathbb{F}_p), & p \equiv 1 (mod\, 3) \\ PSL_2(\mathbb{F}_{p^3}), & p \not\equiv 1 (mod\, 3). \end{cases}$$

In both cases ϕ is an epimorphism, since if $q = p^t$ is a prime power with $t \geq 1$, and $q \neq 2^t (t > 1)$ or 3^2, then $PSL_2(\mathbb{F}_q)$ is generated by $\begin{pmatrix} 1 & 1 \\ 0 & 1 \end{pmatrix}$ and $\begin{pmatrix} 1 & 0 \\ -\xi & 1 \end{pmatrix}$, where ξ generates $\mathbb{F}_q$ over $\mathbb{F}_p$, see [6], Theorem 2.8.4.

Elementary properties of 3-dimensional Hurwitz groups: i) and ii) are identical with the corresponding properties in dimension 2. However, since Γ_4 maps onto the infinite cyclic group $\mathbb{Z}$, property iii) must be modified. For example no abelian group of rank greater than or equal to 2 can be Hurwitz, and the positive answer to the Smith conjecture imposes severe restraints on the solvable groups which can arise. The divisibility condition iv) must also be weakened: for any prime number $\ell \geq 5$, the example above shows the existence of Hurwitz groups G with $\ell \nmid |G|$. Thus $PSL_2(\mathbf{F}_7)$ has order $2^3.3.7$ and $PSL_2(\mathbf{F}_{19})$ has order $2^2.3^2.5.19$. The question of which sporadic simple groups (if any) can be homomorphic images of Γ_4 would seem to be interesting.

We now turn to the problem of realising an arbitrary finite group G first as a group of hyperbolic isometries, and then as the full group $I(M^3)$ for some manifold M^3. We are able to prove the weak analogue of Theorem 2 - this corresponds to the existence of a Fuchsian group with preassigned signature. However as mentioned previously there are problems in extending the notion of "maximal" from Fuchsian to Kleinian groups. We also outline two proofs: the more geometric applies to a wider class of groups, but the algebraic includes an estimate of volume and hence of the index $[I(M^3) : A_n]$.

THEOREM 4. *The alternating group A_n can be realised as a group of hyperbolic isometries for infinitely many values of n.*

First proof ($n =$ odd): Consider the Pretzel knot of type $\underbrace{(3,3,\ldots,3)}$ illustrated below.

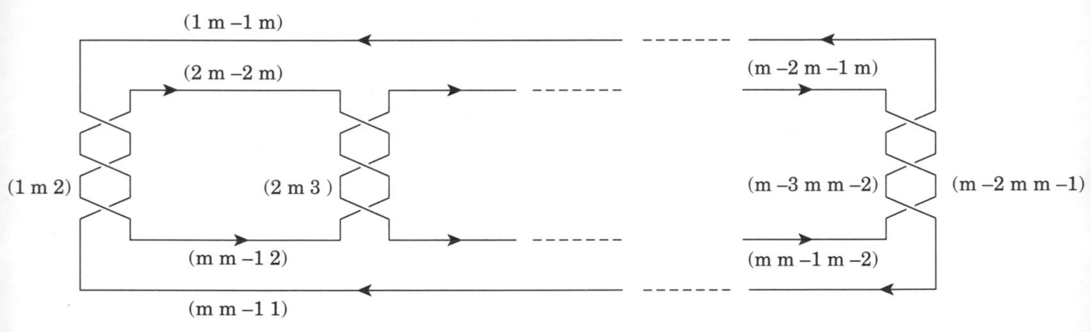

Here $m \geq 5$ and $m = 2n - 1$. With each crossing associate the marked 3-cycle, and use it to map a generator of πK into A_{2n-1}. This gives a homomorphism of πK onto A_{2n-1}, and it remains to show that the complement of K is hyperbolic. It is easy to see that K is not a torus knot, for example by showing that the centre of πK is trivial. On the other hand $S^3 \setminus K$ has a branched double cover, which is Seifert fibred. The only incompressible tori bound neighbourhoods of exceptional fibres over points of S^2, and on applying $\mathbb{Z}/2$ their images in $S^3 \setminus K$ turn out to be singular. Hence K cannot be a satellite knot, and Thurston's theorem provides a hyperbolic structure. (I am grateful to W.B.R. Lickorish for explaining some of these properties of Pretzel knots to me.)

There is a discussion of the case $m = 5$, i.e. of the Pretzel knot $(3,3,3)$, described as a modest example in [18] page 621. As in the case of the figure of eight knot Riley constructs an explicit excellent representation, this time over $\mathbb{Z}(\eta)$, where η is a root of $3 + (h(y))^2 = 0$, $h(y) = y - 3y^2 + 2y^3 - 4y^4 + y^5 - y^6$. Of course by Reid's theorem the index of the image of πK in $PSL_2(\mathbb{Z}(\eta))$ must be infinite.

Second proof ($n = 16t + 1, 16t + 6, 16t + 8$): According to W. Magnus [12, p.153] Drillick proves in his thesis, that for the stated values of n, A_n is a homomorphic image of $PSL_2(\mathbb{Z}(i))$. If L is the Whitehead link, illustrated below, then there is an excellent representation of πL in $PSL_2(\mathbb{Z}(i))$, such that the image has index 12.

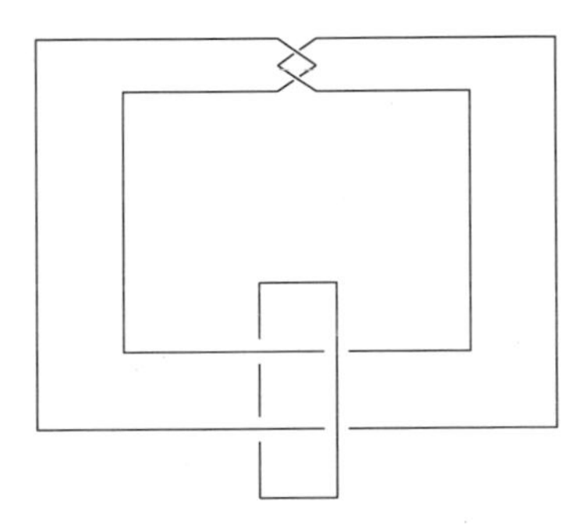

Indeed $S^3 \setminus L$ corresponds to the manifold M_{125} in Weeks' list, and has volume equal to $3.6638\ldots$. The fundamental group Γ_{125} must project onto A_n, since the order $n!$ is always greater than 12. The volume can also be estimated as follows - we include the calculation in view of its potential application to other discrete groups Γ.

Let O be the ring of integers in the imaginary quadratic field k. Write D_k for the discriminant of k over $\mathbb{Q}$ and ζ_k for the Riemann zeta function of k. Then the volume of the quotient orbifold defined by $\Gamma = PSL_2(O)$ equals

$|D_k|^{3/2}\zeta_k(2) \mid 4\pi^2$. With $k = \mathbb{Q}(i)$ the discriminant equals 4, so that $|D|^{3/2} = 8$. The value of $\zeta_k(2)$ is not known in closed form, but

$$\zeta_k(2) = \zeta_{\mathbb{Q}}(2)L(2), \text{ where } L(2) = \prod_p 1/(1 - (\frac{-1}{p})p^{-2}) = 0.9159\ldots$$

Since $\zeta_{\mathbb{Q}}(2) = \frac{\pi^2}{6}$, the orbifold volume equals $8 \cdot \frac{\pi^2}{6} \cdot \frac{1}{4\pi^2} \cdot (0.9159) = 0.3053\ldots$. The subgroup Γ_{125} has index 12 in $PSL_2(\mathbb{Z}(i))$, so the volume of $S^3 \setminus L$ equals $3.6638\ldots$ agreeing with the tabulated value.

Remark: C. Adams has shown recently that the threee smallest possible limit volumes for hyperbolic 3-orbifolds are $0.3053\ldots$, $0.4444\ldots$ and $0.4579\ldots$, see Abstracts presented to the American Math. Soc., October 1989, no. 851-57-15. Note that M_{125} possesses 8 symmetrics, and $3.6638\ldots/8 = 0.4579\ldots$. Confining our attention to manifolds, $v(M_{125})/2$ equals the volume of the non-orientable manifold M_1, triangulated by 2 ideal tetrahedra, and having 2 symmetries. The considerations suggest that $A_n(n \equiv 1, 6 \text{ or } 8(\mathit{mod}\ 16))$ is a subgroup of very small index in the full isometry group of the covering manifold $\tilde{M}$ of M_{125} given by the second proof.

4. AN EXTENSION OF THEOREM 4 TO CLOSED MANIFOLDS

If M is a hyperbolic manifold with one or more cusps, then M can be closed by means of Dehn surgeries, each of which is associated with a pair (r, s) of coprime integers. We choose a basis m, ℓ for $\pi_1 T = H_1 T$, where T is a toral cross-section of the cusp, cut this off, and glue in a solid torus $D^2 \times S^1$ in such a way as to kill the homotopy class $\ell^s m^r$. Then it is known that for almost all pairs (r, s) the resulting manifold with one fewer cusps is still hyperbolic. In the case of $M_4 = S^3 \setminus K$, where K is the figure of eight knot, the only bad pairs are $(4,1)$ and $(3,1)$, and we denote the resulting closed manifold by $M_{(r,s)}$. Furthermore there is a formula for approximating $v(M_{(r,s)})$, which in the case of the figure of eight knot is

$$v(M_{(r,s)}) = v(M_4) - \frac{2\sqrt{3}\pi^2}{r^2 + 12s^2} + 4\sqrt{3}\frac{(r^4 - 72r^2s^2 + 144s^4)\pi^4}{3(r^2 + 12s^2)^4} - \ldots,$$

see [14, §6].

As in the previous section let $\phi : PSL_2(\mathbb{Z}(\omega)) \twoheadrightarrow PSL_2(\mathbb{F}_p)$ be reduction modulo p for some prime $p \equiv 1(\mathit{mod}\ 3)$, and let the image of ω be u, a primitive cube root of 1 in $\mathbb{F}_p$. The fundamental group of the closed manifold is

$$\pi_1 M_{(r,s)} = \{x_1, x_2 : wxw^{-1} = x_2, \ell^s m^r = 1\},$$

where $w = [x_1^{-1}, x_2]$, $m = x_1$, and $\ell = [x_2^{-1}, x_1][x_1^{-1}, x_2]$.

The image of the relator $\ell^s m^r$ in $PSL_2(\mathbb{F}_p)$ equals $(-1)^{r+s+1}\begin{pmatrix}-1 & s(4u+2)+r \\ 0 & -1\end{pmatrix}$, and in order to factorise ϕ through $\pi_1 M_{(r,s)}$ we must satisfy the pair of equations (modulo p):

$$s(4u+2)+r = 0, \text{ and } u^2+u+1 = 0.$$

Assuming that $s \not\equiv 0 (mod\, p)$, these can be rewritten as

$$u = -(2s+r)/4s, \; r^2 = -12s^2, \text{ see [15]}.$$

THEOREM 5. *Let $p \equiv 1 (mod\, 3)$ and $\epsilon > 0$ be an arbitrary positive real number. There exists a closed hyperbolic manifold M with $|v(M) - v(M_4)| < \epsilon$, such that $PSL_2(\mathbb{F}_p)$ is a homomorphic image of $\pi_1 M$. Hence there exists a closed mainfold $\tilde{M}$ with $PSL_2(\mathbb{F}_p) \subseteq I(\tilde{M})$, such that $v(\tilde{M})$ is arbitrarily close to $p(p^2-1)$ $(1.0149\ldots)$.*

Proof: Assume first the $p \equiv 1 (mod\, 12)$, so that -1 is a quadratic residue. In order to replace the equation $r^2 = -12s^2$ by a linear equation it will be enough to show that 3 is a quadratic residue modulo p. Writing $p = 3k+1$ the quadratic reciprocity law implies that $\left(\frac{3}{3k+1}\right) = \left(\frac{3k+1}{3}\right) = 1$, and it follows that -12 is a square in these cases. If on the other hand $p \equiv 7 (mod\, 12)$, the argument is similar, and uses the fact that both -1 and 3 are now non-quadratic residues modulo p. The theorem is now clear, since we can represent the solutions of a pair of linear equations $(mod\, p)$ by (r, s), where r and s are arbitrarily large.

Additional comments - with $d^2 = -12$ the condition $(r, s) = 1$ for a Dehn surgery implies that the smallest relevent pairs are $(\pm d, 1)$ and $(1, \pm d^{-1})$. We tabulate these for the first few primes:

p	$(d,1)$	$(-d,1)$	$(1,d^{-1})$	$(1,-d^{-1})$
7	(4,1)	(3,1)	(1,2)*	(1,5)
13	--	(12,1)*	--	(1,12)
19	(8,1)*	(11,1)	(1,12)	(1,7)
31	(9,1)*	(22,1)	(1,7)	(1,24)
37	(5,1)*	(32,1)	(1,15)	(1,22)
42	(17,1)*	(26,1)	(1,38)	(1,5)

An asterisk against an entry indicates that it corresponds to a manifold of smallest possible volume for this construction. Note that for $p = 7$, the pairs (4,1), (3,1) are excluded by the hyperbolic condition, and that the pair (2,1) leading to the minimal volume obtainable from M_4 in this way fails to satisfy the equation $v = \pm 4s$.

Those prime numbers p such that $p-12$ is a prefect square would seem to form an interesting subclass. Thus we have

COROLLARY 6. *Let $p \equiv 1(\mathit{mod}\ 3)$, $p = 12+d^2$, d odd and not divisible by 3. Then $PSL_2(\mathbf{F}_p)$ is a homomorphic image of $\pi_1(M_{(d,1)})$ with*

$$v(M_4) - \frac{2\sqrt{3}\pi^2}{p} < v(M_{(d,1)}) < v(M_4) - \frac{2\sqrt{3}\pi^2}{p} + \frac{4\sqrt{3}\pi^4}{3p^4}(p^2 - 96p + 1152).$$

As pointed out in the introduction this method can be greatly generalised, and will serve to realise a large class of finite simple groups quite explicitly as hyperbolic isometry groups. A possible alternative method of attack would be to use Thurston's condition (stated in terms of pseudo-Anosov diffeomorphisms) for a surface bundle over S^1 to be hyperbolic. "Vertical" isometries of the total space would then be defined by isometries of the fibre compatible with the structural diffeomorphism. Clearly this approach raises interesting mathematical problems, but is harder than the one used here.

REFERENCES

1. C. Adams, The non-compact hyperbolic 3-manifold of minimal volume, preprint 1989, Williams College, MA 01267.

2. M. Conder, Generators for alternating and symmetric groups, J. London Math. Soc. (1980) 22, 75-86.

3. J. Conway, et al., Atlas of finite groups, Clarendon Press (Oxford) 1985.

4. A. Drillick, The Picard group, Ph.D. Thesis (1971), NYU.

5. H. Glover, D. Sjerve, Representing $PSL_2(p)$ on a Riemann surface of least genus, L'Enseignement Math. (1985) 31, 305-325.

6. D. Gorenstein, Finite groups, Harper & Row (NY), 1968.

7. L. Greenberg, Maximal groups and signatures, Annals of Math. Studies, no.79 (1973), 207-226.

8. A. Hurwitz, Über algebraische Gebilde mit eindeutigen Transformationen in sich, Math. Annalen (1893) 41, 403-442.

9. S. Kobayashi, Transformation groups in differential geometry (EM 70), Springer-Verlag (Heidelberg) 1972.

10. A. Macbeth, Generators of the linear fractional groups, Proc. Symp. Pure Math. (1969) 12, 14-32.

11. W. Magnus, Untersuchungen über einige unendliche diskontinuierliche Gruppen, Math. Armalen (1931) 105, 52-74.

12. W. Magnus, Non-Euclidean tesselations and their groups, Academic Press (NY) 1974.

13. G. Malle, Hurwitz groups and $G_2(q)$, preprint 1989, Heidelberg University, D-6900.

14. W. Neumann, D. Zagier, Volumes of hyperbolic 3-manifolds, Topology (1985) 24, 307-332.

15. A. Nicas, An infinite family of non-Haken hyperbolic 3-manifolds with vanishing Whitehead groups, Math. Proc. Camb. Phil. Soc. (1986) 99, 239-246.

16. A. Reid, Arithmeticity of knot complements, preprint 1989, Ohio State University, 43210.

17. R. Riley, A quadratic parabolic knot, Math. Proc. Camb. Phil. Soc. (1975) 77, 281-288.

18. R. Riley, Applications of a computer implementation of Poincaré's theorem on fundamental polyhedra, Math. of Computation (1983) 40, 607-632.

19. W. Thurston, Geometry and Topology of 3-manifolds, Princeton University notes (1978).

20. C.T.C. Wall, Geometric structures on compact complex analytic surfaces, Topology (1986) 25, 119-153.

21. J. Weeks, Updated manifold census, dated 16.1.1989.

22. A. Woldar, Genus actions on surfaces, preprint 1986, Villanova University, PA 19085.

Cambridge University & ETH Zürich,
January 1990.

Pin Structures on Low–dimensional Manifolds

by
R. C. Kirby[1] and L. R. Taylor[1]

§0. Introduction

Pin structures on vector bundles are the natural generalization of *Spin* structures to the case of non-oriented bundles. $Spin(n)$ is the central $\mathbf{Z}/2\mathbf{Z}$ extension (or double cover) of $SO(n)$ and $Pin^-(n)$ and $Pin^+(n)$ are two different central extensions of $O(n)$, although they are topologically the same. The obstruction to putting a *Spin* structure on a bundle ξ $(= R^n \to E \to B)$ is $w_2(\xi) \epsilon H^2(B; \mathbf{Z}/2\mathbf{Z})$; for Pin^+ it is still $w_2(\xi)$, and for Pin^- it is $w_2(\xi) + w_1^2(\xi)$. In all three cases, the set of structures on ξ is acted on by $H^1(B; \mathbf{Z}/2\mathbf{Z})$ and if we choose a structure, this choice and the action sets up a one–to–one correspondence between the set of structures and the cohomology group.

Perhaps the most useful characterization (Lemma 1.7) of $Pin^\pm$ structures is that Pin^- structures on ξ correspond to *Spin* structures on $\xi \oplus \det \xi$ and Pin^+ to *Spin* structures on $\xi \oplus 3 \det \xi$ where $\det \xi$ is the determinant line bundle. This is useful for a variety of "descent" theorems of the type: a $Pin^\pm$ structure on $\xi \oplus \eta$ descends to a Pin^+ (or Pin^- or *Spin*) structure on ξ when $\dim \eta = 1$ or 2 and various conditions on η are satisfied.

For example, if η is a trivialized line bundle, then $Pin^\pm$ structures descend to ξ (Corollary 1.12), which enables us to define $Pin^\pm$ bordism groups. In the *Spin* case, *Spin* structures on two of ξ, η and $\xi \oplus \eta$ determine a *Spin* structure on the third. This fails, for example, for Pin^- structures on η and $\xi \oplus \eta$ and ξ orientable, but versions of it hold in some cases (Corollary 1.15), adding to the intricacies of the subject.

Another kind of descent theorem puts a $Pin^\pm$ structure on a submanifold which is dual to a characteristic class. Thus, if V^{m-1} is dual to $w_1(T_M)$ and M^m is $Pin^\pm$, then $V \pitchfork V$ gets a $Pin^\pm$ structure and we have a homomorphism of bordism groups (Theorem 2.5),

$$[\cap w_1^2] : \Omega_m^{Pin^\pm} \longrightarrow \Omega_{m-2}^{Pin^\mp}$$

that proved useful in [K–T]. Or, if F^{m-2} is the obstruction to extending a Pin^- structure on $M^m - F$ over M, then F gets a Pin^- structure if M is oriented (Lemma 6.2) or M is not orientable but $F \pitchfork V$ has a trivialized normal bundle in V (Theorem 6.9). These results give generalizations of the Guillou-Marin formula [G-M], Theorem 6.3,

$$2\beta(F) \equiv F \cdot F - \text{sign } M \qquad (\text{mod } 16)$$

[1] Partially supported by the N.S.F.

to any characterized pair (M^4, F^2) with no condition on $H_1(M^4; \mathbf{Z}/2\mathbf{Z})$.

Here, $\beta(F)$ is the $\mathbf{Z}/8\mathbf{Z}$ Brown invariant of a $\mathbf{Z}/4\mathbf{Z}$ quadratic enhancement of the $\mathbf{Z}/2\mathbf{Z}$ intersection form on $H_1(F; \mathbf{Z}/2\mathbf{Z})$; given a Pin^- structure on F, the enhancement counts half-twists, mod 4, in imbedded circles representing elements of $H_1(F; \mathbf{Z}/2\mathbf{Z}Z)$. This is developed in §3, where it is shown that

$$\beta : \Omega_2^{Pin^-} \longrightarrow \mathbf{Z}/8\mathbf{Z}$$

gives the isomorphism in the following table.

$$\begin{array}{llll}
\Omega_1^{Spin} = \mathbf{Z}/2\mathbf{Z} & \Omega_2^{Spin} = \mathbf{Z}/2\mathbf{Z} & \Omega_3^{Spin} = 0 & \Omega_4^{Spin} = \mathbf{Z} \\
\Omega_1^{Pin^-} = \mathbf{Z}/2\mathbf{Z} & \Omega_2^{Pin^-} = \mathbf{Z}/8\mathbf{Z} & \Omega_3^{Pin^-} = 0 & \Omega_4^{Pin^-} = 0 \\
\Omega_1^{Pin^+} = 0 & \Omega_2^{Pin^+} = \mathbf{Z}/2\mathbf{Z} & \Omega_3^{Pin^+} = \mathbf{Z}/2\mathbf{Z} & \Omega_4^{Pin^+} = \mathbf{Z}/16\mathbf{Z}
\end{array}$$

In §2 we calculate the 1 and 2 dimensional groups and show that the non–zero one dimensional groups are generated by the circle with its Lie group framing, S^1_{Lie}, (note the Möbius band is a Pin^+ boundary for S^1_{Lie}); $\mathbf{RP}^2$ generates $\Omega_2^{Pin^-}$; the Klein bottle, the twisted S^1_{Lie} bundle over S^1, generates $\Omega_2^{Pin^+}$; and T^2_{Lie}, the torus with its Lie group framing generates Ω_2^{Spin}. By §5 enough technique exists to calculate the remaining values and show that $\Omega_3^{Pin^+}$ is generated by the twisted T^2 bundle over S^1 with Lie group framing on the fiber torus; $\Omega_4^{Pin^+}$ is generated by $\mathbf{RP}^4$. The Cappell–Shaneson fake RP^4 represents $\pm 9 \in \mathbf{Z}/16\mathbf{Z}$ [Stolz]; the Kummer surface represents $8 \in \mathbf{Z}/16\mathbf{Z}$ and in fact, a *Spin* 4–manifold bounds a Pin^+ 5-manifold iff its index is zero mod 32. The Kummer surface also generates Ω_4^{Spin}.

Section 4 contains a digression on *Spin* structures on 3–manifolds and a geometric interpretation of Turaev's work [Tu] on trilinear intersection forms

$$H_2\left(M^3; \mathbf{Z}/2\mathbf{Z}\right) \otimes H_2\left(M^3; \mathbf{Z}/2\mathbf{Z}\right) \otimes H_2\left(M^3; \mathbf{Z}/2\mathbf{Z}\right) \longrightarrow \mathbf{Z}/2\mathbf{Z}\ .$$

This is used in calculating the μ–invariant: let $\mu(M, \Theta_1)$ be the μ–invariant of M^3 with *Spin* structure Θ_1. The group $H^1\left(M^3; \mathbf{Z}/2\mathbf{Z}\right)$ acts on *Spin* structures, so let $\alpha \in H^1\left(M^3; \mathbf{Z}/2\mathbf{Z}\right)$ determine Θ_2. Then α is dual to an imbedded surface F^2 in M which gains a Pin^- structure from Θ_1 and

$$\mu(\Theta_2) = \mu(\Theta_1) - 2\beta(F) \qquad (\text{mod } 16)$$

Four dimensional characteristic bordism $\Omega_4^!$ is studied in §6 with generalizations of [F-K] and [G-M]. We calculate, in Theorem 6.5, the μ–invariant of circle bundles over surfaces, $S(\eta)$, whose disk bundle, $D(\eta)$, has orientable total space. Fix a *Spin* structure on $S(\eta)$, Θ. Then

$$\mu(S(\eta), \Theta) = \text{sign }(D(\eta)) - \text{ Euler class}(\eta) + 2 \cdot b(F) \qquad (\text{mod } 16)$$

where $b(F) = 0$ if the *Spin* structure Θ extends across $D(\eta)$ and is β of a Pin^- structure on F induced on F from Θ otherwise.

The characteristic bordism groups are calculated geometrically in §7, in particular,

$$\Omega_4^! = \mathbf{Z}/8\mathbf{Z} \oplus \mathbf{Z}/4\mathbf{Z} \oplus \mathbf{Z}/2\mathbf{Z}\ .$$

Just as Robertello was able to use Rochlin's Theorem to describe the Arf invariant of a knot [R], so we can use $\beta : \Omega_2^{Pin^-} \longrightarrow \mathbf{Z}/8\mathbf{Z}$ to give a $\mathbf{Z}/8\mathbf{Z}$ invariant to a characterized link L in a *Spin* 3–manifold M with a given set of even longitudes for L (Definition 8.1). This invariant is a concordance invariant (Corollary 8.4), and if each component of L is torsion in $H_1(M; Z)$, then L has a natural choice of even longitudes (Definition 8.5).

Section 9 contains a brief discussion of the topological case of some of our 4-manifold results. In particular, the formula above must now contain the triangulation obstruction $\kappa(M)$ for an oriented, topological 4–manifold M^4:

$$2\beta(F) \equiv F \cdot F - \text{sign}\ (M) + 8\kappa(M) \qquad (\text{mod } 16)$$

(recall that (M, F) is a characterized pair).

§1. *Pin* Structures and generalities on bundles

The purpose of this section is to define the *Pin* groups and to discuss the notion of a *Pin* structure on a bundle.

Recall that rotations of $\mathbf{R}^n$ are products of reflections across $(n-1)$–planes through the origin, an even number for orientation preserving rotations and an odd number for orientation reversing rotations. These $(n-1)$–planes are not oriented so they can equally well be described by either unit normal vector. Indeed, if $\mathbf{u}$ is the unit vector, and if $\mathbf{x}$ is any point in $\mathbf{R}^n$, then the reflection is given by $x - 2(\mathbf{x}\cdot\mathbf{u})\mathbf{u}$. Thus an element of $O(n)$ can be given as $(\pm\mathbf{v}_1)(\pm\mathbf{v}_2)\cdots(\pm\mathbf{v}_k)$ where each $\mathbf{v}_i$ is a unit vector in $\mathbf{R}^n$ and k is even for $SO(n)$. Then elements of $Pin(n)$, a double cover of $O(n)$, are obtained by choosing an orientation for the $(n-1)$–planes or equivalently choosing one of the two unit normals, so that an element of $Pin(n)$ is $\mathbf{v}_1\cdots\mathbf{v}_k$; if k is even we get elements of $Spin(n)$. With this intuitive description as motivation, we proceed more formally to define *Pin* (see [ABS]).

Let V be a real vector space of dimension n with a positive definite inner product, $(\ ,\ )$. The Clifford algebra, $\mathbf{Cliff}^{\pm}(V)$, is the universal algebra generated by V with the relations

$$\mathbf{v}\mathbf{w} + \mathbf{w}\mathbf{v} = \begin{matrix} 2(\mathbf{v}, \mathbf{w}) & \quad \text{for } \mathbf{Cliff}^+(V) \\ -2(\mathbf{v}, \mathbf{w}) & \quad \text{for } \mathbf{Cliff}^-(V) \end{matrix}$$

If $\mathbf{e}_1, \cdots \mathbf{e}_n$ is an orthonormal basis for V, then the relations imply that $\mathbf{e}_i\mathbf{e}_j = -\mathbf{e}_i\mathbf{e}_j$, $i \neq j$ and $\mathbf{e}_i\mathbf{e}_i = \pm 1$ in $\mathbf{Cliff}^{\pm}(V)$. The elements $\mathbf{e}_I = \mathbf{e}_{i_1}\cdots\mathbf{e}_{i_k}$, $I =$

$\{1 \leq i_1 < i_2 \cdots < i_k \leq n\}$ form a ($\mathbf{e}_I\mathbf{e}_J = 0$, $\mathbf{e}_I\mathbf{e}_I = \pm 1$) basis for $\mathbf{Cliff}^{\pm}(V)$. So dim $\mathbf{Cliff}^{\pm}(V) = 2^n$; note that as vector spaces, $\mathbf{Cliff}^{\pm}(V)$ is isomorphic to the exterior algebra generated by V, but the multiplications are different, e.g. $\mathbf{e}_i\mathbf{e}_i = \pm 1 \neq 0 = \mathbf{e}_i \wedge \mathbf{e}_i$.

Let $Pin^{\pm}(V)$ be the set of elements of $\mathbf{Cliff}^{\pm}(V)$ which can be written in the form $\mathbf{v}_1\mathbf{v}_2 \cdots \mathbf{v}_k$ where each $\mathbf{v}_i$ is a unit vector in V; under multiplication, $Pin^{\pm}(V)$ is a compact Lie group. Those elements $\mathbf{v}_1\mathbf{v}_2 \cdots \mathbf{v}_k \in Pin^{\pm}(V)$ for which k is even form $Spin(V)$.

Define a "transpose" $\mathbf{e}_I^t = \mathbf{e}_{i_k} \cdots \mathbf{e}_{i_1} = (-1)^{k-1}\mathbf{e}_I$ and an algebra homomorphism $\alpha(\mathbf{e}_I) = (-1)^k\mathbf{e}_I = (-1)^{|I|}\mathbf{e}_I$ and extend linearly to $\mathbf{Cliff}^{\pm}(V)$. We have a $\mathbf{Z}/2\mathbf{Z}$–grading on $\mathbf{Cliff}^{\pm}(V)$: $\mathbf{Cliff}^{\pm}(V)_0$ is the $+1$ eigenspace of α and $\mathbf{Cliff}^{\pm}(V)_1$ is the -1 eigenspace. For $w \in \mathbf{Cliff}^{\pm}(V)$, define an automorphism $\rho(w)\colon \mathbf{Cliff}^{\pm}(V) \to \mathbf{Cliff}^{\pm}(V)$ by

$$\rho(w)(v) = \begin{cases} wvw^t & \text{for } \mathbf{Cliff}^-(V) \\ \alpha(w)vw^t & \text{for } \mathbf{Cliff}^+(V) \end{cases}$$

We can define a norm in the Clifford algebra, $N\colon \mathbf{Cliff}^{\pm} \to \mathbf{R}^+$ by $N(x) = \alpha(x)x$ for all $x \in \mathbf{Cliff}^{\pm}(V)$. Then we can define $Pin^{\pm}(V)$ to be $\{w \in \mathbf{Cliff}^{\pm}(V) \mid \rho(w)(V) = V \text{ and } N(w) = 1\}$. Hence if $w \in Pin^{\pm}(v)$, $\rho(w)$ is an automorphism of V so ρ is a representation $\rho\colon Pin^{\pm}(V) \to O(V)$ and by restriction $\rho\colon Spin(V) \to SO(V)$.

It is easy to verify that $\rho(w)$ acts on V by reflection across the hyperplane $w^{\perp}$, e.g. for $Pin^-(V)$,

$$\rho(\mathbf{e}_1)\mathbf{e}_i = \mathbf{e}_1\mathbf{e}_i\mathbf{e}_1 = \begin{cases} -\mathbf{e}_1^2\mathbf{e}_i = \mathbf{e}_i & i \neq 1 \\ \mathbf{e}_1^2\mathbf{e}_i = -\mathbf{e}_i & i = 1 \end{cases}$$

If r and I are basepoints in the components of $O(V)$, where r is reflection across $\mathbf{e}_1^{\perp}$, then $\rho^{-1}\{r, I\} = \{\pm\mathbf{e}_1, \pm 1\}$ and

$$\rho^{-1}\{r, I\} \cong \begin{cases} \mathbf{Z}/2\mathbf{Z} \oplus \mathbf{Z}/2\mathbf{Z} & \text{for } Pin^+(V) \\ \mathbf{Z}/4\mathbf{Z} & \text{for } Pin^-(V) \end{cases}.$$

The $\mathbf{Z}/2\mathbf{Z} = \{-1, 1\} \in Pin^{\pm}$ is central and $Pin^{\pm}(V)/\{\pm 1\} = O(V)$. If $n > 1$, this $\mathbf{Z}/2\mathbf{Z}$ is the center of $Pin^{\pm}(V)$ and, since $O(V)$ has a non–trivial center, for $n > 1$, the $\mathbf{Z}/2\mathbf{Z}$ central extensions $Pin^{\pm} \to O(V)$ are non–trivial.

Thus $Pin^{\pm}(V)$ is a double cover of $O(V)$. As spaces, $Pin^{\pm}(V) = Spin(V) \amalg Spin(V)$ but the group structure is different in the two cases. We can think of $-1 \in \rho^{-1}(I)$ as rotation of V (about any axis) by 2π and $+1 \in \rho^{-1}(I)$ as the identity. More precisely, an arc in $Pin^{\pm}(V)$ from 1 to -1 maps by ρ to a loop in $O(V)$ which generates $\pi_1(O(V))$; in fact, for $\theta \in [0, \pi]$, the arc $\theta \to \pm\mathbf{e}_1 \cdot (\cos\theta\mathbf{e}_1 + \sin\theta\mathbf{e}_2)$ is one such. Even better, we may think of $Pin^{\pm}$ as scheme for distinguishing an odd number of full twists from an even number.

We use $Pin^{\pm}(n)$ to denote $Pin^{\pm}(V)$ where V is $\mathbf{R}^n$.

Remark. The tangent bundle of $\mathbf{RP}^2$, $T_{\mathbf{RP}^2}$, has a $Pin^-(2)$–structure.

We can "see" the $Pin^-(2)$ structure on $T_{\mathbf{RP}^2}$ as follows: decompose $\mathbf{RP}^2$ into a 2–cell, B^2, and a Möbius band, MB, with core circle $\mathbf{RP}^1$. Then $T_{\mathbf{RP}^2}|_{MB}$ can be described using two coordinate charts, U_1 and U_2, with local trivializations $(\mathbf{e}_1, \mathbf{e}_2)$, in which $\mathbf{e}_1$ is parallel to $\mathbf{RP}^1$ and $\mathbf{e}_2$ is normal, and with transition function $U_1 \cap U_2 \to Pin^-(2)$ which sends the two components of $U_1 \cap U_2$ to 1 and $\mathbf{e}_2$. Then $T_{\mathbf{RP}^2}|_{\partial MB}$ is a trivial $\mathbf{R}^2$–bundle over $S^1 = \partial MB$ which is trivialized by the transition function 1 and $\mathbf{e}_2^2 = -1$. Now $\mathbf{e}_1$ would be tangent to S^1 but the $\mathbf{e}_2^2 = -1$ adds a rotation by 2π as $S^1 = \partial MB$ is traversed. But this trivialization on $T_{\mathbf{RP}^2}|_{S^1}$ is exactly the one which extends over the 2–cell B^2. Thus $\mathbf{RP}^2$ is Pin^-. Note that this process fails if $\mathbf{e}_2^2 = +1$, and, in fact, $\mathbf{RP}^2$ does not support a Pin^+ structure (see Lemma 1.3 below).

We now review the theory of G bundles, for G a topological group, and the theory of H structures on a G bundle. A principal G bundle is a space E with a left G action, $E \times G \to E$ such that no point in E is fixed by any non–identity element of G. We let $B = E/G$ be the orbit space and $p\colon E \to B$ be the projection. We call B the *base* of the bundle and say that E is a *bundle over* B. We also require a local triviality condition. Explicitly, we require a numerable cover, $\{\mathcal{U}_i\}$, of B and G maps $r_i\colon \mathcal{U}_i \times G \to E$ such that the composite $\mathcal{U}_i \times G \xrightarrow{r_i} E \xrightarrow{p} B$ is just projection onto $\mathcal{U}_i$ followed by inclusion into B. Such a collection is called an *atlas* for the bundle and it is convenient to describe bundles in terms of some atlas. The functions $r_j^{-1} \circ r_i$ are G maps, $\mathcal{U}_i \cap \mathcal{U}_j \times G \to \mathcal{U}_i \cap \mathcal{U}_j \times G$, which commute with the projection. Hence they can be given as *transition* functions $g_{ij}\colon \mathcal{U}_i \cap \mathcal{U}_j \to G$. Note $g_{ii} = id$, $g_{ij}^{-1} = g_{ji}$ and $g_{ik} = g_{ij} \circ g_{jk}$ on $\mathcal{U}_i \cap \mathcal{U}_j \cap \mathcal{U}_k$. Conversely, given any numerable cover of a space B and a set of maps satisfying these three conditions, we can find a principal G bundle and an atlas for it so the base space is B and the transitions functions are our given functions.

Suppose E_0 and E_1 are two G bundles over B_0 and B_1 respectively. Let $f\colon E_0 \to E_1$ be a map. A bundle map covering f is a G map $F\colon E_0 \to E_1$ so that $p_1 \circ F = f \circ p_0$, where p_i is the projection in the i-th bundle. We say two bundles over B are *equivalent* iff there exists a bundle map between them covering the identity.

Given a bundle over B, say E, with atlas $\mathcal{U}_i$ and g_{ij}, and a map $f\colon B_0 \to B$, the *pull–back of E along f* is the bundle over B_0 with numerable cover $f^{-1}(\mathcal{U}_i)$ and transition functions $g_{ij} \circ f$. The pull–backs of equivalent bundles are equivalent. A bundle map between E_0 and E_1 covering $f\colon B_0 \to B_1$ is equivalent to a bundle equivalence between E_0 and the pull–back of E_1 along f. Hence we mostly discuss the case of bundle equivalence.

Given any atlas for a bundle, say $\mathcal{U}_i; g_{ij}$, and a subcover $\mathcal{V}_\alpha$ of $\mathcal{U}_i$ we can restrict the g_{ij} to get a new family of transition functions $g_{\alpha\beta}$. Clearly these two atlases represent the same bundle. Given two numerable covers, it is possible to find a third numerable cover which refines them both, so it is never any loss of generality when considering two bundles over the same base to assume the transition functions are defined on a common cover.

A bundle equivalence between bundles given by transition functions g_{ij} and g'_{ij} for the same cover is given by maps $h_i : \mathcal{U}_i \to G$ such that, for all i and j and all $u \in \mathcal{U}_i \cap \mathcal{U}_j$, $g'_{ij}(u) = h_i(u)\, g_{ij}(u) \left(h_j(u)\right)^{-1}$.

Given a continuous homomorphism $\psi: H \to G$, we can form a principal G bundle from a principal H bundle by applying ψ to any atlas for the H bundle. If $p: E \to B$ is the H bundle, we let $p_\psi: E \times_H G \to B$ denote the associated G bundle. Equivalent H bundles go to equivalent G bundles. We say that a G bundle, $p: E \to B$, had an H structure provided that there exists an H bundle, $p_1: E_1 \to B$ so that the associated G bundle, $(p_1)_\psi: E_1 \times_H G \to B$ is equivalent to the G bundle. More correctly one should say that we have a ψ structure on our G bundle, but we won't. An *H structure* for a G bundle, $p: E \to B$ consists of a pair: an H bundle, $p_1: E_1 \to B$, and a G equivalence, γ from $(p_1)_\psi: E_1 \times_H G \to B$ to the original G bundle, $p: E \to B$. Two structures $p_1: E_1 \to B$, γ_1 and $p_2: E_2 \to B$, γ_2 on $p: E \to B$ are *equivalent* if there exists an equivalence of H bundles $f: E_1 \to E_2$ such that, if f_ψ denotes the corresponding equivalence of G bundles, $\gamma_1 = \gamma_2 \circ f_\psi$.

We assume the reader is familiar with this next result.

Theorem 1.1. *For any topological group, G, there exists a space B_G such that equivalence classes of G bundles over B are in 1–1 correspondence with homotopy classes of maps $B \to B_G$. (A map $B \to B_G$ corresponding to a bundle is called a classifying map for the bundle.) Given $\psi: H \to G$ we get an induced map $B\psi: B_H \to B_G$. If this map is not a fibration, we may make it into one without changing B_G or the homotopy type of B_H, so assume $B\psi$ is a Hurewicz fibration. Given a G bundle with a classifying map $B \to B_G$, H structures on this bundle are in 1–1 correspondence with lifts of the classifying map for the G bundle to B_H.*

Example. Let $p: E \to B$ be a trivial $O(n)$ bundle, and suppose the atlas has one open set, namely B, and one transition function, the identity. One $SO(n)$ structure on this bundle consists of the same transition function but thought of as taking values in $SO(n)$ together with the bundle equivalence which maps B to the identity in $O(n)$. Another $SO(n)$ structure is obtained by using the same transition functions but taking as the bundle equivalence a map B to $O(n)$ which lands in the orientation reversing component of $O(n)$. Indeed any map $B \to O(n)$ gives an $SO(n)$ structure on our bundle. It is not difficult to see that any two maps into the same component of $O(n)$ give equivalent structures and that two maps into different components give structures that are not equivalent as structures. Clearly the $SO(n)$ bundle in all cases is the same. One gets from here to the more traditional notion of orientation for the associated vector bundle as follows. Since the transition functions are in $O(n)$, $O(n)$ acts on the vector space fibre. But for matrices to act on a vector space a basis needs to be chosen. This basis orients the $SO(n)$ bundle: in the first case the equivalence orients the underlying $O(n)$ bundle one way and in the second case the equivalence orients the bundle the other way.

Finally recall that an $O(n)$ bundle has an orientation iff the first Stiefel–Whitney class, w_1 of the bundle vanishes. If there is an $SO(n)$ structure then $H^0(B; \mathbf{Z}/2\mathbf{Z})$ acts in a simply transitive manner on the set of structures.

The Lie group $Spin(n)$ comes equipped with a standard double cover map $Spin(n) \to SO(n)$, and this is the map ψ we mean when we speak of an $SO(n)$ bundle, or an oriented vector bundle, having a *Spin* structure. There is a fibration sequence $B_{Spin(n)} \to B_{SO(n)} \to K(\mathbf{Z}/2\mathbf{Z}, 2)$, so the obstruction to the existence of a *Spin* structure is a 2–dimensional cohomology class which is known to be the second Stiefel–Whitney class w_2. If the set of *Spin* structures is non–empty, then $H^1(B; \mathbf{Z}/2\mathbf{Z})$ acts on it in a simply transitive manner.

The action can be seen explicitly as follows. Fix one *Spin* structure, say g_{ij}. An element in $H^1(B; \mathbf{Z}/2\mathbf{Z})$ can be represented by a Cech cocycle: i.e. a collection of maps $c_{ij}: \mathcal{U}_i \cap \mathcal{U}_j \to \pm 1$ satisfying the same conditions as the transition functions for a bundle. The new *Spin* structure consists of the transition functions $g_{ij} \cdot c_{ij}$ with the same $SO(n)$ bundle equivalence, where we think of ± 1 as a subgroup of $Spin(n)$ and $\cdot$ denotes group multiplication. It is not hard to check that cohomologous cocycles give equivalent structures.

We now explore the relation between *Spin* structures on an oriented vector bundle and framings of that bundle. A framing of a bundle is the same thing as an H structure where H is the trivial subgroup. Hence H is naturally a subgroup of $Spin(n)$ and an equivalence class of framings of a bundle gives rise to an equivalence class of *Spin* structures. Consider first the case $n = 1$. Recall $SO(1)$ is trivial and $Spin(1) = \mathbf{Z}/2\mathbf{Z}$. Hence an $SO(1)$ bundle already has a unique trivialization, and hence a "canonical" *Spin* structure. There are often other *Spin* structures, but, none of these come from framings. In case $n = 2$, $Spin(2) = S^1$, $SO(2) = S^1$ and the map is the double cover. If an $SO(2)$ bundle is trivial, framings are acted on simply transitively by $H^1(B; \mathbf{Z})$. The corresponding *Spin* structures are equivalent iff the class in $H^1(B; \mathbf{Z}/2\mathbf{Z})$ is trivial. If B is a circle the bundle is trivial iff it has a *Spin* structure and both *Spin* structures come from framings. The *Spin* structure determines the framing up to an action by an even element in $\mathbf{Z}$, so we often say that the *Spin* structure determines an *even* framing. If $n > 2$ and B is still a circle, then the bundle is framed iff it has a *Spin* structure and now framings and *Spin* structures are in 1–1 correspondence.

Of course, given any *Spin* structure on a bundle over B, and any map $f: S^1 \to B$, we can pull the bundle back via f and apply the above discussion. Since *Spin* structures on the bundle are in 1–1 correspondence with $H^1(B; \mathbf{Z}/2\mathbf{Z})$, which is detected by mapping in circles, we can recover the *Spin* structure by describing how the bundle is framed when restricted to each circle (with a little care if $n = 1$ or 2). Moreover, if an $SO(n)$ bundle over a CW complex is trivial when restricted to the 2–skeleton, then w_2 vanishes, so the bundle has a *Spin* structure. If $n \neq 2$ and the bundle has a *Spin* structure then, restricted to the 2– skeleton, it is trivial. If $n = 2$ this last remark is false as the tangent bundle to S^2 shows.

Finally, we need to discuss stabilization. All our groups come in families indexed by the natural numbers and there are inclusions of one in the next. An example is the family $O(n)$ with $O(n) \to O(n+1)$ by adding a 1 in the bottom right, and all our other families have similar patterns. This is of course a special case of our general discussion of H structures on G bundles. Given a vector bundle, ξ, and an oriented line bundle, ϵ^1, the $O(n)$ transition functions for ξ extend naturally to a set of $O(n+1)$ transition functions for $\xi \oplus \epsilon^1$ using the above homomorphism, and any of our structures on ξ will extend naturally to a similar structure on $\xi \oplus \epsilon^1$. We call the structure on $\xi \oplus \epsilon^1$ the *stabilization* of the structure on ξ.

A particular case of great interest to us is the relation between tangent bundles in a manifold with boundary. Suppose M is a codimension 0 subset of the boundary of W. We can consider the tangent bundle of W, say T_W, restricted to M. It is naturally identified with $T_M \oplus \nu_{M \subset W}$ where ν denotes the normal bundle. This normal bundle is framed by the "inward–pointing" normal, so we can compare structures on M with structures on W using stabilization.

Since both $Pin^{\pm}(n)$ are Lie groups and have homomorphisms into $O(n)$, the above discussion applies.

Remarks. *With this definition it is clear that, if there is a $Pin^{\pm}$ structure on a bundle ξ over a space B then $H^1(B; \mathbf{Z}/2\mathbf{Z})$ acts on the set of $Pin^{\pm}$ structures in a simply transitive manner. It is also clear that the obstruction to existence of such a structure must be a 2–dimensional cohomology class in $H^2(B_{O(n)}; \mathbf{Z}/2\mathbf{Z})$ that restricts to $w_2 \in H^2(B_{SO(n)}; \mathbf{Z}/2\mathbf{Z})$ and hence is either $w_2(\xi)$ or $w_2(\xi) + w_1^2(\xi)$. Here w_i denotes the i–th Stiefel–Whitney class of the bundle.*

We sort out the obstructions next.

Lemma 1.2. *Let λ be a line bundle over a CW complex B. Then λ has a Pin^+ structure and $\lambda \oplus \lambda \oplus \lambda$ has a Pin^- structure.*

Proof: Since $Pin^+(1) \to O(1)$ is just a projection, $\mathbf{Z}/2\mathbf{Z} \oplus \mathbf{Z}/2\mathbf{Z} \to \mathbf{Z}/2\mathbf{Z}$, there is a group homomorphism, $O(1) \to Pin^+(1)$, splitting the projection. If we compose transition functions for λ with this homomorphism, we get a set of Pin^+ transition functions for λ. If we have an equivalent $O(1)$ bundle, the two $Pin^+(1)$ bundles are also equivalent.

Transition functions for 3λ are given by taking transition functions for λ and composing with the homomorphism $O(1) \to O(3)$ which sends ± 1 to the matrix $\begin{pmatrix} \pm 1 & 0 & 0 \\ 0 & \pm 1 & 0 \\ 0 & 0 & \pm 1 \end{pmatrix}$. It is easy to check that this homomorphism lifts through a homomorphism $O(1) \to Pin^-(3)$. If we have an equivalent $O(1)$ bundle, the two $Pin^-(3)$ bundles are also equivalent. ∎

Addendum to 1.2. Notice that we have proved a bit more. The homomorphisms we chose are not unique, but can be chosen once and for all. Hence a line bundle has a "canonical" Pin^+ structure and 3 times a line bundle has a "canonical" Pin^- structure.

Remark. There are two choices for the homomorphisms above. If we choose the other then the two "canonical" Pin^+ structures on a line bundle differ by the action of w_1 of the line bundle, with a similar remark for the Pin^- case.

Lemma 1.3. *The obstruction to lifting an $O(n)$–bundle to a $Pin^+(n)$–bundle is w_2, and to a $Pin^-(n)$–bundle is $w_2 + w_1^2$. If $\xi \oplus \lambda$ = trivial bundle, then ξ has a Pin^- structure iff λ has a Pin^+ structure.*

Proof: A line bundle has a Pin^+ structure by Lemma 1.2, so $w_2 = 0$, but there are examples, e.g. the canonical bundle over $\mathbf{RP}^2$, for which $w_1^2 \neq 0$. Hence w_2 is the obstruction to a bundle having a Pin^+ structure.

For 3 times a line bundle, $w_2 = w_1^2$, so we can find examples, e.g. 3 times the canonical bundle over $\mathbf{RP}^2$, for which $w_2 + w_1^2 = 0$ but $w_2 \neq 0$. Hence $w_2 + w_1^2$ is the obstruction to having a Pin^- structure.

The remaining claim is an easy characteristic class calculation. ∎

The fact that the tangent bundle and normal bundles have different structures can lead to some confusion. In the rest of this paper, when we say a manifold has a $Pin^\pm$ structure, we mean that the *tangent bundle* to the manifold has a $Pin^\pm$ structure. As an example of the possibilities of confusion, the *Pin* bordism theory calculated by Anderson, Brown and Peterson, [ABP2], is Pin^- bordism. They do the calculation by computing the stable homotopy of a Thom spectrum, which as usual is the Thom spectrum for the *normal* bundles of the manifolds. The key fact that makes their calculation work is that w_2 vanishes, but this is w_2 of the normal bundle, so the tangent bundle has a Pin^- structure and we call this Pin^- bordism.

We remark that a $Pin^\pm$ structure is equivalent to a stable $Pin^\pm$ structure and similarly for *Spin*. This can be seen by observing that

$$\begin{array}{ccc} Pin^\pm(n) & \longrightarrow & Pin^\pm(n+1) \\ \downarrow & & \downarrow \\ O(n) & \longrightarrow & O(n+1) \end{array}$$

commutes and is a pull–back of groups, with a similar diagram in the *Spin* case.

In order to be able to carefully discuss structures on bundles, we introduce the following notation and definitions. Given a vector bundle, ξ, let $\mathcal{P}in^\pm(\xi)$ denote the set of $Pin^\pm$ structures on it. If ξ is an oriented vector bundle, let $\mathcal{S}pin(\xi)$ denote the set of *Spin* structures on it. Throughout this paper we will be writing down functions between sets of $Pin^{\pm}$ or *Spin* structures. All these sets, if non-empty are acted on, simply transitively, by $H^1(B; \mathbf{Z}/2\mathbf{Z})$ where B is the base of the bundle.

Definition 1.4. We say that a function between two sets of structures on bundles over bases B_1 and B_2 respectively is *natural* provided there is a homomorphism $H^1(B_1; \mathbf{Z}/2\mathbf{Z}) \to H^1(B_2; \mathbf{Z}/2\mathbf{Z})$ so that the resulting map is equivariant.

One example of this concept is the following construction.

Construction 1.5. Let $\hat{f}: \xi_1 \to \xi_2$ be a bundle map covering $f: B_1 \to B_2$. Given a cover and transition functions for B_2 and ξ_2, we can use f and $\hat{f}$ to construct a cover and transition functions for B_1 and ξ_1. This construction induces a natural function

$$\hat{f}^*: \mathcal{P}in^{\pm}(\xi_2) \to \mathcal{P}in^{\pm}(\xi_1)$$

with a similar map for *Spin* structures if we use $\hat{f}$ to pull back the orientation.

There are two examples of this construction we will use frequently. The first is to consider an open subset $U \subset M$ of a manifold M: here the derivative of the inclusion is a bundle map so Construction 1.5 gives us a natural restriction of structures. The second is to consider a codimension 0 immersion between two manifolds, say $f: N \to M$. Again the derivative is a bundle map so we get a natural restriction of structures.

We can also formally discuss stabilization.

Lemma 1.6. *Let ξ be a vector bundle, and let ϵ^1 be a trivial line bundle, both over a connected space B. There are natural one to one correspondences*

$$\mathcal{S}_r(\xi): \mathcal{P}in^{\pm}(\xi) \to \mathcal{P}in^{\pm}(\xi \oplus \bigoplus_{i=1}^{r} \epsilon^1) \ .$$

If ξ is oriented there is a natural one to one correspondence

$$\mathcal{S}_r^+(\xi): \mathcal{S}pin^{\pm}(\xi) \to \mathcal{S}pin^{\pm}(\xi \oplus \bigoplus_{i=1}^{r} \epsilon^1) \ .$$

Given a bundle map $\hat{f}: \xi_1 \to \xi_2$, there is another bundle map $(\widehat{f \oplus \bigoplus_{i=1}^{r} 1}): \xi_1 \oplus \bigoplus_{i=1}^{r} \epsilon^1 \to \xi_2 \oplus \bigoplus_{i=1}^{r} \epsilon^1$. The obvious squares involving these bundle maps and the stabilization maps commute.

We would like a result that relates $Pin^{\pm}$ structures on bundles to the geometry of the bundle restricted over the 1–skeleton mimicking the framing condition for the *Spin* case. We settle for the next result. Let ξ^n be an n–plane bundle over a CW–complex X, and let $\det \xi$ be the determinant bundle of ξ^n.

Lemma 1.7. *There exist natural bijections*

$$\Psi_{4k+1}(\xi)\colon \mathcal{P}in^-(\xi) \to \mathcal{S}pin(\xi \oplus (4k+1)\det\xi)$$
$$\Psi_{4k+3}(\xi)\colon \mathcal{P}in^+(\xi) \to \mathcal{S}pin(\xi \oplus (4k+3)\det\xi)$$
$$\Psi_{4k+2}(\xi)\colon \mathcal{P}in^\pm(\xi) \to \mathcal{P}in^\mp(\xi \oplus (4k+2)\det\xi)$$
$$\Psi_{4k}(\xi)\colon \mathcal{P}in^\pm(\xi) \to \mathcal{P}in^\pm(\xi \oplus (4k)\det\xi)$$

and

$$\Psi^+_{4k}(\xi)\colon \mathcal{S}pin(\xi) \to \mathcal{S}pin(\xi \oplus (4k)\det\xi)\ .$$

A bundle map $\hat{f}\colon \xi_1 \to \xi_2$ defines a bundle map $\det\xi_1 \to \det\xi_2$. Using this map between determinant bundles, all the squares involving the Ψ maps commute.

Proof: It follows from Lemma 1.3 that the existence of a structure of the correct sort on ξ is equivalent to the existence of a structure of the correct sort on $\xi \oplus r\det\xi$.

Let us begin by recalling the transition functions for the various bundles. There are homomorphisms $\delta_r\colon O(n) \to O(n+r)$ defined by sending an $n \times n$ matrix A to the $(n+r)\times(n+r)$ matrix which is A in the first $m \times m$ locations, $\det A$ in the remaining r diagonal locations, and zero elsewhere.

If $\mathcal{U}_i$, $g_{ij}\colon \mathcal{U}_i \cap \mathcal{U}_j \to O(n)$ is a family of transition functions for ξ, then $\delta_r \circ g_{ij}$ is a family of transition functions for $\xi \oplus r\det\xi$.

Next, we describe a function from the set of structures on ξ to the set of structures on $\xi \oplus r\det\xi$.

Begin with the case in which ξ has a Pin^- structure with transition functions $G_{ij}\colon \mathcal{U}_i \cap \mathcal{U}_j \to Pin^-(n)$ lifting the given set g_{ij} into $O(n)$. Pick an element e in the Clifford algebra for $\mathbf{R}^n \oplus \mathbf{R}^1$ so that $e^2 = -1$ and e maps to reflection through $\mathbf{R}^n$ under the canonical map to $O(n+1)$. There are two such choices but choose one once and for all. Define H_{ij} into $Pin^-(n+1)$ by $H_{ij}(u) = i(G_{ij}(u)) \cdot x_{ij}(u)$ where i denotes the natural inclusion of $Pin^-(n)$ into $Pin^-(n+1)$ and $x_{ij}(u)$ is e if $\det g_{ij}(u) = -1$ and 1 otherwise.

It is clear that the H_{ij} land in $Spin(n+1)$, but what needs to be checked is that they are a set of transition functions for our bundle. Clearly they lift the transition functions for the underlying $SO(n+1)$ bundle, so we need to consider the cocycle relation. This says that $H_{ij}(u)H_{jk}(u)H_{ki}(u) = 1$. If we replace the H's by G's, we do have the relation, so let us compute $H_{ij}(u)H_{jk}(u)H_{ki}(u) = G_{ij}(u)x_{ij}(u)G_{jk}(u)x_{jk}(u)G_{ki}(u)x_{ki}(u)$. Any x commutes past a G if the x associated to the G is 1 and it goes past with a sign switch if the x associated to the G is e. Also note that either none or two of the x's in our product are e. We leave it to the reader to work through the cases to see that the cocycle relation always holds and to note that the key point is that $e^2 = -1$.

Next, consider the case in which ξ has a Pin^+ structure, and let G_{ij} continue to denote the transition functions. Let e_1, e_2 and e_3 denote elements in the Pin^+

Clifford algebra for $\mathbf{R}^n \oplus \mathbf{R}^3$: each e_i covers reflection in a hyperplane perpendicular to one of the three standard basis vectors for the $\mathbf{R}^3$ factor. Define H_{ij} as above except replace e by $e_1e_2e_3$. The proof goes just as before after we note that $(e_1e_2e_3)^2 = -1$.

For the case in which $r = 2$ and ξ may have either a Pin^+ or a Pin^- structure, choose e_1 and e_2; note that $(e_1e_2)^2 = -1$ and proceed as above.

The last natural bijection is also easy. If g_{ij} are transition functions for ξ it is easy to choose the cover so that there are lifts G_{ij} of our functions to $Pin^-(n)$ (or $Pin^+(n)$ if the reader prefers), but the cocycle relation may not be satisfied. We can define new functions H_{ij} into $Spin(4n)$ by just juxtaposing 4 copies of G_{ij} thought of as acting on four copies of the same space. These functions can easily be checked to satisfy the cocycle condition.

Now that we have defined our functions, the results of the theorem are easy. The reader should check that the functions we defined are $H^1(\ ;\mathbf{Z}/2\mathbf{Z})$ equivariant and hence induce 1–1 transformations. ∎

Remark 1.8. We did make a choice in the proof of 1.7. The choice was global and so the lemma holds, but it is interesting to contemplate the effect of making the other choice. It is not too hard to work out that if we continue to use 1, but replace e by $-e$, the new *Spin* structure will differ from the old one by the action of $w_1(\xi)$. The same result holds if we switch an odd number of the e_1, e_2, e_3 in the Pin^+ case or an one of e_1, e_2 in the $r = 2$ case.

For later use, we need a version of Lemma 1.7 in which the line bundles are merely isomorphic to the determinant bundle. To be able to describe the effect of changing our choices, we need the following discussion.

There is a well–known operation on an oriented vector bundle known as "reversing the orientation". Explicitly, suppose that we have transition functions, g_{ij}, defined into $SO(n)$ based on a numerable cover $\{\mathcal{U}_i\}$. Then we choose maps $h_i: \mathcal{U}_i \to O(n) - SO(n)$ and let the bundle with the "opposite orientation" have transition functions $h_i \circ g_{ij} \circ h_j^{-1}$ and use the maps h_i to get the $O(n)$ equivalence with the original bundle. The choice of the h_i is far from unique, but any two choices yield equivalent $SO(n)$ bundles. In the same fashion, given a $Spin(n)$ bundle, we can consider the *opposite Spin structure.* Proceed just as above using $Spin(n)$ for $SO(n)$ and $Pin^+(n)$ or $Pin^-(n)$ for $O(n)$.

Note that a *Spin* structure and its opposite are equivalent Pin^+ or Pin^- structures. Conversely, given a $Pin^\pm$ structure on a vector bundle which happens to be orientable, then there are two compatible *Spin* structures which are the opposites of each other. We summarize the above discussion as

Lemma 1.9. *If ξ is an oriented vector bundle, then there is a natural one to one correspondence, called reversing the spin structure,*

$$\mathcal{R}_\xi: Spin(\xi) \to Spin(-\xi)$$

where $-\xi$ denotes ξ with the orientation reversed. We have that $\mathcal{R}_\xi \circ \mathcal{R}_{-\xi}$ is the identity. Finally, given a bundle map $\hat{f}$ as in Construction 1.5, the obvious square commutes.

Proof: We described the transformation above, and it is not hard to see that it is $H^1(;\mathbf{Z}/2\mathbf{Z})$ equivariant. It is also easy to check that the composition formula holds. ∎

In practice, we can rarely identify our bundles with the accuracy demanded by Lemma 1.7 or Lemma 1.6, so we discuss the effect of a bundle automorphism on the sets of structures. Suppose we have a bundle $\chi = \xi \oplus \bigoplus_{i=1}^{r} \lambda$, where λ is a line bundle. We will study the case λ is trivial (so called "stabilization") and the case λ is isomorphic to $\det \xi$. Let γ be a bundle automorphism of χ which is the sum of the identity on ξ and some automorphism of $\bigoplus_{i=1}^{r} \lambda$. The transition functions for $\bigoplus_{i=1}^{r} \lambda$ are either the identity or minus the identity, both of which are central in $O(r)$ so γ is equivalent to a collection of maps $\gamma: B \to O(r)$, where B is the base of the bundle. The bundle automorphism induces a natural automorphism of $Pin^\pm$ structures on χ, described in the proof of

Lemma 1.10. *Let the base of the bundle, B, be path connected. The map induced by γ on structures, denoted γ^*, is the identity if γ lands in $SO(r)$. Otherwise it reverses the Spin structure in the Spin case and acts via $w_1(\xi)$ in the $Pin^\pm$ case if λ is trivial and by $r \cdot w_1(\xi)$ if λ is isomorphic to $\det \xi$.*

Proof: To fix notation, choose transition functions for a structure on ξ (either *Spin* or $Pin^\pm$). Pick transition functions for λ using the same cover. If λ is trivial, take the identity for the transition functions and if λ is the determinant bundle take the determinant of the transition functions for ξ. The new structure induced by γ has transition functions $\tilde{\gamma}(u) o_{ij}(u) \tilde{\gamma}^{-1}(u)$ where o_{ij} denotes the old transition functions and $\tilde{\gamma}(u)$ denotes a lift of $\gamma(u)$ to $Pin^\pm(r)$ and then into $Pin^\pm(n+r)$ where ξ has dimension n. There may be no continuous choice of $\tilde{\gamma}$, but since the two lifts yield the same conjugation, the new transition functions remain continuous. The element $o_{ij}(u) Pin^\pm(n+r)$ has the form x with x involving only the first n basis vectors in the Clifford algebra if $\det o_{ij}(u) = 1$ or if λ is trivial: otherwise $x e_{n+1} \cdots e_{n+r}$ with x as before.

Recall $\tilde{\gamma} x = (-1)^{\alpha(x)\alpha(\gamma)} x \tilde{\gamma}$ and $\tilde{\gamma} e_{n+1} \cdots e_{n+r} = (-1)^{\alpha(\gamma)(r-1)} e_{n+1} \cdots e_{n+r} \tilde{\gamma}$ where α on $Pin^\pm$ is the restriction of the mod 2 grading from the Clifford algebra and α on $O(r)$ is 1 iff the element is in $SO(r)$. The result now follows for $Pin^\pm$ structures. The result for *Spin* structures is now clear. If γ takes values in $SO(r)$ then the bundle map preserves the orientation and the underlying Pin^- structure, hence the *Spin*structure. If γ takes values in $O(r) - SO(r)$, compose the map induced by γ with the reverse *Spin* structure map. The reverse *Spin* structure map is induced by any constant map $B \to O(r) - SO(r)$. Hence the composite of these two maps is induced by a map $B \to SO(r)$ and hence is the identity. ∎

There are a couple of further compatibility questions involving the functions we have been discussing. Given an $SO(n)$ bundle ξ and an oriented trivial line bundle ϵ^1, we get a natural $SO(n+r)$ bundle $\xi \oplus r\epsilon^1$ and an isomorphism $-\xi \oplus r\epsilon^1 \cong -(\xi \oplus r\epsilon^1)$.

Lemma 1.11. *With the above identifications, stabilization followed by reversing the Spin structure agrees with reversing the Spin structure and then stabilizing: i.e.* $\mathcal{R}_{\xi\oplus r\epsilon}\circ\mathcal{S}_r^+(\xi) = \mathcal{S}_r^+(-\xi)\circ\mathcal{R}(\xi)$.

Proof: Left to the reader. ∎

Let M^m be $Pin^\pm$ and let V^{m-1} be a codimension 1 manifold of M with normal line bundle ν. We wish to apply Lemma 1.7 to the problem of constructing a "natural" structure on V. If there is a natural map from structures on M to structures on V, we say that V *inherits* a structure from the structure on M. Of course, the homomorphism $H^1(M;\mathbf{Z}/2\mathbf{Z}) \to H^1(V;\mathbf{Z}/2\mathbf{Z})$ implicit in the use of "natural" is just the one induced by the inclusion.

Corollary 1.12. *If ν is trivialized then V inherits a $Pin^\pm$ structure from a $Pin^\pm$ structure on M. If M and V are oriented then V inherits a Spin structure from a Spin structure on M.*

Proof: When ν is trivialized the result follows from Lemma 1.6. If M and V are oriented, then we can trivialize (i.e. orient) ν so that the orientation on $T_V \oplus \nu$ agrees with the orientation on $T_M|_V$. ∎

A case much like Corollary 1.12 occurs when M is a manifold with boundary, $V = \partial M$. In this case, the normal bundle, ν, is trivialized by the geometry, namely the preferred direction is inward. Just as in Corollary 1.12, we put ν last getting $T_M|_{\partial M} = T_{\partial M} \oplus \nu$. On orientations this gives the convention "inward normal last" which we adopt for orienting boundaries. Furthermore, a *Spin* or $Pin^\pm$ structure on M now induces one on ∂M, so we have a bordism theory of *Spin* manifolds and of $Pin^\pm$ manifolds.

In the *Spin* case, the inverse in the bordism group is formed by taking the manifold, M, with *Spin* structure on T_M, and reversing the *Spin* structure. In either the Pin^+ or the Pin^- case, the inverse in bordism is formed by acting on the given structure by $w_1(M)$. Having to switch the $Pin^\pm$ structure to form the inverse is what prevents $\Omega_*^{Pin^\pm}$ from being a $\mathbf{Z}/2\mathbf{Z}$ vector space like ordinary unoriented bordism. The explicit formula for the inverse does imply

Corollary 1.13. *The image of $\Omega_r^{Spin}(X)$ in $\Omega_r^{Pin^\pm}(X)$ has exponent 2 for any CW complex X, or even any spectrum.*

The "inward normal last" rule has some consequences. Suppose we have a manifold with boundary M, ∂M, and a structure on $M \times \mathbf{R}^1$. We can first restrict to the boundary, which is $(\partial M) \times \mathbf{R}^1$, and then do the codimension 1 restriction, or else we can do the codimension 1 restriction to M and then restrict to the boundary.

Lemma 1.14. *The two natural functions described above,*

$$\mathcal{P}in^{\pm}(M \times \mathbf{R}^1) \to \mathcal{P}in^{\pm}(\partial M) \ ,$$

differ by the action of $w_1(M)$. The same map between Spin structures reverses the Spin structure.

Proof: By considering restriction maps it is easy to see that it suffices to prove the result for $M = (\partial M) \times [0, \infty)$, and here the functions are bijections. Consider the inverse from structures on ∂M to structures on $\partial M \times \mathbf{R}^1 \times [0, \infty)$. The two different functions differ by a bundle automorphism which interchanges the last two trivial factors. By Lemma 1.10, this has the effect claimed. ∎

In the not necessarily trivial case we also have a "restriction of structure" result.

Corollary 1.15. *If ν is not necessarily trivial, then V inherits a structure from one on M in three of the four cases below:*

	Pin$^+$	*Pin*$^-$
V orientable $\nu = \det T_M$	*Spin*	*None*
V not necessarily orientable $\nu = \det T_V$	*Pin*$^-$	*Pin*$^-$

Proof: In the northwest case, $T_V \oplus \nu = T_M|_V$ has a *Pin*$^+$ structure, so $T_M \overset{3}{\oplus} \det T_M$ has a *Spin* structure. But $T_M \overset{3}{\oplus} \det T_M|_V = T_V \oplus \nu \oplus \overset{3}{\oplus} \det T_M|_V = T_V \overset{4}{\oplus} \det T_M|_V$ so T_V and hence V acquires a *Spin* structure. However, there is a choice in the above equation: we have had to identify ν with $\det T_M|_V$. When we say that the ν and $\det T_M$ are equal, we mean that we have fixed a choice.

A similar argument works in the southeast case: $T_V \oplus \det T_V$ is naturally oriented, so an identification of ν with $\det T_V$ gives $T_V \oplus \nu = T_M|_V$. Since M has a *Pin*$^-$ structure, V gets a *Pin*$^-$structure.

In the southwest case, consider $E \subset M$, a tubular neighborhood of V. Since ν and $\det T_V$ are identified, and since $T_V \oplus \det T_V$ is naturally oriented, E is oriented and hence the *Pin*$^+$ structure reduces uniquely to a *Spin* structure. From here the argument is the same as in the last paragraph.

Lastly, consider the northeast case. If we let $V = \mathbf{RP}^5 \subset \mathbf{RP}^6 = M$, we see that M has a *Pin*$^-$ structure; ν and $\det T_M$ are isomorphic; V is orientable but does not have any *Spin* structures at all. ∎

Remark. If we just assume that the line bundles in the table are isomorphic, which is surely the more usual situation, then we no longer get a well-defined structure. The new structure is obtained from the old one by first reversing orientation in the *Spin* case, and then acting by $w_1(\nu)$. A similar remark applies to Corollary 1.12.

§2. Pin^- structures on low-dimensional manifolds and further generalities.

We begin this section by recalling some well-known characteristic class formulas. Every 1-dimensional manifold is orientable and has *Spin* and $Pin^\pm$ structures. It is easy to parlay this into a proof that $\Omega_0^{Spin} \cong \mathbf{Z}$ and $\Omega_0^{Pin^\pm} \cong \mathbf{Z}/2\mathbf{Z}$, with the isomorphism being given by the number of points (for *Spin*) and the number of points mod 2 for $Pin^\pm$. Using the Wu relations, [M–S, p. 132], we see that every surface and every 3-manifold has a Pin^- structure, and hence oriented 2 and 3-manifolds have *Spin* structures. We can also say that a 2 or 3-manifold has a Pin^+ structure iff $w_1^2 = 0$. For surfaces this translates into having even Euler characteristic or into being an unoriented boundary.

We next give a more detailed discussion of structures on S^1. The tangent bundle to S^1 is trivial and 1-dimensional, hence a trivialization is the same thing as an orientation. Since $H^1\left(S^1; \mathbf{Z}/2\mathbf{Z}\right) \cong \mathbf{Z}/2\mathbf{Z}$, there are two *Spin* structures on the circle. Since the tangent bundle to S^1 does not extend to a non-zero vector field over the 2-disk, the two *Spin* structures on an oriented S^1 can be described as follows: one of them is the *Spin* structure coming from the framing given by the orientation (this is called the *Lie group framing* or the *Lie group Spin structure*) and the other one is the one induced by the unique *Spin* structure on the 2-disk restricted to S^1.

Theorem 2.1. *The group $\Omega_1^{Spin} \cong \mathbf{Z}/2\mathbf{Z}$, generated by the Lie group Spin structure on the circle; $\Omega_1^{Pin^-} \cong \mathbf{Z}/2\mathbf{Z}$ and the natural map $\Omega_1^{Spin} \to \Omega_1^{Pin^-}$ is an isomorphism; $\Omega_1^{Pin^+} = 0$.*

Proof: Since the 2-disk has an orientation reversing involution, the restriction of this involution to the boundary gives an equivalence between S^1 with Lie group *Spin* structure and S^1 with the orientation reversed and the Lie group *Spin* structure. Hence Ω_1^{Spin} and $\Omega_1^{Pin^\pm}$ are each 0 or $\mathbf{Z}/2\mathbf{Z}$. Suppose S^1 is the boundary of an oriented surface $\hat{F}$. It is easy to check that all *Spin* structures on $\hat{F}$ induce the same *Spin* structure on S^1. If we let F denote $\hat{F} \cup B^2$ then F also has a *Spin* structure, and it is easy to see that any *Spin* structure on $\hat{F}$ extends (uniquely) to one on F. In particular, the *Spin* structure induced on S^1 is the one which extends over the 2-disk, so S^1 with the Lie group *Spin* structure does not bound.

The proof for the Pin^- case is identical because any surface has a Pin^- structure.

In the Pin^+ case however, $\mathbf{RP}^2$ does not have a Pin^+ structure. On the other hand, $\mathbf{RP}^2 - \operatorname{int} B^2$ (which is the Möbius band) does have a Pin^+ structure. The induced Pin^+ structure on the boundary must therefore be one which does not extend over the 2–disk, and hence the circle with the Lie group Pin^+ structure does bound. ▪

In dimension 4, the generic manifold supports neither a *Spin* nor a $Pin^\pm$ structure. A substitute which works fairly well is to consider a 4–manifold with a submanifold dual to w_2 or $w_2+w_1^2$. We will also have need to consider submanifolds dual to w_1. A general discussion of these concepts does not seem out of place here.

Let M be a paracompact manifold, with or without boundary. Let a be a cohomology class in $H^i(M;\mathbf{Z}/2\mathbf{Z})$. We say that a codimension i submanifold of M, say $W \subset M$, is dual to a iff the embedding of W in M is proper and the boundary of M intersects W precisely in the boundary of W. The fundamental class of W is a class in $H^{l.f.}_{n-i}(W, \partial W;\mathbf{Z}/2\mathbf{Z})$, where $H^{l.f.}$ denotes homology with locally finite chains. With the conditions we have imposed on our embedding, this class maps under the inclusion to an element in $H^{l.f.}_{n-i}(M, \partial M;\mathbf{Z}/2\mathbf{Z})$. Under Poincaré duality, $H^{l.f.}_{n-i}(M, \partial M;\mathbf{Z}/2\mathbf{Z})$ is isomorphic to $H^i(M;\mathbf{Z}/2\mathbf{Z})$ and we require that the image of the fundamental class of W map under this isomorphism to a. Specifically, in $H^{l.f.}_{n-i}(M, \partial M;\mathbf{Z}/2\mathbf{Z})$, we have the equation $a \cap [M, \partial M] = i_*[W, \partial W]$.

A cohomology class in $H^n(B;A)$, is given by a homotopy class of maps, $B \to K(A,n)$, where $K(A,n)$ is the Eilenberg–MacLane space with $\pi_n \cong A$. If $TO(n)$ denotes the Thom space of the universal bundle over $BO(n)$, then the Thom class gives a map $TO(n) \to K(\mathbf{Z}/2\mathbf{Z}, n)$. If M is a manifold, the Pontrjagin–Thom construction shows that $a \in H^n(M;\mathbf{Z}/2\mathbf{Z})$ is dual to a submanifold iff the map $M \to K(\mathbf{Z}/2\mathbf{Z}, n)$ representing a lifts to a map $M \to TO(n)$. Similar remarks hold if $A = \mathbf{Z}$ with $BO(n)$ replaces by $BSO(n)$. The submanifold, V, is obtained by transversality, so the normal bundle is identified with the universal bundle over $BO(n)$ or $BSO(n)$ and the Thom class pulls back to a. Hence there is a map $(M, M-V) \to (TO(n), *)$ which is a monomorphism on $H^n(;\mathbf{Z}/2\mathbf{Z})$ by excision. The Thom isomorphism theorem shows $H^n(M, M-V;\mathbf{Z}/2\mathbf{Z}) \cong H^0(V;\mathbf{Z}/2\mathbf{Z})$ so $H^n(M, M-V;\mathbf{Z}/2\mathbf{Z})$ is naturally isomorphic to a direct product of $\mathbf{Z}/2\mathbf{Z}$'s and the Thom class in $H^n(TO(n), *;\mathbf{Z}/2\mathbf{Z})$ restricts to the product of the generators. It follows that a restricted to $M - V$ is 0. It also follows that a restricted to V is the Euler class of the normal bundle.

Since $TO(1) = \mathbf{RP}^\infty = K(\mathbf{Z}/2\mathbf{Z}, 1)$ all 1–dimensional mod 2 cohomology classes have dual submanifolds. Since $TSO(1) = S^1 = K(\mathbf{Z}, 1)$ all 1–dimensional integral homology classes have dual submanifolds with oriented normal bundles. This holds even if M is not orientable, in which case the submanifold need not be orientable either. Since $TSO(2) = \mathbf{CP}^\infty = K(\mathbf{Z}, 2)$, any 2–dimensional integral cohomology class has a dual submanifold with oriented normal bundle. A case of interest to us is $TO(2)$. The map $TO(2) \to K(\mathbf{Z}/2\mathbf{Z}, 2)$ is not an equivalence, and not all 2–dimensional mod 2 cohomology classes have duals. As long as the manifold has dimension ≤ 4, duals can be constructed directly, but these techniques fail in dimensions 5 or more. A more detailed analysis of the map $TO(2) \to K(\mathbf{Z}/2\mathbf{Z}, 2)$ also shows the same thing: there are no obstructions to doing the lift until one gets to dimension 5 and then there are. It is amusing to note that the obstruction to realizing a class a in a 5–manifold is $Sq^2 Sq^1 a + a Sq^1 a \in H^5(M; \mathbf{Z}/2\mathbf{Z}) / Sq^1(H^4(M; \mathbf{Z}/2\mathbf{Z}))$: in particular, if M is not orientable, then any class can be realized.

In our case we want to consider duals to w_1, w_2 and $w_2 + w_1^2$. We begin with w_1. This is an example for which the above discussion shows that we always have a dual, say $V^{m-1} \subset M^m$. We want to use the fact that we have a dual to w_1. The first question we want to consider is when is an arbitrary codimension 1 submanifold dual to w_1. The answer is supplied by

Lemma 2.2. *A codimension 1 submanifold $V \subset M$ is dual to $w_1(M)$ iff there exists an orientation on $M - V$ which does not extend across any component of V. The set of such orientations is acted on simply transitively by $H^0(M; \mathbf{Z}/2\mathbf{Z})$.*

Remark. We say that an orientation on $N - X$ does not extend across X if there is no orientation on N which restricts to the given one on $N - X$. We can take $N = (M - V) \cup V_0$ and $X = V_0$, where V_0 is a component of V. By varying V_0 over the path components of V we get a definition of an orientation on $M - V$ which does not extend across any component (= path component) of V. A similar definition applies to the case of a *Spin* or $Pin^{\pm}$ structure on $M - V$ which does not extend across any component of V.

Proof: Suppose that $M - V$ is orientable and fix an orientation. If ν_i denotes the normal bundle to the component V_i of V, let $\big(D(\nu_i), S(\nu_i)\big)$ represent the disk sphere bundle pair. Each $S(\nu_i)$ is oriented by our fixed orientation on $M - V$ since $M - \amalg D(\nu_i) \subset M - V$ is a codimension 0 submanifold (hence oriented) and $\amalg S(\nu_i)$ can be naturally added as a boundary. Define $b \in H^1(M, M - V; \mathbf{Z}/2\mathbf{Z}) \cong \oplus H^1(D(\nu_i), S(\nu_i); \mathbf{Z}/2\mathbf{Z}) \cong \oplus \mathbf{Z}/2\mathbf{Z}$ on each summand as 1 if the orientation on $S(\nu_i)$ extends across $D(\nu_i)$ and -1 if it does not. The class b hits $w_1(M)$ in $H^1(M; \mathbf{Z}/2\mathbf{Z})$. This can be easily checked by considering any embedded circle in M and making it transverse to the V_i's subject to the further condition that if it intersects V_i at a point then it just enters $S(\nu_i)$ at one point and runs downs a fibre and out the other end. The tangent bundle of M restricted to this circle is oriented iff it crosses the V_i in an even number of points iff $\langle i^*(b), j_*[S^1]\rangle = 1$, where $i^*(b)$ is the image of b in $H^1(M; \mathbf{Z}/2\mathbf{Z})$ and $j_*[S^1]$ is the image of the fundamental class of the circle in $H_1(M; \mathbf{Z}/2\mathbf{Z})$. Since $w_1(M)$ also has this property, $i^*(b) = w_1(M)$ as claimed. If we act on this orientation by $c \in H^0(M - V; \mathbf{Z}/2\mathbf{Z})$, the new element in $H^1(M, M - V; \mathbf{Z}/2\mathbf{Z})$ is just $b + \delta^*(c)$, where $\delta^*(c)$ is the image of c under the coboundary $H^0(M - V; \mathbf{Z}/2\mathbf{Z}) \to H^1(M, M - V; \mathbf{Z}/2\mathbf{Z})$.

Now suppose that $M - V$ has an orientation which does not extend across any component of V. The b for this orientation has a -1 in each summand, and is hence the image of the Thom class. Therefore V is dual to $w_1(M)$.

Next suppose that V is dual to $w_1(M)$. Then $w_1(M)$ restricts 0 to $M - V$, and hence $M - V$ is orientable. Fix one such orientation and consider the corresponding b. Since both b and the image of the Thom class hit w_1, we can find $c \in H^0(M - V; \mathbf{Z}/2\mathbf{Z})$ so that $b + \delta^*(c)$ is the image of the Thom class. If we alter the given orientation on $M - V$ by c, we get a new one which does not extend across any component of V. ∎

There is also a "descent of structure" result here.

Proposition 2.3. *Given M^m, the Poincaré dual to $w_1(M)$ is an orientable $(m-1)$–dimensional manifold V^{m-1}. There is an orientation on $M - V$ which does not extend across any component of V and this orients the boundary of a tubular neighborhood of V. This boundary is a double cover of V and the covering translation is an orientation preserving free involution. In particular, V is oriented. Recall that $\alpha \in H^0(M; \mathbf{Z}/2\mathbf{Z})$ acts simply transitively on the orientations of $M - V$ which do not extend across any component of V. Hence α acts on the set of orientations of V by taking the image of α in $H^0(V; \mathbf{Z}/2\mathbf{Z})$ and letting this class act as it usually does.*

Remark. If V has more components than M, not all orientations on V can arise from this construction.

Proof: Suppose there is a loop λ in V which reverses orientation in V. If the normal line bundle ν to V in M is trivial when restricted to λ, then λ reverses orientation in M also, so $\lambda \bullet V \equiv 1 \pmod 2$; but $\lambda \bullet V = 0$ since ν is trivial over λ, a contradiction. If $\nu|_\lambda$ is nontrivial, then λ preserves orientation in M so $\lambda \bullet V \equiv 0 \pmod 2$; but $\lambda \bullet V = 1$ since ν is nontrivial, again a contradiction. So orientation reversing loops λ cannot exist.

Another proof that V is orientable: As we saw above $w_1(\nu) = i^*(w_1(M))$, where $i: V \subset M$. Since $T_M|_V = T_V \oplus \nu$, it follows easily from the Whitney sum formula that $w_1(V) = 0$.

We now continue with the proof of the proposition. Let E be a tubular neighborhood of V and recall that $H^1(E, \partial E; \mathbf{Z}/2\mathbf{Z})$ is $H^0(V; \mathbf{Z}/2\mathbf{Z})$ by the Thom isomorphism theorem. By Lemma 2.2 each component of ∂E can be oriented so that the orientation does not extend across E. Clearly ∂E is a double cover of V classified by $i^*(w_1(M))$. Since V is orientable, the covering translation must be orientation preserving and we can orient V so that the projection map is degree 1. It is easy to check the effect of changing the orientation on $M - V$ which does not extend across any component of V. ∎

We continue this discussion for the 2–dimensional cohomology classes w_2 and $w_2 + w_1^2$. Again we need a lemma which enables us to tell if a codimension 2 submanifold is dual to one of these classes. We have

Theorem 2.4. *Let M be a paracompact manifold, with or without boundary. Let F be a codimension 2 submanifold of M with finitely many components and with*

$\partial M \cap F = \partial F$. Then F is dual to $w_2 + w_1^2$ iff there is a Pin^- structure on $M - F$ which does not extend across any component of F. Furthermore $H^1(M; \mathbf{Z}/2\mathbf{Z})$ acts simply transitively on the set of Pin^- structures which do not extend across any component of F. There are similar results for Pin^+ structures and $Spin$ structures.

Proof: The proof is rather similar to the proof of the previous result. First, let F be a codimension 2 submanifold of M with $i: F \to M$ denoting the inclusion. Let $(D(\nu_i), S(\nu_i))$ denote the disk, sphere bundle tubular neighborhoods to the components of F. Suppose $M - F$ has a Pin^- structure. (The proof for Pin^+ or $Spin$ structures is sufficiently similar that we leave it to the reader.) From Lemma 1.6, each $S(\nu_i)$ inherits a Pin^- structure. Define $b \in H^2(M, M - F; \mathbf{Z}/2\mathbf{Z}) \cong \oplus H^2(D(\nu_i), S(\nu_i); \mathbf{Z}/2\mathbf{Z}) \cong \oplus \mathbf{Z}/2\mathbf{Z}$ on each summand as 1 if the Pin^- structure on $S(\nu_i)$ extends across $D(\nu_i)$ and -1 if it does not. The class b hits $w_2(M)$ in $H^2(M; \mathbf{Z}/2\mathbf{Z})$. To see this, let $j: N \to M$ be an embedded surface which either misses an F_i or hits it in a collection of fibre disks. As before $\langle i^*(b), j_*[N] \rangle$ is 1 if $T_M|_N$ has a Pin^- structure and is -1 if it does not, since a bundle over a surface with a Pin^- structure over $N - \amalg D^2$ such that the Pin^- structure does not extend over the disks has a Pin^- structure iff there are an even number of such disks. Since $w_2(M)$ has the same property, $i^*(b) = w_2(M)$.

Now $H^1(M - F; \mathbf{Z}/2\mathbf{Z})$ acts simply transitively on the Pin^- structures on $M - F$ and, for $c \in H^1(M - F; \mathbf{Z}/2\mathbf{Z})$, the new b one gets is $b + \delta^*(c)$. The proof is now sufficiently close to the finish of the proof of Lemma 2.2 that we leave it to the reader to finish. ∎

There is also a "descent of structure" result in this case, but it is sufficiently complicated that we postpone the discussion until §6.

There are two cases in which we can show a "descent of structure" result for $Pin^{\pm}$ structures. As above, given M we can find a submanifold V dual to $w_1(M)$. We can then form $V \pitchfork V$ which is the submanifold obtained by making V transverse to itself. If ν denotes the normal bundle to V in M, then the normal bundle to $V \pitchfork V$ in V is naturally identified with $\nu|_{V \pitchfork V}$ and hence the normal bundle to $V \pitchfork V$ in M is naturally identified with $\nu|_{V \pitchfork V} \oplus \nu|_{V \pitchfork V}$. Since V is orientable, 2.3, $\nu|_{V \pitchfork V}$ is isomorphic to $\det T_M|_{V \pitchfork V}$. Hence by Lemma 1.7, a $Pin^{\pm}$ structure on M induces one on $V \pitchfork V$ after we identify $\nu|_{V \pitchfork V}$ with $\det T_M|_{V \pitchfork V}$. If we choose the other identification, the structure on $V \pitchfork V$ changes by twice $w_1(M)$ restricted to $V \pitchfork V$: i.e. the final structure on $V \pitchfork V$ is independent of the identification.

Theorem 2.5. *The function above*

$$[\cap w_1^2]: \mathcal{P}in^{\pm}(M) \to \mathcal{P}in^{\mp}(V \pitchfork V)$$

is a natural function using the map, $H^1(M; \mathbf{Z}/2\mathbf{Z}) \to H^1(V \pitchfork V; \mathbf{Z}/2\mathbf{Z})$, induced by the inclusion. If $V_1 \pitchfork V_1$ is another choice then there is a dual to w_1, $W \subset M \times [0,1]$ which is V at one end and V_1 at the other, so that $W \pitchfork W$ can be constructed *as a $Pin^{\mp}$ bordism between the two $Pin^{\mp}$ structures. The map $[\cap w_1^2]$ induces a homomorphism of bordism theories*

$$[\cap w_1^2]: \Omega_m^{Pin^{\pm}}(X) \to \Omega_{m-2}^{Pin^{\mp}}(X)$$

for any CW complex or spectrum X.

Proof: The naturality result follows easily from the naturality result in Lemma 1.7. The first bordism result follows easily once we recall that $TO(1) \cong K(\mathbf{Z}/2\mathbf{Z}, 1)$ so 1–dimensional cohomology classes in M are the same as codimension 1 submanifolds up to bordism in $M \times [0,1]$. The bordism result is also not hard to prove. ∎

For another example of "descent of structure", we consider the following: given any manifold, M^m, the dual to $w_1(M)$ is a codimension 1 submanifold V^{m-1}. Since V is orientable, Proposition 2.3, we are in the northwest situation of Corollary 1.15 and V receives a pair of *Spin* structures. Let $\left(\Omega_m^{Pin^+}\right)_0$ denote the subgroup of $\Omega_m^{Pin^+}$ consisting of those elements so that the two *Spin* structures on V are bordant. It is not hard to see that if the two structures are bordant for one representative in $\Omega_m^{Pin^+}$, then they are for any representative. Moreover, it is easy to check that the induced map is a homomorphism:

Lemma 2.6. *There is a well–defined homomorphism*

$$[\cap w_1]: \left(\Omega_m^{Pin^+}\right)_0 \to \Omega_{m-1}^{Spin} \quad .$$

Remark. It is not difficult to see that $\left(\Omega_m^{Pin^+}\right)_0$ contains the kernel of the map $[\cap w_1^2]$ since any such element has a representative for which the normal bundle to V is trivial. For such a V, we see a *Spin* bordism of $2 \cdot V$ to zero, so V and $-V$ represent the same element in *Spin* bordism. Moreover, the cohomology class by which we need to change the *Spin* structure is the zero class.

We conclude this section with some results we will need later which state that different ways of inducing structures are the same.

The first relates structures (*Spin* or $Pin^{\pm}$) and immersions. Given an immersion $f: N \to M$ the derivative gives a bundle map between the tangent bundles and so we can use it to pull structures on M back to N. The induced map on structures, denoted f^*, is natural in the technical sense defined earlier. Suppose we have an embedding $M_0 \times \mathbf{R}^1 \subset M$. Let $N_0 = f^{-1}(M_0)$ and note that there is an embedding $N_0 \times \mathbf{R}^1 \subset N$ so that f restricted to $N_0 \times \mathbf{R}^1$ is $g \times$ id where $g: N_0 \to M_0$ is also an immersion.

Lemma 2.7. *The following diagram commutes*

$$\begin{array}{ccc} \mathcal{P}in^{\pm}(N) & \xrightarrow{f^*} & \mathcal{P}in^{\pm}(M) \\ \mathcal{S}'_N \downarrow & & \downarrow \mathcal{S}'_M \\ \mathcal{P}in^{\pm}(N_0) & \xrightarrow{g^*} & \mathcal{P}in^{\pm}(M_0) \end{array}$$

where we orient $\mathbf{R}^1$ and Lemma 1.6 gives us the natural map $\mathcal{S}'_M$ as the composite $\mathcal{P}in^{\pm}(M) \to \mathcal{P}in^{\pm}(M_0 \times \mathbf{R}^1) \xrightarrow{\mathcal{S}} \mathcal{P}in^{\pm}(M_0)$ with a similar definition for $\mathcal{S}'_N$. There is a similar result for Spin structures.

Proof: We can easily reduce to the case $M = M_0 \times \mathbf{R}^1$. The required result can now be checked by choosing transition functions on M_0 and extending to transition functions for all the other bundles in sight, The two bundle we want to be isomorphic will be identical. ∎

The next result relates double covers and Pin^+ structures. Let M be a manifold with a *Spin* structure, and let $x: \pi_1(M) \to \mathbf{Z}/2\mathbf{Z}$ be a homomorphism (equivalently, $x \in H^1(M; \mathbf{Z}/2\mathbf{Z})$). Let E be the total space of the induced line bundle over M. By Lemma 1.7, E has a natural Pin^+ structure induced from the *Spin* structure on M. Hence ∂E receives a Pin^+ structure. Furthermore, ∂E is orientable and we orient it by requiring the covering map $\pi: \partial E \to M$ to be degree 1. The Pin^+ structure and the orientation give a *Spin* structure on ∂E. We can also use the immersion π to pull the *Spin* structure on M back to one on ∂E.

Lemma 2.8. *The two Spin structures on ∂E are the same.*

Proof: Begin with the 1–dimensional case. Here we are discussing *Spin* structures on the circle. Suppose that the line bundle is non–trivial. Thinking of the circle as the boundary of E, we see that it has the Lie *Spin* structure from Theorem 2.1. Thinking of it as the connected double cover we also see that it has the Lie group *Spin* structure, so the result is true in dimension 1. The case in which the line bundle is trivial is even easier.

The proof proceeds by induction on dimension. Suppose we know the result in dimension $m-1$ and let M have dimension $m > 1$. It suffices to show that the two *Spin* structures on ∂E agree when restricted to embedded circles. We can span $H^1(M; \mathbf{Z}/2\mathbf{Z})$ by embedded circles, S^1_i, $i = 1, \cdots, r$, where all the circles except the first lift to disjoint circles in the double cover. The first double covers itself if the line bundle is non–trivial and lifts to disjoint circles otherwise. The group $H_1(\partial E; \mathbf{Z}/2\mathbf{Z})$ is spanned by the collection of connected components of the covers from the circles in M.

Let M_0 be the boundary of the tubular neighborhood of such a circle and let $\tilde{M}_0$ be a connected component of the corresponding double cover. It suffices to prove that the two *Spin* structures on ∂E agree when restricted to $\tilde{M}_0$. We can restrict the line bundle to M_0 and consider the resulting total space E_0. First note that E_0 has trivial normal bundle in E and that it suffices to show that the two *Spin* structures on ∂E agree when restricted to ∂E_0.

Consider first the *Spin* structure induced by the double cover map. This map is an immersion, so Lemma 2.7 shows that inducing the structure on ∂E and then restricting to ∂E_0 is the same as first restricting the structure to M_0 and then inducing via the double cover map $\partial E_0 \to M_0$.

Next consider the *Spin* structure induced by restricting the Pin^+ structure to the boundary. We can restrict the Pin^+ structure on E to E_0 and then restrict to ∂E_0 or else restrict to the boundary and then to ∂E_0. These are not obviously the same: if we let ν_1 be the normal vector to E_0 in E, restricted to ∂E_0, and let ν_2 be the normal bundle to ∂E in E, again restricted to ∂E_0. We have a *Spin* structure on $T_E|_{\partial E_0}$, and in the two cases we identify this bundle with $T_{\partial E_0} \oplus \nu_1 \oplus \nu_2$ in one case and with $T_{\partial E_0} \oplus \nu_2 \oplus \nu_1$ in the other. By Lemma 1.10, these two ways of getting the *Spin* structure via boundaries agree up to a reverse of *Spin* structure. But we are using the orientation of M to keep track of all the other orientations, so the structures turn out to agree.

Our inductive hypothesis applies over M_0 and we conclude that the two *Spin* structures on ∂E_0 agree. ∎

The other result relates double covers and the Ψ_2. Let M be a manifold and let E' be the total space of the bundle $\det T_M \oplus \det T_M$ over M. There is a natural one to one function $\Psi_2: \mathcal{P}in^{\pm}(M) \to \mathcal{P}in^{\mp}(E')$. Let $E \subset E'$ be the total space of the first copy of $\det T_M$: note $\partial E \to M$ is a 2 sheeted cover. The embedding $\partial E \subset E'$ has a normal bundle which we see as two copies of the trivial bundle, which happens to be $\det T_{\partial E}$. This gives a natural function $\Psi_2^!: \mathcal{P}in^{\mp}(E') \to \mathcal{P}in^{\pm}(\partial E)$.

Theorem 2.9. *The $Pin^{\pm}$ structure defined above on ∂E is the same as the one induced by the double cover map.*

Proof: We begin by proving that certain diagrams commute. To fix notation, let $M_0 \times \mathbf{R}^1 \subset M$. Let E_0 denote the total space of $\det T_{M_0} \oplus \det T_{M_0}$ and observe that we can embed $E_0 \times \mathbf{R}^1$ in E. We can arrange the embedding so that on 0 sections it is our given embedding, and so that $(\partial E_0) \times \mathbf{R}^1$ is embedded in ∂E. We begin with

$$\begin{array}{ccc} \mathcal{P}in^{\pm}(M) & \xrightarrow{\Psi_2} & \mathcal{P}^{\mp}(E') \\ {\scriptstyle L_1}\downarrow & & \downarrow{\scriptstyle L_2} \\ \mathcal{P}in^{\pm}(M_0) & \xrightarrow{\Psi_2} & \mathcal{P}^{\mp}(E_0') \end{array}$$

where L_1 is just $\mathcal{S}^{-1}$ followed by the restriction map induced by the embedding of $M_0 \times \mathbf{R}^1$ in M and L_2 is defined similarly but using the embedding of $E_0 \times \mathbf{R}^1$ in E. This diagram commutes by Lemma 1.10. We can then restrict this structure to

∂E and then further to $(\partial E_0) \times \mathbf{R}^1$. Since stabilization commutes with restriction we see

$$\begin{array}{ccc} \mathcal{P}in^{\pm}(M) & \longrightarrow & \mathcal{P}in^{\pm}(\partial E) \\ {\scriptstyle L_3}\downarrow & & \downarrow{\scriptstyle L_4} \\ \mathcal{P}in^{\pm}(M_0) & \longrightarrow & \mathcal{P}in^{\pm}(\partial E_0) \end{array}$$

commutes, where L_3 is defined by restricting from M to $M_0 \times \mathbf{R}^1$ followed by the inverse stabilization map and L_4 is defined by restricting from ∂E to $(\partial E_0) \times \mathbf{R}^1$ followed by the inverse stabilization map.

The proof now proceeds much like the last one. First we check the result for S^1. Applying the last diagram to the 2–disk with boundary S^1 shows the result for the structure which bounds. Apply the Pin^+ diagram to the Möbius band to see the result for the Lie Pin^+ structure. The result now holds for any Pin^+ structure on S^1. Hence it holds for *Spin* structures and hence for Pin^- structures.

For M of dimension at least 2 we induct on the dimension. But just like the proof of the preceding result, this follows from the commutativity of our second diagram. ∎

§3. Pin^- structures on surfaces, quadratic forms and Brown's arf invariant.

In this section we want to recall an algebraic way of describing Pin^- structures due to Brown [Br].

Definition 3.1. A function $q\colon H_1(F;\mathbf{Z}/2\mathbf{Z}) \to \mathbf{Z}/4\mathbf{Z}$ is called a *quadratic enhancement* of the intersection form provided it satisfies $q(x+y) = q(x)+q(y)+2\cdot x{\bullet}y$ for all $x, y \in H_1(F;\mathbf{Z}/2\mathbf{Z})$ (here $2\cdot$ denotes the inclusion $\mathbf{Z}/2\mathbf{Z} \subset \mathbf{Z}/4\mathbf{Z}$ and $\bullet$ denotes intersection number.

The main technical result of this section is

Theorem 3.2. *There is a canonical 1–1 correspondence between Pin^- structures on a surface F and quadratic enhancements of the intersection form.*

Discussion. One sometimes says that there is a 1–1 correspondence between Pin^- structures on F and $H^1(F;\mathbf{Z}/2\mathbf{Z})$, but this is non–canonical. Canonically, there is an action of $H^1(F;\mathbf{Z}/2\mathbf{Z})$ on the set of Pin^- structures which is simply transitive. Once a base point has been selected, the action gives a 1–1 correspondence between $H^1(F;\mathbf{Z}/2\mathbf{Z})$ and the set of Pin^- structures.

Note also that $H^1(F;\mathbf{Z}/2\mathbf{Z})$ acts on the set of quadratic enhancements, by $q \times \gamma$ goes to q_γ defined by

$$q_\gamma(y) = q(y) + 2\cdot\gamma(y) \tag{3.3}$$

and note that this action is simply transitive. The 1–1 correspondence in Theorem 3.2 is equivariant with respect to these actions. Indeed, the proof of Theorem 3.2 will be to fix a Pin^- structure on F and use it to write down a quadratic enhancement. This gives a transformation from the set of Pin^- structures to the set of quadratic enhancements. We will check that it is equivariant for the $H^1(F;\mathbf{Z}/2\mathbf{Z})$ action and this will prove the theorem.

Before describing the enhancement, we prove a lemma that produces enhancements from functions on embeddings. Specifically

Lemma 3.4. *Let $\hat{q}$ be a function which assigns an element in $\mathbf{Z}/4\mathbf{Z}$ to each embedded disjoint union of circles in a surface F subject to the following conditions:*

(a) $\hat{q}$ is additive on disjoint union; if L_1 and L_2 are two embedded collections of circles such that $L_1 \amalg L_2$ is also an embedding then $\hat{q}(L_1 \amalg L_2) = \hat{q}(L_1) + \hat{q}(L_2)$

(b) if L_1 and L_2 are embedded collections of circles which cross transversely at r points, then we can get a third embedded collection, L_3, by replacing each crossing: we require $\hat{q}(L_3) = \hat{q}(L_1) + \hat{q}(L_2) + 2 \cdot r$

(c) if L is a single embedded circle which bounds a disk in F, then $\hat{q}(L) = 0$.

Then $\hat{q}(L)$ depends only on the underlying homology class of L, and the induced function $q\colon H_1(F;\mathbf{Z}/2\mathbf{Z}) \to \mathbf{Z}/4\mathbf{Z}$ is a quadratic enhancement.

Proof: The first step is to show how given L, we may replace it with a single embedded circle K such that the L and K represent the same homology class in $H_1(F;\mathbf{Z}/2\mathbf{Z})$ and have the same $\hat{q}$. If L has more than one component, it is possible to draw an arc between two different components. A small regular neighborhood of this arc is a disk, and let K_1 be its boundary circle. By (c), $\hat{q}(K_1) = 0$. The circle K_1 has two pairs of intersection points with L. Apply (b): the new embedding consists of a new collection L_1 which has one fewer components that L, and two small circles K_2 and K_3, each of which bounds a disk. Condition (b) says that $\hat{q}(L_1 \amalg K_2 \amalg K_3) = \hat{q}(L) + \hat{q}(K_1) = \hat{q}(L)$. From (a) and (c) we see that $\hat{q}(L_1 \amalg K_2 \amalg K_3) = \hat{q}(L_1)$, so $\hat{q}(L) = \hat{q}(L_1)$, and L and L_1 represent the same homology class. Continue until there is only one component left.

Next we prove isotopy invariance of $\hat{q}$ in several steps. First, suppose $A \subset F$ is an embedded annulus with boundary $K_0 \amalg K_1$ and core C. We want to show $\hat{q}(K_0) = \hat{q}(K_1) = \hat{q}(C)$. Draw an arc from K_0 to C and let K_3 be a circle bounding a regular neighborhood of this arc. Apply condition (b): the result is two circles, each of which bounds a disk. From conditions (a) and (c) we see $\hat{q}(C) = \hat{q}(K_0)$. A similar proof establishes the rest. We can also show that $\hat{q}(C)$ must be even. Let C_1 be a copy of C pushed off itself in the annular structure. Then $\hat{q}(C) = \hat{q}(C_1)$ since they are both $\hat{q}(K_0)$. Let $L = C \amalg C_1$. Then $\hat{q}(L) = 2\hat{q}(C)$ by (a). On the other hand, just as above, we can use (b) to transform L into a picture with two circles bounding disks, so by (a) and (c) we see $\hat{q}(L) = 0$ and the result follows. Hence any curve in F with trivial normal bundle has even $\hat{q}$. Finally, suppose that C_1 is embedded in A and represents the same element in mod 2 homology as C.

We can find a third curve C_2 which also represents the same element in mod 2 homology and which intersects both C_1 and C transversely. Consider say C_2 and C. Apply (b): r is even as are both $\hat{q}(C)$ and $\hat{q}(C_2)$. Hence $\hat{q}(C) = \hat{q}(C_2)$. Similarly $\hat{q}(C_1) = \hat{q}(C_2)$ and we have our result.

Next suppose that $M \subset F$ is a Möbius band with core C_0. We can push C_0 to get another copy, C_1 intersecting C_0 transversely in one point. We can push off another copy C_2 which intersects C_0 and C_1 transversely in a single point and all three points are distinct. Applying (b) to pairs of these circles, we get $\hat{q}(C_i) + \hat{q}(C_j) = 2$ for $0 \leq i, j \leq 2$, $i \neq j$. Adding all three equations we see $2(\hat{q}(C_0) + \hat{q}(C_1) + \hat{q}(C_2)) = 2$, so at least one $\hat{q}(C_i)$ must be odd. But then returning to the individual equations we see that $\hat{q}(C_0) = \hat{q}(C_1) = \hat{q}(C_2)$, so we see that $\hat{q}(C)$ must be odd whenever the normal bundle to C is non–trivial. Let C_1 be any embedded circle in M which represents the core in mod 2 homology. It is possible to find a third embedded circle, C_2 which also represents the core and intersects C_0 and C_1 transversely. Since $\hat{q}(C_i)$ must be odd, it is not hard to use (b) to show that $\hat{q}(C_0) = \hat{q}(C_1)$.

To show isotopy invariance proceed as follows. Let K be a circle with a neighborhood W. Any isotopy of K will remain for a small interval inside W and the image K_t will continue to represent the core in mod 2 homology. By the above discussion $\hat{q}$ will be constant on K_t, the circle at time t. Hence, the subset of $t \in [0,1]$ for which $\hat{q}(K_t) = \hat{q}(K)$ is an open set. Likewise the set of $t \in [0,1]$ for which $\hat{q}(K_t) \neq \hat{q}(K)$ is an open set, so we have isotopy invariance for a single circle. By part (a), the result for general isotopies follows as above.

Next we prove homology invariance. Suppose L_1 and L_2 represent the same element in homology. By isotopy invariance, we may assume that they intersect transversely. Let L_3 be the result of applying condition (b). $\hat{q}(L_3) = \hat{q}(L_1) + \hat{q}(L_2) + 2 \cdot r$, and L_3 is null–homologous. If we can prove $\hat{q}(L_3) = 0$ then we are done. As we saw above, it is no loss of generality to assume that L_3 is connected, and since it is null–homologous, it has trivial normal bundle, so $\hat{q}(L_3)$ is even. Also, since L_3 is null–homologous, there exists a 2–manifold with boundary a single circle, say W, and an embedding $W \subset F$ so that $\partial W = L_3$. If W is a disk we are done by (c), so we work by induction on the Euler characteristic of W. If W is not a disk then we can write $W = W_1 \cup V$ where $\partial V = \partial_0 V \amalg \partial_1 V = S^1 \amalg S^1$, V is either a twice punctured torus or a punctured Möbius band, and W_1 has larger Euler characteristic than W. We are done if we can show $\hat{q}(\partial_0 V) = \hat{q}(\partial_1 V)$. We begin with the toral case. Using (b) and (c) as usual, we can see that $\hat{q}(\partial_0 V) = \hat{q}(S_a) + \hat{q}(S_b)$ where S_a and S_b are two meridian circles, one on either side of the hole. Likewise $\hat{q}(\partial_1 V) = \hat{q}(S_a) + \hat{q}(S_b)$ so we are done with this case. In the Möbius band case we can again use (b) and (c) and see that $\hat{q}(\partial_0 V) + \hat{q}(\partial_1 V) = 0$. Since they are both even, again they are equal.

This shows that $\hat{q}$ induces a function $q \colon H_1(F; \mathbf{Z}/2\mathbf{Z}) \to \mathbf{Z}/4\mathbf{Z}$, and (b) translates immediately into the relation $q(x + y) = q(x) + q(y) + 2 \cdot x \bullet y$. ∎

Now we describe our function. Let λ be a line bundle over F with $w_1(\lambda) = w_1(F)$ and let $E(\lambda)$ denote its total space. From Lemma 1.7, a *Spin* structure on $E(\lambda)$ gives a Pin^- structure on F. Let K be an embedded circle in F, and let τ denote the tangent bundle of $E(\lambda)$ restricted to K. A *Spin* structure on $E(\lambda)$ yields a trivialization of τ. It is also true that $\tau = T_{S^1} \oplus \nu_{K\subset F} \oplus \nu_{F\subset E(\lambda)}$, where ν denotes normal bundle. Note all three of these bundles are line bundles. Pick a point $p \in K$ and orient each of the line bundles at p so that the orientation on τ agrees with that coming from the *Spin* structure. Since T_{S^1} is trivial, the orientation picks out a trivialization, and hence $\nu_{K\subset F} \oplus \nu_{F\subset E}$ acquires a preferred even framing. (Note that framings of a 2–plane bundle correspond to $\mathbf{Z}$, while those of a 3–plane bundle correspond to $\mathbf{Z}/2\mathbf{Z}$. Hence the framing of the 3–plane bundle picks out a set of framing of the 2–plane bundle, a set we call *even*.)

Definition 3.5. Choose an odd framing on $\nu_{K\subset F} \oplus \nu_{F\subset E}$ and using it, count the number (mod 4) of right half twists that $\nu_{K\subset F}$ makes in a complete traverse of K. This is $\hat{q}(K)$. Given a disjoint union of circles, Lemma 3.4 (a) gives the value of $\hat{q}$ in terms of the individual components.

We first need to check that $\hat{q}$ really only depends on the embedded curve and not on the choice of p or the local orientations made at p or on the choice of odd framing. It is easy to see that the actual choice of framing within its homotopy class is irrelevant because we get the same count in either frame. If we choose a new odd framing the new count of right half twists will change by a multiple of 4, so the specific choice of odd framing is irrelevant. If we move p to a new point, we can move around K in the direction of the orientation and transport the local orientations as we go. If we make these choices at our new point, nothing changes so the choice of point is irrelevant. Since we must keep the same orientation on τ, we are only free to change orientations in pairs. If we keep the same orientation on K, the odd framing on the normal bundle remains the same and so we get the same count. Finally, suppose we switch the orientation on K. We can keep the same framing on the normal bundle provided we switch the order of the two frame vectors. If we do this and traverse K in the old positive direction we get the same count as before, except with a minus sign. Fortunately, we are now required to traverse K in the other direction which introduces another minus sign, so the net result is the same count as before. Hence $\hat{q}$ only depends on the embedded curve.

Since $\hat{q}$ satisfies Lemma 3.4 (a) by definition, we next show that it satisfies conditions (b) and (c) also. We begin with (c). In this case, all three line bundles are trivial, hence framed after our choice of p and the local orientations. However, this stable framing of the circle is the Lie group one, so it is not the stable framing of the circle which extends over the disk, Theorem 2.1. Since the framing from the *Spin* structure does extend over the disk, the framing constructed above is an odd framing, and $\hat{q}$ is clearly 0 for these choices. To show (b), consider a small disk neighborhood of a crossing. It is not hard to check that in the framing coming from that of the disk, we can remove the crossing without changing the count. However, this is the even framing and we are supposed to do the counting using the odd framing. This introduces a full twist, and so we get a contribution of 2 for each crossing. This is (b).

Thanks to Lemma 3.4 we have described a function from the set of Pin^- structures on F to the set of quadratic enhancements on the intersection form on $H_1(F;\mathbf{Z}/2\mathbf{Z})$. Suppose now we change the Pin^- structure by $\gamma \in H^1(F;\mathbf{Z}/2\mathbf{Z})$. The effect of this change is to reverse even and odd framings on K for which $\gamma(K) = -1$ and to leave things alone for K for which $\gamma(K) = 1$. The effect on the resulting q is to add 2 to $q(x)$ if $\gamma(x) = -1$ and add nothing to it if $\gamma(x) = 1$. But this is just q_γ.

This completes the proof of Theorem 3.2.

Next we describe an invariant due to Brown, [Br], associated to any quadratic enhancement q. Given q, form the Gauss sum

$$\Lambda_q = \sum_{x \in H_1(F;\mathbf{Z}/2\mathbf{Z})} e^{2\pi i q(x)/4} \quad .$$

This complex number has absolute value $\sqrt{|H_1(F;\mathbf{Z}/2\mathbf{Z})|}$ and there exists an element $\beta(q) \in \mathbf{Z}/8\mathbf{Z}$ such that $\Lambda_q = \sqrt{|H_1(F;\mathbf{Z}/2\mathbf{Z})|}\ e^{2\pi i\beta(q)/8}$.

Hence we can think of β as a function from Pin^- structures on surfaces to $\mathbf{Z}/8\mathbf{Z}$. It also follows from Brown's work, that β is an invariant of Pin^- bordism: two surfaces with Pin^- structures that are Pin^- bordant have the same β.

Lemma 3.6. *The homomorphism*

$$\beta\colon \Omega_2^{Pin^-} \to \mathbf{Z}/8\mathbf{Z}$$

is an isomorphism. The composite $\Omega_2^{Pin^-} \xrightarrow{\beta} \mathbf{Z}/8\mathbf{Z} \to \mathbf{Z}/2\mathbf{Z}$ *is the* mod 2 *Euler characteristic and hence determines the unoriented bordism class of the surface.*

Proof: Brown proves that β induces an isomorphism between Witt equivalence classes of quadratic forms and $\mathbf{Z}/8\mathbf{Z}$. One homomorphism from the Witt group is the dimension mod 2 of the underlying vector space. Since this is just the mod 2 Euler characteristic of our surface, the second result follows.

Hence, if $\beta(F) = 0$, the manifold is an unoriented boundary, say of W^3. There is an obstruction in $H^2(W, \partial W;\mathbf{Z}/2\mathbf{Z})$ to extending the Pin^- structure on F across W. If this obstruction is 0 we are done, so assume otherwise. There is a dual circle, $K \subset W - F$ and the Pin^- structure on F extends across $W - K$. The boundary of a neighborhood of K is either a torus or a Klein bottle, so if $\beta(F) = 0$, F is Pin^- bordant to a torus or a Klein bottle with β still 0. Moreover, since the Pin^-

structure is not supposed to extend across the neighborhood of K, one of the non-zero classes in H_1 has a non-zero q. For the Klein bottle, two of the non-zero classes have odd square and the other has even square. It is the class with even square that must have a non-trivial q on it to prevent the Pin^- structure from extending across the disk bundle. But the Klein bottle with this sort of enhancement has non-zero β, so the boundary of K must be a torus. For the torus, q must vanish on the remaining classes in H_1 in order to have $\beta = 0$ and it is easy to find a Pin^- boundary for it. ▪

Exercise. Show that $\mathbf{RP}^2$ with its two Pin^- structures has $\beta = \pm 1 \in \mathbf{Z}/8\mathbf{Z}$.

The relation between Pin^- structures and quadratic enhancements is pervasive in low-dimensional topology. In [Ro], [F–K] and [G–M] enhancements were produced on characteristic surfaces in order to generalize Rochlin's theorem. In §6, we will show how to find an enhancement without the use of membranes. This gives some generalizations of the previous work. In the next section we will study surfaces embedded in "spun" 3-manifolds. An interesting theory that we do not pursue is Brown's idea of studying immersions of a surface in $\mathbf{R}^3$. Since $\mathbf{R}^3$ has a unique *Spin* structure, an immersion pulls back a *Spin* structure onto the total space of a line bundle over the surface with oriented total space.

Another point we wish to investigate is the behavior of β under change of Pin^- structure. Hence fix a quadratic form $q: V \to \mathbf{Z}/4\mathbf{Z}$: i.e. V is a $\mathbf{Z}/2\mathbf{Z}$-vector space; $q(rx) = r^2 q(x)$ for all $x \in V$ and $r \in \mathbf{Z}$; and $q(x+y) - q(x) - q(y)$ is always even and gives rise to a non-singular bilinear pairing $\lambda: V \times V \to \mathbf{Z}/2\mathbf{Z}$.

Given $a \in V$, define q_a by $q_a(x) = q(x) + 2 \cdot \lambda(a, x)$.

Lemma 3.7. *With notation as above,* $\beta(q_a) = \beta(q) + 2 \cdot q(a)$.

Proof: There is a rank 1 form (1) consisting of a $\mathbf{Z}/2\mathbf{Z}$ vector space with one generator, x, for which $q(x) = 1$. There is a similar form (-1). It is easy to check the formula by hand for these two cases. Or, having checked it for (1) and $a = x$ and $a = 0$, argue as follows. Given any form q, there is another form $-q$ defined on the same vector space by $(-q)(x) = -q(x)$. It is easy to check that $\beta(-q) = -\beta(q)$. If the formula holds for q and a, it is easily checked for $-q$ and a after we note $(-q)_a = -(q_a)$.

Given two forms q_1 on V_1 and q_2 on V_2, we can form the *orthogonal sum* $q_1 \perp q_2$ on $V_1 \oplus V_2$ by the formula $(q_1 \perp q_2)(v_1, v_2) = q_1(v_1) + q_2(v_2)$. Brown checks that $\beta(q_1 \perp q_2) = \beta(q_1) + \beta(q_2)$. If $a_i \in V_i$, note $(q_1 \perp q_2)_{(a_1,a_2)} = (q_1)_{a_1} \perp (q_2)_{a_2}$, so if the formula holds for the two pieces, it holds for the orthogonal sum. Moreover, if it holds for the sum and one of the pieces, it holds for the other piece.

Finally, note that if $a = 0$, the formula is true.

Now use Brown, [Br, Theorem 2.2 (viii)] to see that it suffices to prove the formula for a form isomorphic to $m(1) + n(-1)$ and any a and this follows from the above discussion. ∎

Next we present a "geometric" calculation of the *Spin* and Pin^+ bordism groups in dimension 2.

Proposition 3.8. *Any Spin structure induces a unique Pin^- structure, so β is defined just as above for surfaces with a Spin structure. We have β defines an isomorphism $\Omega_2^{Spin} \to \mathbf{Z}/2\mathbf{Z}$. Any surface with odd Euler characteristic with any Pin^- structure is a generator for $\Omega_2^{Pin^-}$ and the 2–torus with the Lie group Spin structure is a generator for Ω_2^{Spin}.*

Proof: The proof is almost identical to that of Lemma 3.6. The surface F bounds an oriented 3–manifold W and by considering the obstruction to extending the *Spin* structure we see that F is *Spin* bordant to a torus with the same *Spin* structure as in the proof of Lemma 3.6. Just note that the boundary constructed there is actually a *Spin* boundary. It is a fact from Brown that β restricted to even forms only takes on the values 0 and 4. The results about the generators are straightforward. ∎

The Pin^+ case is more interesting. We have already seen that the only way a surface can have a Pin^+ structure is for w_1^2 to be 0. Hence the $[\cap w_1^2]$ map must also be 0, so the $[\cap w_1]$ map is defined on all of $\Omega_2^{Pin^+}$.

Proposition 3.9. *The homomorphism $[\cap w_1]\colon \Omega_2^{Pin^+} \to \Omega_1^{Spin} \cong \mathbf{Z}/2\mathbf{Z}$ is an isomorphism. A generator is given by the Klein bottle in half of its four Pin^+ structures.*

Proof: A surface, F, has a Pin^+ structure iff $w_2(F) = 0$ iff F is an unoriented boundary, say $F = \partial W$. The obstruction to the Pin^+ structure on F extending to W is given by a relative 2–dimensional cohomology class, so its dual is a 1–dimensional absolute homology class. We can assume that it is a single circle, and so F is Pin^+ bordant to either a torus or a Klein bottle, and the Pin^+ structure has the property that it does not extend over the corresponding 2–disk bundle over S^1.

Since S^1 with either Pin^+ structure is a Pin^+ boundary it is not hard to see that the torus with any Pin^+ structure is a Pin^+ boundary. There are two Pin^+ structures on the Klein bottle which do not extend over the disk bundle. If one cuts the Klein bottle open along the dual to w_1 and glues in two copies of the Möbius band, one sees a Pin^+ bordism between these two Pin^+ structures. Hence $\Omega_2^{Pin^+}$ has at most two elements. On the other hand it is not hard to see that the Klein bottle with the Pin^+ structures which do not extend over the disk bundle hit the non–zero element in Ω_1^{Spin} under $[\cap w_1]$. ∎

For future convenience let us discuss another way to "see" structures on the torus and the Klein bottle. We begin with the torus, T^2.

Example 3.10. We can write T^2 as the union of two open sets $U_i = S^1 \times (-1, 1)$ so that $U_1 \cap U_2$ is two disjoint copies of $S^1 \times (-1, 1)$, say $U_1 \cap U_2 = V_{12} \amalg \overline{V}_{12}$.

We can frame $S^1 \times (-1,1)$ using the product structure and the framings of the two 1–dimensional manifolds, S^1 and $(-1,1)$. If we form an $SO(2)$ bundle over T^2 with transition function g_{12} defined by $g_{12}(U_1 \cap U_2) = 1$ then we get the tangent bundle. If we think of 1 as the identity of $Spin(2)$ then the same transition functions give a *Spin* structure on T^2. This *Spin* structure is the Lie group one: clearly the copy of S^1 in the $S^1 \times (-1,1)$'s receives the Lie group structure, and it is not difficult to start with a framing of $(-1,1)$ and transport it around the torus to get the Lie group structure on this circle. If we take as $Spin(2)$ transition functions h_{12} defined by $h_{12}(V_{12}) = 1$ and $h_{12}(\overline{V}_{12}) = -1 \in Spin(2)$, then we get a *Spin* structure whose enhancement is 0 on the obvious S^1 and 2 on the circle formed by gluing the two intervals.

Example 3.11. We can write the Klein bottle, K^2, as the union of two open sets $U_i = S^1 \times (-1,1)$ so that $U_1 \cap U_2$ is two disjoint copies of $S^1 \times (-1,1)$, say $U_1 \cap U_2 = W_{12} \amalg \overline{W}_{12}$. We can frame $S^1 \times (-1,1)$ using the product structure and the framings of the two 1–dimensional manifolds, S^1 and $(-1,1)$. If we form an $O(2)$ bundle over K^2 with transition function g_{12} defined by $g_{12}(W_1) = 1$ and $g_{12}(\overline{W}_{12}) = \begin{pmatrix} -1 & 0 \\ 0 & 1 \end{pmatrix} \in O(2)$ then we get the tangent bundle (we are writing the S^1 tangent vector first). If we define $h_{12}(W_1) = 1$ and $h_{12}(\overline{W}_{12}) = e_1 \in Pin(2)$, we get a *Pin* structure on the tangent bundle. The copy of S^1 in the $S^1 \times (-1,1)$'s receives the Lie group structure, so if we are describing a Pin^- structure, then we get the bordism generator.

We conclude this section with two amusing results that we will need later.

Theorem 3.12. *Let F be a surface with a Spin structure. Let $q\colon H_1(F;\mathbf{Z}/2\mathbf{Z}) \to \mathbf{Z}/2\mathbf{Z}$ denote the induced quadratic enhancement. Let $x \in H^1(F;\mathbf{Z}/2\mathbf{Z})$. Corresponding to x there is a double cover of F, $\tilde{F}$ which has an induced Spin structure. There is also a dual homology class a and*

$$[\tilde{F}] = q(a) \in \mathbf{Z}/2\mathbf{Z} \quad .$$

Proof: We can write F as $T^2 \# F_1$ where T^2 is a 2 torus and a is contained in T^2. Then $\tilde{F} = T_1^2 \# F_1 \# F_1$, where T_1^2 is a double cover of T^2 given by $x \in H^1(T^2;\mathbf{Z}/2\mathbf{Z})$. Note $\langle x, a\rangle = 1$ not -1,so a lifts to 2 disjoint parallel circles. Moreover, $H_1(T_1^2;\mathbf{Z}/2\mathbf{Z})$ is generated by one component of the cover of a, say $\tilde{a}$, and another circle, say $\tilde{b}$ which double covers a circle, say b in T^2.

Note $[\tilde{F}] = [T_1^2] + 2[F_1]$, so $[\tilde{F}] = [T_1^2]$. The enhancement $\tilde{q}\colon H_1(T_1^2;\mathbf{Z}/2\mathbf{Z})$ satisfies $\tilde{q}(\tilde{a}) = q(a)$ and $\tilde{q}(\tilde{b}) = -1$. Hence the *Spin* bordism class of T_1^2 in $\mathbf{Z}/2\mathbf{Z}$ is given by $q(a)$. ∎

The second result is the following. Given any surface, F, we can take the orientation cover, $\tilde{F}$, and orient $\tilde{F}$ so that the orientation does not extend across any component of the total space of the associated line bundle. Given a $Pin^{\pm}$ structure on F, we can induce a $Spin$ structure on $\tilde{F}$.

Lemma 3.13. *The orientation double cover map induces homomorphisms*

$$\Omega_2^{Pin^{\pm}} \to \Omega_2^{Spin}$$

which are independent of the orientation on the double cover. The Pin^- map is trivial, and the Pin^+ map is an isomorphism.

Proof: If we switch to orientation on $\tilde{F}$, we get the reverse of the $Spin$ structure we originally had. Since $\Omega_2^{Spin} \cong \mathbf{Z}/2\mathbf{Z}$ this shows that the answer is independent of orientation. By applying the construction to a bordism between two surfaces we see that the maps are well–defined on the bordism groups. Since addition is disjoint union, the maps are clearly homomorphisms.

In the Pin^- case, $\mathbf{RP}^2$ is a generator of the bordism group. The oriented cover is S^2 which has a unique $Spin$ structure and is a $Spin$ boundary. This shows the Pin^- map is trivial.

In the Pin^+ case, a generator is given by the Klein bottle. Consider the transition functions that we gave for this Pin^+ structure in Example 3.11. This give us a set of transition functions for the torus which double covers the Klein bottle. We get 4 open sets, but it is not difficult to amalgamate three of the cylinders into one. The new transition function, h_{12}, takes the value 1 on one component of the overlap and the value e_1^2 on the other. Since $e_1 \in Pin^+(2)$, $e_1^2 = 1$ so we get the Lie group structure on T^2 by Example 3.10. ∎

Remark. If we started with a non–bounding Pin^- structure on the Klein bottle, then the above proof would show that the double cover has $Spin$ transition functions given by 1 on one component of the overlap and -1 on the other, and, as we saw, this $Spin$ structure bounds (as Lemma 3.13 requires).

§4. *Spin* structures on 3–manifolds.

Let M^3 be a closed 3–manifold with a given $Spin$ structure. We begin by generalizing some of the basic ideas in the calculus of framed links in S^3.

Given any embedded circle $k: S^1 \to M^3$, the normal bundle is trivial, and therefore has a countable number of framings. If the homology class represented by k is torsion, we can give a somewhat more geometric description of these framings. Recall that there is a non–singular linking form

$$\ell : tor H_1(M;\mathbf{Z}) \otimes tor H_1(M;\mathbf{Z}) \to \mathbf{Q}/\mathbf{Z}\ .$$

Let $x \in H_1(M;\mathbf{Z})$ be the class represented by k, and assume that x is torsion.

Lemma 4.1. *The framings on the normal bundle to k are in one–to–one correspondence with rational numbers q such that the class of q in $\mathbf{Q}/\mathbf{Z}$ is $\ell(x,x)$.*

Proof: We describe the correspondence. A framing on the normal bundle of k is equivalent to a choice of longitude in the torus which bounds a tubular neighborhood of k. Suppose $r \in \mathbf{Z}$ is chosen so that $r \cdot x = 0$ in $H_1(M;\mathbf{Z})$. Take r copies of the longitude in the boundary torus and let F be an oriented surface which bounds these r circles. Count the intersection of F and k with signs as usual. If one gets $p \in \mathbf{Z}$, then assign the rational number $\frac{p}{r}$ to this framing. It is a standard argument that $\frac{p}{r}$ is well–defined once the framing is fixed. It is also easy to see that $\frac{p}{r}$ mod $\mathbf{Z}$ is $\ell(x,x)$, and that if we choose a new framing which turns through t full right twists with respect to our original framing, then the new rational number that we get is $\frac{p}{r} + t$. ∎

A *Spin* structure on M gives a *Spin* structure on the normal bundle to k as follows. Restriction gives a *Spin* structure on the tangent bundle to S^1 plus the normal bundle. Choose the *Spin* structure on the normal bundle so that this *Spin* structure plus the one on S^1 which makes S^1 into a *Spin* boundary gives the restricted *Spin* structure.

Definition 4.2. We call the above framings *even*.

If x as above is torsion and M is spun, then the *Spin* structure picks out half of the rational numbers for which the longitude gives a framing compatible with the *Spin* structure on the normal bundle. Given one of these rational numbers, say q, the remaining ones are of the form $q + 2t$ for t an integer. Hence we can define a new element in $\mathbf{Q}/\mathbf{Z}$, namely $\frac{q}{2}$. This gives a map

$$\gamma\colon tor H_1(M;\mathbf{Z}) \to \mathbf{Q}/\mathbf{Z}$$

which is a *quadratic enhancement of the linking form* i.e.

$$\begin{aligned} \gamma(x+y) &= \gamma(x) + \gamma(y) + \ell(x,y) \\ \gamma(rx) &= r^2 \cdot \gamma(x) \text{ for any integer } r\ . \end{aligned}$$

Suppose now that x is zero in $H_1(M;\mathbf{Z}/2\mathbf{Z})$, but not necessarily torsion in $H_1(M;\mathbf{Z})$. Then any *Spin* structure on M induces the same *Spin* structure in a neighborhood of k, and hence the notion of even framing is independent of *Spin* structure for these classes.

Theorem 4.3. *A knot k which is* mod 2 *trivial as above, bounds a surface which does not intersect k. This surface selects a longitude for the normal bundle to k, and this longitude represents an even framing.*

Proof: Let E be a tubular neighborhood for k with boundary T^2. (This T^2 is often called the *peripheral* torus.) We can select a basis for $H_1(T^2;\mathbf{Z}/2\mathbf{Z})$ as follows. One element, the *meridian*, is the unique non–trivial element in the kernel of the map $H_1(T^2; \mathbf{Z}/2\mathbf{Z}) \to H_1(E; \mathbf{Z}/2\mathbf{Z})$. One calculates that the sequence $H_1(T^2; \mathbf{Z}/2\mathbf{Z}) \to H_1(M - k; \mathbf{Z}/2\mathbf{Z}) \to H_1(M; \mathbf{Z}/2\mathbf{Z})$ is exact, and that the image of $H_1(T^2; \mathbf{Z}/2\mathbf{Z})$ in $H_1(M - k; \mathbf{Z}/2\mathbf{Z})$ is 1–dimensional and generated by the meridian. Hence there is a unique non–trivial element, the mod 2 *longitude*, in the kernel of $H_1(T^2; \mathbf{Z}/2\mathbf{Z}) \to H_1(M - k; \mathbf{Z}/2\mathbf{Z})$. An even longitude for k is an element $\ell \in H_1(T^2; \mathbf{Z})$ which reduces in mod 2 homology to the mod 2 longitude.

Fix an even longitude, ℓ. It follows that there is an embedded surface, $F^2 \subset M$ such that $\partial F = k$. This surface can be chosen to intersect T^2 transversely in the even longitude. The southeast corner of Corollary 1.15 assigns a Pin^- structure to F. Restricted to k, the normal bundle to F in M is trivial, so the surface frames the normal bundle to k in M. Hence the *Spin* structure on M restricted to k is seen as the *Spin* structure on the circle coming from the restriction of the Pin^- structure on F plus the *Spin* structure on the normal bundle coming from the framing. We saw in the proof of Theorem 2.1 that, regardless of the Pin^- structure on F, the boundary circle receives the non–Lie structure. This is the definition of the even framing. ∎

Remarks 4.4.

(i) In S^3 with its unique *Spin* structure, the framing on k designated by an even number in the framed link calculus is an even framing in the above sense.

(ii) If the class x has odd order, then $\ell(x, x) = \frac{p}{r}$ with r odd. There are then two sorts of representatives in $\mathbf{Q}$ for $\ell(x, x)$: the p is even for half the representatives and odd for the other half. The framings that the *Spin* structure will call even are the ones with even numerator.

(iii) If we change the *Spin* structure on M by a class $\alpha \in H^1(M; \mathbf{Z}/2\mathbf{Z})$ the even framings on a circle change iff α evaluates non–trivially on the fundamental class of the circle.

(iv) If we attach a handle to a knot in a 3–manifold, M^3, we get a 4–manifold W with $H_2(W, M; \mathbf{Z}) = \mathbf{Z}$. If our knot in M^3 is torsion, we get a unique (up to sign) class $x \in H_2(W; \mathbf{Q})$ which hits our relative class. If we attach a handle with framing $q \in \mathbf{Q}$ from Lemma 4.1, then x intersects itself with a value of q. Hence the signature of W is sign (q), where sign $(q) = 1$ if $q > 0$; -1 if $q < 0$ and 0 if $q = 0$.

By Corollary 1.15 the surface F we used in the proof of Theorem 4.3 inherits a Pin^- structure from one on M. This suggests trying to define a knot invariant in this situation. Indeed, for knots in S^3, this is one way to define Robertello's Arf invariant, [R]. The situation in general is more complicated and needs results from §6, so we carry out the discussion in §8.

An invariant of a 3–manifold with a *Spin* structure is the μ–invariant. We discuss in Theorem 5.1 the classical result that $\Omega_3^{Spin} = 0$. It follows that any

3–manifold, M^3, is the boundary of a *Spin* 4–manifold, W.

Definition 4.5. The signature of W, reduced mod 16, is the *μ–invariant* of the manifold M with its *Spin* structure. It follows from Rochlin's theorem that $\mu(M)$ is well–defined once the *Spin* structure on M is fixed.

Remark. Some authors stick to $\mathbf{Z}/2\mathbf{Z}$ homology spheres so that there is a unique *Spin* structure and hence a μ invariant that depends only on the manifold.

We now turn to a geometric interpretation of some work of Turaev [Tu]. Intersection defines a symmetric trilinear product

$$\tau: H_2(M;\mathbf{Z}/2\mathbf{Z}) \times H_2(M;\mathbf{Z}/2\mathbf{Z}) \times H_2(M;\mathbf{Z}/2\mathbf{Z}) \to \mathbf{Z}/2\mathbf{Z}$$

We introduce a symmetric bilinear form

$$\lambda: H_2(M;\mathbf{Z}/2\mathbf{Z}) \times H_2(M;\mathbf{Z}/2\mathbf{Z}) \to \mathbf{Z}/2\mathbf{Z}$$

which is defined as follows. Let F_x and F_y be embedded surfaces representing two classes x and y in $H_2(M;\mathbf{Z}/2\mathbf{Z})$. To define $\lambda(x,y)$ put the two surfaces in general position. The intersection will be a collection of embedded circles. The normal bundle of each circle in M has a sub–line bundle, ξ_x, given by the inward normal to the surface F_x. Define $\lambda(x,y)$ to be the number of circles with non–trivial ξ_x.

Here is an equivalent definition of λ. Any codimension 1 submanifold of a manifold is mod 2 dual to a 1–dimensional cohomology class in the manifold. If this cohomology class is pulled–back to the submanifold, it becomes w_1 of the normal bundle to the embedding. Hence, if x^* and y^* are the Poincaré duals to x and y, $\lambda(x,y) = x^* \cup x^* \cup y^*[M]$, where $[M]$ is the fundamental class of the 3–manifold. This follows because $x^* \cup y^* \cap [M]$ is the homology class represented by the intersection circles, and to count the number with non–trivial ξx we just evaluate w_1 of the normal bundle on these circles. But $w_1 = x^*$ so we are done. We can also prove symmetry using this definition. Since M is orientable, $0 = w_1(M)x^*y^* = Sq^1(x^*y^*) = (x^*)^2y^* + x^*(y^*)^2$.

Yet another definition of λ is

$$\lambda(x,y) = \tau(x,x,y) \ .$$

Hence λ is symmetric and bilinear.

Given a *Spin* structure on M, we can enhance λ to a function

$$f: H_2(M;\mathbf{Z}/2\mathbf{Z}) \times H_2(M;\mathbf{Z}/2\mathbf{Z}) \to \mathbf{Z}/4\mathbf{Z} \ .$$

To begin, we define f on embedded surfaces F_x and F_y in M as above, but now use the *Spin* structure to put even framings on the intersection circles and then count the number of half twists in each ξ_x. (Since the collection of circles is embedded, there is no correction term needed to account for intersections.) Note if we defined ξ_y in the obvious manner and counted half twists in it instead of in ξ_x, we would get the same number, so f is symmetric.

Here is another description of $f(F_x, F_y)$. In M^3, F_y is dual to a cohomology class, $\alpha \in H^1(M; \mathbf{Z}/2\mathbf{Z})$, and we could take α and restrict it to F_x, getting $\alpha_x \in H^1(F_x; \mathbf{Z}/2\mathbf{Z})$. The Poincaré dual of α_x in F_x is just the class represented by our collection of circles, which we will denote by $\hat{y}$. Associated to our Pin^- structure on F_x, there is a quadratic enhancement ψ_x. Note

$$f(F_x, F_y) = \psi_x(\hat{y}) \ . \tag{4.6}$$

In particular, note $f(F_x, F_y)$ only depends on the homology class of F_y, and hence by symmetry also only on the homology class of F_x.

Once we see the pairing is well–defined, it is easy to see that $f(x, 0) = f(0, x) = 0$ for all $x \in H_2(M; \mathbf{Z}/2\mathbf{Z})$. We have lost bilinearity and gained

$$f(x, y + z) = f(x, y) + f(x, z) + 2\tau(x, y, z) \ . \tag{4.7}$$

Proof: With notation as above, we apply formula 4.6. We need to show $\psi_x(\widehat{y + x}) = \psi_x(\hat{y}) + \psi_x(\hat{z}) + 2\tau(x, y, z)$, which is just the quadratic enhancement property of ψ_x and the identification of $\hat{y} \bullet \hat{z}$ in F_x with $\tau(x, y, z)$. ∎

If we change the *Spin* structure on M by $\alpha \in H^1(M; \mathbf{Z}/2\mathbf{Z})$, then we change f as follows. Let f_α denote the new pairing and let $a \in H_2(M; \mathbf{Z}/2\mathbf{Z})$ be the Poincaré dual to α. Then

$$f_\alpha(x, y) = f(x, y) + 2\tau(x, y, a) \ ,$$

or

$$f_\alpha(x, y) = f(x, y + a) - f(x, a) \ .$$

Proof: We prove the first formula. Using 4.6 we see that the first formula is equivalent to $\psi_\alpha(\hat{y}) = \psi(\hat{y}) + 2\tau(x, y, a)$, which follows easily from formula 3.3. ∎

Finally, we have a function

$$\beta \colon H_2(M; \mathbf{Z}/2\mathbf{Z}) \to \mathbf{Z}/8\mathbf{Z} \ . \tag{4.8}$$

We define β by taking an embedded surface representing x, using the *Spin* structure on M to get a Pin^- structure on F_x, taking the underlying Pin^- bordism class, and using our explicit identification of this group with $\mathbf{Z}/8\mathbf{Z}$.

We need to see why this is independent of the choice of embedded surface. Given two such surfaces, there is a bordism in $M \times [0,1]$ between them. Let $W \subset M \times [0,1]$ be a 3–manifold with the two boundary components representing the same element in $H_2(M; \mathbf{Z}/2\mathbf{Z})$. Since $M \times [0,1]$ is spun, we get a Pin^- structure on W which is our given Pin^- structure at the two ends. Since Brown's $\mathbf{Z}/8\mathbf{Z}$ is a Pin^- bordism invariant, we are done. It further follows that $\beta(0) = 0$.

Reduced mod 2 $\beta(x)$ is just the mod 2 Euler class of an embedded surface representing x, and hence β is additive mod 2. We have

$$\beta(x+y) = \beta(x) + \beta(y) + 2f(x,y) \quad . \tag{4.9}$$

which we will prove in a minute. It follows that $f(x,x) = -\beta(x)$ reduced mod 4. Note that, mod 4, $\beta(x+y) = \beta(x) + \beta(y) + 2\tau(x,x,y)$.

How does β change when we change the *Spin* structure by $\alpha \in H^1(M;\mathbf{Z}/2\mathbf{Z})$? The principle is easy. Given a surface, F, restrict α to F and consider it to be a change in Pin^- structure on F. Compute the Brown invariant for this new Pin^- structure, and this is the value of the new β on F. It follows from Lemma 3.7 that

$$\beta_\alpha(x) = \beta(x) + 2f(x,a) \tag{4.10}$$

with notation as above.

Given the theorem below, we now prove formula 4.9. From this theorem we get: $u - u_\alpha = 2\beta(a)$ and $u - u_{\alpha_1} = 2\beta(a_1)$. Also $u_\alpha - u_{\alpha_1} = 2\beta_\alpha(a_1 - a)$. Hence $\beta_\alpha(a_1 - a) = \beta(a_1) - \beta(a)$. Set $a_1 = x + a$ and use formula 4.10. ∎

The main result concerning β is

Theorem 4.11. *Let M be a spun 3–manifold with resulting function β and μ–invariant u in $\mathbf{Z}/16\mathbf{Z}$. Let $\alpha \in H^1(M;\mathbf{Z}/2\mathbf{Z})$ be used to change the Spin structure, and let u_α be the new μ–invariant. Then*

$$u - u_\alpha = 2\beta(a) \qquad \pmod{16}$$

where $a \in H_2(M;\mathbf{Z}/2\mathbf{Z})$ is the Poincaré dual to α.

Proof: The proof is just the Guillou–Marin formula, [G–M, Theoreme, p. 98], or our discussion of it in §6, 6.4. On $M \times [0,1]$ put the original *Spin* structure on $M \times 0$ and put the altered one on $M \times 1$. We can cap this off to a closed 4–manifold by adding *Spin* manifolds that the two copies of M bound to either end. The resulting 4–manifold has index $u_\alpha - u$. Let F be a surface in M representing a. Then $F \times 1/2$ is a dual to w_2 for the 4–manifold. Since F is in a product, $F \bullet F = 0$ and the enhancement used in the Guillou–Marin formula is the same as the one we put on F to calculate β. By formula 6.4, $u - u_\alpha = 2\beta(a)$. ∎

As a corollary we get a result of Turaev, [Tu]

Corollary 4.12. *The quadratic enhancement of the linking form gives the μ–invariant* mod 8 *via the Milgram Gauss sum formula.*

Proof: This was proved in [Ta] for rational homology spheres. Pick a basis for the torsion free part of H_1 and do surgery on this basis. The resulting bordism, W, has signature 0; both boundary components have isomorphic torsion subgroups of H_1; and the top boundary component has no torsion free part. Put a *Spin* structure on the bordism, which puts a *Spin* structure at both ends. The two enhancements on the linking forms are equal, and they stay equal if we change both *Spin* structures by an element in $H^1(W; \mathbf{Z}/2\mathbf{Z})$. Any *Spin* structure on M can be obtained from our initial one by acting on it by an element of the form $x + y$, where x comes from $H^1(W; \mathbf{Z}/2\mathbf{Z})$ and y comes from $H^1(M; \mathbf{Z})$. But acting by this second sort of element does not change the mod 8 μ–invariant or the quadratic enhancement of the linking form. ▪

§5. Geometric calculations of $\Omega_{3,4}^{Pin^\pm}$.

We begin this section with a calculation for the 3–dimensional *Spin*, Pin^- and Pin^+ bordism groups.

Theorem 5.1. $\Omega_3^{Spin} \cong 0$; $\Omega_3^{Pin^-} \cong 0$ *and* $[\cap w_1]: \Omega_3^{Pin^+} \to \Omega_2^{Spin} \cong \mathbf{Z}/2\mathbf{Z}$ *is an isomorphism.*

Proof: The *Spin* bordism result is classical: [ABP1], [Ka] or [Ki].

Given a non–orientable $Pin^\pm$ manifold M^3, we will try to find a $Pin^\pm$ bordism to an orientable manifold which then $Pin^\pm$ bounds by the *Spin* case. The dual to $w_1(M)$ is an orientable surface F by Proposition 2.3. The first step is to reduce to the case when F has trivial normal bundle. If not, consider F intersected transversely with itself. It can be arranged that this is a single circle C, which is dual in F to $w_1(M)$ pulled back to F. The normal bundle to C in M is $\nu_{F\subset M}|_C \oplus \nu_{F\subset M}|_C$ which is also $\nu_{C\subset F} \oplus \nu_{C\subset F}$ which is trivialized. Hence the $Pin^\pm$ structure on M induces a $Pin^\mp$ structure on C. Suppose C with this structure bounds Y^2; let E denote the total space of $\zeta \oplus \zeta$ over Y, where ζ is the determinant line bundle for Y. Note that inside ∂E there is a copy of $(\partial Y) \times B^2$, and E has a $Pin^\pm$ structure extending the one on $(\partial Y) \times B^2$. We can form $M \times [0,1] \cup E$ by gluing $(\partial Y^2) \times B^2$ to $C \times B^2 \times 1$ where $C \times B^2$ is the trivialized disk bundle to C above. Clearly the $Pin^\pm$ structure extends across the bordism, and the "top" is a new $Pin^\pm$ manifold M_1 with a new dual surface F_1 with trivial normal bundle.

In the Pin^- case, C has a Pin^+ structure which bounds ($\Omega_1^{Pin^+} = 0$, Theorem 2.1) so we have achieved the (M_1, F_1) case. In the Pin^+ case an argument is needed to see that we never get C representing the non–zero element in $\Omega_1^{Pin^-} = \mathbf{Z}/2\mathbf{Z}$, i.e. C does not get the Lie group *Spin* structure.

To show this, let V be a dual to w_1 and let E be a tubular neighborhood of V. By the discussion just before Lemma 2.7, since E as a Pin^+ structure, there is an inherited *Spin* structure on V (in fact there are two which differ by the action of $x \in H^1(V; \mathbf{Z}/2\mathbf{Z})$, where x denotes the restriction of w_1 to V). Note x also describes the double cover $\partial E \to V$. The boundary, ∂E, also inherits a Pin^+ structure and we saw, Lemma 2.7, that, if we orient ∂E and V so that the covering map is degree 1, the *Spin* structure on ∂E is the same as the one induced by the covering map. The *Spin* structure on ∂E bounds the *Spin* manifold which is the closure of $M - E$, so if C is the dual to x and q is the quadratic enhancement on $H_1(V; \mathbf{Z}/2\mathbf{Z})$, $q(C) = 0$ by Theorem 3.12. Recall that the normal bundle to V in M, when restricted to C is trivial. Hence the framing on C as a circle in V is the same as the Pin^- structure on C as V intersect V in a Pin^+ manifold. Hence C has the non–Lie group *Spin* structure and hence represents 0 in Ω_1^{Spin}.

Hence we may now assume that F has trivial normal bundle in M. Therefore F inherits a $Pin^\pm$ structure from the one on M, and hence, after choosing an orientation, F has a *Spin* structure. If the *Spin* structure on F is a boundary then it is easy as above to construct a $Pin^\pm$ bordism to an oriented manifold. In the Pin^+ case we are entitled to assume that the surface bounds because that is what the invariant $[\cap w_1]$ is measuring. In the Pin^- case, the Klein bottle $\times S^1$ with the Lie group framing is an example for which the F has the non–bounding *Spin* structure. But if we add this manifold to our original M, for the new manifold, F will bound and we are done.

We have now proved that $[\cap w_1]$ is injective in the Pin^+ case and that $\Omega_3^{Pin^-}$ is generated by $K \times S^1$, where K is the Klein bottle and the Pin^- structure comes from some structure on the surface and the Lie group *Spin* structure on S^1. In some Pin^- structures, K bounds and hence so does $K \times S^1$. In the others, K is Pin^- bordant to two copies of $\mathbf{RP}^2$, so $K \times S^1$ is bordant to two copies of $\mathbf{RP}^2 \times S^1$. Hence, if we can prove that $[\cap w_1]$ is onto and that $\mathbf{RP}^2 \times S^1$ bounds, we are done.

If we take the generator of $\Omega_2^{Pin^+}$ and cross it with S^1 with the Lie group *Spin* structure, we get a 3–manifold with $[\cap w_1]$ being the 2–torus with Lie group *Spin* structure so by Proposition 3.8, $[\cap w_1]$ is onto.

Consider $\mathbf{RP}^2$ in $\mathbf{RP}^4$: it is the dual to $w_1^2 + w_2$ so there is a Pin^- structure on $\mathbf{RP}^4 - \mathbf{RP}^2$ which restricts to the Lie group structure on the normal circle to $\mathbf{RP}^2$. An easy calculation of Stiefel–Whitney classes shows that the normal bundle ν of $\mathbf{RP}^2$ in $\mathbf{RP}^4$ is orientable but $w_2(\nu) \neq 0$. So we take the pairwise connected sum $(\mathbf{RP}^4,\ \mathbf{RP}^2)\#(\mathbf{CP}^2,\ \mathbf{CP}^1)$ and then the normal bundle of $\mathbf{RP}^2 = \mathbf{RP}^2\#\mathbf{CP}^1$ in $\mathbf{RP}^4\#\mathbf{CP}^2$ has $w_1 = w_2 = 0$. For a bundle over $\mathbf{RP}^2$ this means that the bundle is trivial, so its normal circle bundle is $\mathbf{RP}^2 \times S^1$. The two Pin^- structures on $\mathbf{RP}^4\#\mathbf{CP}^2 - \mathbf{RP}^2$ bound two Pin^- structures on $\mathbf{RP}^2 \times S^1$ which have the Lie group structure on S^1. Since this is all the Pin^- structures that there are with the Lie group *Spin* structure on the S^1, we are done. ∎

Next we turn to the 4–dimensional case. The result is

Theorem 5.2. *The group* $\Omega_4^{Spin} \cong \mathbf{Z}$ *generated by the Kummer surface;* $\Omega_4^{Pin^-} = 0$*; and the group* $\Omega_4^{Pin^+} \cong \mathbf{Z}/16\mathbf{Z}$ *generated by* $\mathbf{RP}^4$.

Proof: The *Spin* result may be found in [Ki, p. 64, Corollary]. Our first lemma determines the image of Ω_4^{Spin} in the $Pin^\pm$ bordism groups.

Lemma 5.3. *The Kummer surface bounds a Pin^- manifold hence so does any 4–dimensional Spin manifold. Twice the Kummer surface bounds a Pin^+ manifold, but the Kummer surface itself does not. Hence a 4–dimensional Spin manifold Pin^+ bounds iff its signature is divisible by 32.*

Proof: The Enriques surface, E, [Ha], is a complex surface with $\pi_1(E) \cong \mathbf{Z}/2\mathbf{Z}$ with $w_2(E) \neq 0$. Habegger shows that $H^2(M;\mathbf{Z}) \cong \mathbf{Z}^{10} \oplus \mathbf{Z}/2\mathbf{Z}$ and $w_2(M)$ is the image of the non–zero torsion class in $H^2(M;\mathbf{Z})$, see paragraph 2 after the Proposition on p. 23 of [Ha]. If $y \in H^1(E;\mathbf{Z}/2\mathbf{Z})$ is a generator, then from the universal coefficient theorem, $y^2 = w_2(W)$. If L is the total space of the line bundle over E with $w_1 = y$, then it is easy to calculate that L is Pin^- (but not Pin^+), and ∂E is the Kummer surface. This proves the Kummer surface bounds a Pin^- manifold. Since $\Omega_4^{Spin} \cong \mathbf{Z}$ generated by the Kummer surface, this proves any *Spin* 4–manifold bounds as a Pin^- manifold.

Let M^4 is a *Spin* manifold and let W^5 be a Pin^- manifold with $\partial W = M$ as Pin^- manifolds. Consider the obstruction to putting a Pin^+ structure on W extending the one on M^4. The obstruction is $w_2(W) = w_1^2(W)$, so the dual class is represented by a 3–manifold formed as the intersection to a dual to w_1 pushed off itself. As usual, this 3–manifold has a natural Pin^+ structure and it is easy to see that we get a well–defined element in $\Omega_3^{Pin^+} \cong \mathbf{Z}/2\mathbf{Z}$. If this element is 0, then we can glue on the trivializing bordism and extend its normal bundle to get a new Pin^- manifold W_1 which still bounds M and has no obstruction to extending the *Spin* structure on the boundary to a Pin^+ structure on the interior. Hence, if our element in $\Omega_3^{Pin^+}$ is 0, M bounds. From this it is easy to see that twice the Kummer surface bounds. Hence any 4–dimensional *Spin* manifold with index divisible by 32 bounds a Pin^+ manifold.

Suppose that W is a Pin^+ manifold with $\partial W = M$ orientable. Let $V \subset W$ be a dual to w_1 contained in the interior of W. Let E be a tubular neighborhood of V with boundary ∂E. As usual, ∂E is orientable and the covering translation is orientation preserving. Since V is orientable with a normal line bundle, if we fix an orientation, *Spin* structures on V correspond to Pin^+ structures on E. Since W is a Pin^+ manifold, E has an induced Pin^+ structure and V acquires an induced *Spin* structure. The bordism between M and ∂E is an oriented Pin^+ bordism, so M and ∂E have the same signature. But ∂E is the double cover of V so has signature twice the signature of V. Since V is *Spin*, the signature of V is divisible by 16, so the signature of M is divisible by 32. This shows that the Kummer surface does not bound a Pin^+ manifold and indeed that any 4–dimensional *Spin* manifold of index congruent to 16 mod 32 does not bound a Pin^+ manifold. ∎

Since $\Omega_4^{Spin} \cong \mathbf{Z}$ generated by the Kummer surface this lemma calculates the image of Ω_4^{Spin} in $\Omega_4^{Pin^\pm}$ and our next goal is to try to produce a $Pin^\pm$ bordism from any $Pin^\pm$ manifold to an orientable one.

To this end let M be a 4–manifold with V^3 a dual to w_1. Consider the dual to w_1 intersected with itself. It is a surface $F \subset M$ and the normal bundle is two copies of the same line bundle. Indeed, the transversality condition gives an isomorphism between the two bundles. This line bundle is also abstractly isomorphic to the determinant line bundle for F. A $Pin^{\pm}$ structure on F gives rise to a $Pin^{\mp}$ structure on the total space of the normal bundle of F in M by Lemma 1.7. Hence we can use the $Pin^{\pm}$ structure on M to put a $Pin^{\mp}$ structure on F and it is not hard to check that we get a homomorphism $\Omega_4^{Pin^{\pm}} \to \Omega_2^{Pin^{\mp}}$. If F bounds in this structure, one can easily see a $Pin^{\pm}$ bordism to an new 4–manifold M_1 in which the dual to w_1 has trivial normal bundle. This puts a $Pin^{\pm}$ structure on V_1. By orienting V_1 we get a $Spin$ manifold and if V_1 bounds in this $Spin$ structure, M_1 $Pin^{\pm}$ bounds an orientable manifold.

Consider the Pin^- case. Any element in the kernel of the map $[\cap w_1^2]: \Omega_4^{Pin^-} \to \Omega_2^{Pin^+}$ is Pin^- bordant to a Pin^- manifold whose dual to w_1, say V, has trivial normal bundle. Orienting this normal bundle gives a Pin^- structure on V, and since $\Omega_3^{Spin} = 0$, we can further Pin^- bord our element to an orientable representative. It then follows from Lemma 5.3 that the map $[\cap w_1^2]$ is injective.

To show that this map is trivial, which proves $\Omega_4^{Pin^-} = 0$, proceed as follows. Let $V \subset M$ be a dual to $w_1(M)$ and let F^2 denote the transverse intersection of V with itself. Since the normal bundle to F in M is 2 copies of the determinant line bundle for F, F acquires a Pin^+ structure from the Pin^- structure on M. Let $E \subset V$ be a tubular neighborhood for F in V. Theorem 2.9 applies to this situation to show that the Pin^+ structure on ∂E induced by the double cover map $\partial E \to F$ is the same as the Pin^+ structure induced on $\partial E \subset M$ from the fact that its normal bundle is exhibited as the sum of 2 copies of its determinant line bundle. Since the normal bundle to V in M is trivial on $V - F$, $V - F$ has a $Spin$ structure which restricts to the given one on ∂E. By Lemma 3.13, the oriented cover map $\Omega_2^{Pin^+} \to \Omega_2^{Spin}$ is an isomorphism, so F is a Pin^+ boundary, which finishes the Pin^- case.

So consider the Pin^+ case. This time our homomorphism goes from $\Omega_4^{Pin^+}$ to $\Omega_2^{Pin^-} \cong \mathbf{Z}/8\mathbf{Z}$ and the example of $\mathbf{RP}^4$ shows that it is onto. Just as in the Pin^- case, any element in the kernel of this homomorphism is Pin^+ bordant to an orientable manifold. This together with Lemma 5.3 shows that $0 \to \mathbf{Z}/2\mathbf{Z} \to \Omega_4^{Pin^+} \to \mathbf{Z}/8\mathbf{Z} \to 0$ is exact.

To settle the extension requires more work. Given a Pin^+ structure on a 4–manifold M, we can choose a dual to w_1, say $V \subset M$, and an orientation on $M - V$ which does not extend across any component of V. We can consider the bordism group of such structures, say G_4. There is an epimorphism $G_4 \to \Omega_4^{Pin^+}$ defined by just forgetting the dual to w_1 and the orientation. There is another homomorphism $G_4 \to \mathbf{Q}/32\mathbf{Z}$ defined as follows. Let E be a tubular neighborhood of V with boundary ∂E. The covering translation on ∂E is orientation preserving, so V is also oriented. The normal bundle to ∂E in M is a trivial line bundle, oriented by inward normal last, where inward is with respect to the associated disk bundle. Hence ∂E acquires a *Spin* structure, and hence a μ invariant in $\mathbf{Z}/16\mathbf{Z}$. The manifold ∂E is a 3–manifold with an orientation preserving free involution on it, hence there is an associated Atiyah–Singer α invariant, $\alpha(\partial E) \in \mathbf{Q}$. Define $\psi(M, V) = \sigma(M - \text{int } V) + \alpha(\partial E) - 2\mu(V) \in \mathbf{Q}/32\mathbf{Z}$. It is not hard to check that ψ depends only on the class of (M, V) in G_4 and defines a homomorphism. We can make choices so that $\psi(\mathbf{RP}^4, \mathbf{RP}^3) = +2$. Hence $\psi(8(\mathbf{RP}^4, \mathbf{RP}^3)) = 16$ with these choices. The Pin^+ bordism of 8 copies of $\mathbf{RP}^4$ to an oriented manifold is seen to extend to a bordism preserving the dual to w_1 and orientation data. This oriented, hence *Spin* manifold has index congruent to 16 mod 32, and so we have constructed a Pin^+ bordism (with some extra structure which we ignore) from 8 copies of $\mathbf{RP}^4$ to a *Spin* manifold which is Pin^+ bordant to the Kummer surface. This shows $\Omega_4^{Pin^+} \cong \mathbf{Z}/16\mathbf{Z}$. ∎

§6. 4–dimensional characteristic bordism.

The purpose of this section is to study the relations between 4–manifolds and embedded surfaces dual to $w_2 + w_1^2$.

Definition 6.1. A pair (M, F) with the embedding of F in M proper and the boundary of M intersecting F precisely in the boundary of F is called a *characteristic pair* if F is dual to $w_2 + w_1^2$. A characteristic pair is called *characterized* provided we have fixed a Pin^- structure on $M - F$ which does not extend across any component of F. The characterizations of a characteristic pair are in one to one correspondence with $H^1(M; \mathbf{Z}/2\mathbf{Z})$.

We begin by discussing the oriented case.

Lemma 6.2. *Let M be an oriented manifold with a codimension 2 submanifold F which is dual to w_2. There exists a function*

$$\mathcal{C}har(M, F) \to \mathcal{P}in^-(F) \quad .$$

The group $H^1(M; \mathbf{Z}/2\mathbf{Z})$ acts on $\mathcal{C}har(M, F)$, the group $H^1(F; \mathbf{Z}/2\mathbf{Z})$ acts on $\mathcal{P}in^-(F)$ and the map is equivariant with respect to the map induced on $H^1(\,; \mathbf{Z}/2\mathbf{Z})$ by the inclusion $F \subset M$.

Remark. Later in this section we will define this function in a more general situation.

Proof: There is an obvious restriction map from characteristic structures on (M, F) to those on (E, F), where E is the total space of the normal bundle to F in M, denoted ν. Hence it suffices to do the case $M = E$. In this case we expect our function to be a bijection. After restricting to the case $M = E$ it is no further restriction to assume that F is connected since we may work one component at a time.

We begin with the case that F has the homotopy type of a circle. In this case ν has a section, so choose one and write $\nu = \lambda \oplus \epsilon^1$. Orient ϵ^1 and use it to embed F in ∂E. The normal bundle to ∂E in E is oriented; E is oriented; so ∂E is oriented. The normal bundle to the embedding of F in ∂E is λ so the orientation on E plus the orientation of ϵ^1 pick out a preferred isomorphism between λ and $\det T_F$. From Corollary 1.15, there is a Pin^- structure on F induced from the one on ∂E.

We want to see that this Pin^- structure is independent of the section we chose. It is not difficult to work out the effect of reorienting the section: there is none.

Suppose the bundle is trivial. We divide into two cases depending on the dimension of E. In the 1–dimensional case, we may proceed as follows. The manifold F is a circle and since the bundle has oriented total space, it must be trivial. Hence $\partial E = T^2$ and $H_1\left(T^2; \mathbf{Z}/2\mathbf{Z}\right)$ has one preferred generator, the image of the fibre, otherwise known as a meridian, denoted m. Let x denote another generator. Since the *Spin* structure is not to extend over the disk, the enhancement associated to the *Spin* structure on T^2, say q, satisfies $q(m) = 2$. The *Spin* structure on the embedded base is determined by q of the image, which is either x or $x + m$. Check $q(x) = q(x + m)$.

In the higher dimensional case, there is an S^1 embedded in F and the normal bundle to this embedding is trivial. Over the S^1 in F there is an embedded T^2 in ∂E and the bundle projection, p, identifies the normal bundle to T^2 in ∂E with the normal bundle to S^1 in F. Fix a *Spin* structure on one of these normal bundles and use p to put a *Spin* structure on the other. The *Spin* structure on ∂E restricts to one on T^2 and it is not hard to check that the Pin^- structure we want to put on F using the section is determined by using the section over S^1 and checking what happens in T^2. We saw this was independent of section so we are done with the trivial case.

Now we turn to the non–trivial case, still assuming that F is the total space of a bundle over S^1. The minimal dimension for such an F is 2 since the bundle, ν, is non–trivial. In this case F is just a Möbius band. Since E is oriented, the bundle we have over F is isomorphic to $\det \nu \oplus \epsilon^1$. Sitting over our copy of S^1 in F is the Klein bottle, K^2, and the normal bundle to K^2 in ∂E is just the pull–back of ν. One can sort out orientations and check that there is an induced Pin^- structure on K^2 so that the Pin^- structure that we want to put on F is determined by the enhancement of the section applied to S^1 as a longitude of K^2. This calculation is just like the torus case. In the higher dimensional case, ν is a non–trivial line bundle plus a trivial bundle so we can reduce to the dimension 2 case just as above.

Now we turn to the case of a general F.

Since we have done the circle case, we may as well assume that the dimension of F is at least 2. If the dimension of F is 2, then we can find a section of our bundle over $F - pt$. The embedding of $F - pt$ in ∂E gives a Pin^- structure on $F - pt$ and this extends uniquely to a Pin^- structure on F. This argument even works if F

has a boundary and we take as the function on the boundary the function we have already defined. Now if we restrict this structure on F to a neighborhood of an embedded circle, we get our previous structure. Since this structure is independent of the section, the structure on all of F is also independent of the section since Pin^- structures can be detected by restricting to circles.

The higher dimensional case is a bit more complicated. We can define our function by choosing a set of disjointly embedded circles and taking a tubular neighborhood to get U, with $H_1(U;\mathbf{Z}/2\mathbf{Z}) \to H_1(F;\mathbf{Z}/2\mathbf{Z})$ an isomorphism. We then use our initial results to put a Pin^- structure on U and then extend it uniquely to all of F. Now let V be a tubular neighborhood of a circle in F. We can restrict the Pin^- structure on F to V, or we can use our "choose a section, embed in ∂E and induce" technique. There is an embedded surface, W^2, in F which has the core circle for V as one boundary component and some of the cores of U as the others. Let X be a tubular neighborhood of W in F. The bundle restricted to X has a section so we can induce a Pin^- structure on X using the section. This shows that the two Pin^- structures defined above on V agree. It is not hard from this result to see that the Pin^- structure on F is independent of the choice of U. ∎

Remarks. Notice that the proof shows that the Pin^- structure on a codimension 0 subset of F, say X, only depends on the Pin^- structure on the circle bundle lying over X. It is not hard to check that our function commutes with taking boundary, we get a well-defined homomorphism, β, from the rth Guillou–Marin bordism group to $\Omega_{r-2}^{Pin^-}$.

Theorem 6.3. *Let M^4 be an oriented 4–manifold, and suppose we have a characteristic structure on the pair (M,F). The following formula holds:*

$$2\cdot\beta(F) = F{\bullet}F - \operatorname{sign}(M) \pmod{16} \tag{6.4}$$

where the Pin^- structure on F is the one induced by the characteristic structure on (M,F) via 6.2.

Proof: By the Guillou–Marin calculation, their bordism group in dimension 4 is $\mathbf{Z}\oplus\mathbf{Z}$, generated by $(S^4,\mathbf{RP}^2)$ and $(\mathbf{CP}^2,S^2)$. The formula is trivial to verify for $(\mathbf{CP}^2,S^2)$. For $(S^4,\mathbf{RP}^2)$ we must verify that $\mathbf{RP}^2{\bullet}\mathbf{RP}^2=2$ implies that the resulting q is 1 on the generator. Now $\mathbf{RP}^2$ has two sorts of embeddings in S^4. There is a "right–handed" one, which has $\mathbf{RP}^2{\bullet}\mathbf{RP}^2=2$, and a "left–hand" one which has $\mathbf{RP}^2{\bullet}\mathbf{RP}^2=-2$. The "right–handed" one can be constructed by taking a 'right–handed" Möbius strip in the equatorial S^3 and capping it off with a ball in the northern hemisphere. For our vector field, use the north–pointing normal. The "even" framing on the bundle to ν_k, the core of the Möbius band, is the one given by the 0–framing in S^3. Hence we may count half twists in S^3, where the right–hand Möbius band half twists once. ∎

It would be nice to check that the Pin^- structure we put on the characterized surface agrees with those of Guillou–Marin and Freedman–Kirby. For the Freedman–Kirby case we take an embedded curve k in F and cap it off by an orientable surface, B, in M. We start B off in the same direction as our normal vector field, so then the normal bundle to B in M, when restricted to the boundary circle, will be the 2–plane bundle around k we are to consider. The Guillou–Marin case is similar except that B need not be orientable. Since B is a punctured surface, the normal bundle to B in M splits off a trivial line bundle and so is a trivial bundle plus the determinant line bundle for the tangent bundle. Having chosen one section, the others are classified by $H^1(B; \mathbf{Z}^{w_1})$, where $\mathbf{Z}^{w_1}$ denotes $\mathbf{Z}$ coefficients twisted by w_1 of the normal bundle. When restricted to the boundary circle, this gives a well–defined "even" framing of the normal bundle.

If B does not intersect F except along ∂B, Theorem 4.3 shows that the framing on ∂B is the even one in the sense of Definition 4.2. We can assume in general that B intersects F transversally away from ∂B. The surface $\hat{B} = B - \amalg D^2$ lies in $M - F$ and each circle from the transverse intersection has the non–bounding *Spin* structure. Hence, in general, the framing on ∂B is even iff the mod 2 intersection number of F and B is even. Moreover, the number of half right twists mod 4 is just the obstruction to extending the section given by the normal to k in F over all of B. This shows that our enhancement and those of Freedman–Kirby and Guillou–Marin agree when both are defined.

The enhancement above is defined more generally since we do not need the membranes to select the Pin^- structure and hence do not need the condition that $H_1(F; \mathbf{Z}/2\mathbf{Z}) \to H_1(M; \mathbf{Z}/2\mathbf{Z})$ should be 0. One nice application of this is to compute the μ–invariant of circle bundles over surfaces when the associated disk bundle is orientable.

Any $O(2)$–bundle, η, over a 2 complex, X, is determined by $w_1(\eta)$ and the Euler class, $\chi(\eta) \in H^2(X; \mathbf{Z}^{w_1})$, where $\mathbf{Z}^{w_1}$ denotes $\mathbf{Z}$ coefficients twisted by $w_1(\eta)$. In our case, X is a surface which we will denote by F; the bundle η has the same w_1 as the surface; and the Euler class is in $H^2(F; \mathbf{Z}^{w_1}) \cong \mathbf{Z}$. Let $S(\eta)$ denote the circle bundle. One way to fix the isomorphism is to orient the total space of η and then $F{\bullet}F = \chi(\eta)$. The signature of the disk bundle is also easy to compute. We denote it by $\sigma(\eta)$ since we will see it depends only on η; indeed it can be computed from $w_1(\eta)$ and $\chi(\eta)$. If $w_1(\eta) = 0$ then $\sigma(\eta) = \text{sign}\ \chi(\eta)$ (± 1 or 0 depending on $\chi(\eta)$): if $w_1(\eta) \neq 0$ then $\sigma(\eta) = 0$. By Lemma 6.2, *Spin* structures on $S(\eta)$ which do not extend across the disk bundle are in 1–1 correspondence with Pin^- structures on F.

Theorem 6.5. *With notation as above fix a Spin structure on $S(\eta)$. Let $b(F) = 0$ if this structure extends across the disc bundle and let $b(F) = \beta(F)$ if it does not and the Pin^- structure on F is induced via the function in Lemma 6.2. We have*

$$\mu(S(\eta)) = \sigma(\eta) - \chi(\eta) + 2 \cdot b(F) \qquad (\text{mod } 16) \ . \tag{6.6}$$

Proof: The result follows easily from 6.4. ▪

We want to describe a homomorphism from various characteristic bordism groups into the Pin^- bordism group in two dimensions less. Roughly the homomorphism is described as follows. We have a characteristic pair (M, F) and we will see that, with certain hypotheses, F is a Pin^- manifold. We then use the characterization of the pair to pick out a Pin^- structure on F. The homomorphism then just sends (M, F) to the Pin^- bordism class of F.

To describe our hypotheses, consider the following commutative square

$$\begin{array}{ccc} F & \longrightarrow & B_{O(2)} \\ \downarrow & & \downarrow \\ M & \longrightarrow & \mathrm{TO}(2) \end{array}$$

Let $U \in H^2(\mathrm{TO}(2); \mathbf{Z}/2\mathbf{Z})$ denote the Thom class and recall that U pulls back to w_2 in $H^2\left(B_{O(2)}; \mathbf{Z}/2\mathbf{Z}\right)$. The 2–plane bundle classified by ν is just the normal bundle to the embedding $i: F \subset M$, and $f^*(U) \in H^2(M; \mathbf{Z}/2\mathbf{Z})$ is the class dual to F. Let a denote the class dual to F. Then we see that $i^*(a) = w_2(\nu_{F\subset M})$, where $\nu_{F\subset M}$ is the normal bundle to the embedding. Let us apply this last equation to our characteristic situation. The class a is $w_2(M) + w_1^2(M)$ and we have the bundle equation $i^*(T_M) = T_F \oplus \nu_{F\subset M}$. Now $i^*w_1(M) = w_1(F) + w_1(\nu)$ and $i^*w_2(M) = w_2(F) + w_2(\nu) + w_1(F) \cdot w_1(\nu)$. Hence $i^*(w_2(M) + w_1^2(M)) = w_2(F) + w_2(\nu) + w_1(F) \cdot w_1(\nu) + w_1^2(F) + w_1^2(\nu)$ and using our equation for $w_2(\nu)$ we see that $w_2(F) + w_1^2(F) = w_1(\nu) \cdot i^*w_1(M)$. Hence F is Pin^- iff the right hand product vanishes or

Lemma 6.7. *The surface F has a Pin^- structure iff*

$$\big(w_1(F) + w_1(\eta)\big) \cup w_1(\eta) = 0 \quad .$$

To study $w_1(\nu) \cdot i^*w_1(M)$ we may equally study $w_1(\nu) \cap (i^*w_1(M) \cap [F, \partial F])$. The term $i^*w_1(M) \cap [F, \partial F]$ can be described as the image of the fundamental class of the manifold obtained by transversally intersecting F and a manifold V in M dual to w_1. Hence, the product $w_1(\nu) \cdot i^*w_1(M)$ vanishes if the normal bundle to $F \cap V \subset V$ is orientable. This suggests studying the following situation.

Definition 6.8 . Let M be a manifold with a proper, codimension 2 submanifold F (proper means that $\partial M \cap F = \partial F$ and that every compact set in M meets F in a compact set). A *characteristic structure* on the pair (M, F) is a collection consisting of

a) a proper submanifold V dual to $w_1(M)$ which intersects F transversely

b) an orientation on $M - V$ which does not extend across any component of V

c) a Pin^- structure on $M - F$ that does not extend across any component of F (so F is dual to $w_2 + w_1^2$)

d) an orientation for the normal bundle of $V \cap F$ in V.

Let $\mathcal{C}har^-(M, F)$ be the set of characteristic structures on (M, F).

The next goal of this section is to prove a "reduction of structure " result, the Pin^- Structure Correspondence Theorem.

Theorem 6.9. *There exists a function*

$$\Psi : \mathcal{C}har^-(M, F) \to \mathcal{P}in^-(F)$$

which is natural in the following sense. If we change the Pin^- structure on $M - F$ which does not extend across any component of F by acting on it with $a \in H^1(M; \mathbf{Z}/2\mathbf{Z})$, then we change Ψ of the structure by acting on it with $i^*(a) \in H^1(F; \mathbf{Z}/2\mathbf{Z})$, where $i: F \subset M$ is the inclusion. If X denotes a collection of components of $F \cap V$, then the dual to X is a class in $x \in H^1(F; \mathbf{Z}/2\mathbf{Z})$. If we switch the orientation to the normal bundle of $F \cap V$ in F over X and not over the other components, then we alter Ψ by acting with x. If we change the orientation on $M - V$ which does not extend across any component of V, we do not change Ψ of the Pin^- structure. Finally, if $M_1 \subset M$ is a codimension 1 submanifold with trivialized normal bundle such that F and V intersect M_1 transversely (including the case $M_1 = \partial M$), then the characteristic structure on M restricts to one on M_1. The Pin^- structure we get on $F_1 = M_1 \cap F$ is the restriction of the one we got on F.

Remark. The observation that characteristic structures restrict to boundaries allows us to define bordism groups: let $\Omega_r^!$ denote the bordism group of characteristic structures.

Reduction 6.10. Given a closed manifold M with a characteristic structure, let $E \subset M$ denote the total space of the normal bundle of F in M. The associated circle bundle, ∂E, is embedded in M with trivial normal bundle and without loss of generality we may assume that V intersects ∂E transversally. Hence E acquires the above data by restriction.

This reduces the general case to the following local problem. We may deal with one component at a time now and so we must describe how to put a Pin^- structure on a connected Pin^- manifold F, given that we have a 2–disc bundle over F with total space E; a Pin^- structure on ∂E which does not extend to all of E; a codimension 1 submanifold V which is dual to $w_1(E)$ and intersects F transversally; an orientation on $E - V$ which does not extend across any component of V; and an orientation for the normal bundle of $F \cap V$ in V. We must also check that the Pin^-

structure that we get on F is independent of our choice of tubular neighborhood. Note for reassurance that Pin^- structures on F are in one to one correspondence with Pin^- structures on ∂E which do not extend to E.

Let us consider the following situation. We have a circle bundle $p: \partial E \to F$ over F with associated disc bundle ξ. We let E denote the total space of ξ. We have a codimension 1 submanifold, V, of E which is dual to $w_1(E)$ and which intersects F transversally. We are given an orientation on $E - V$ which does not extend across any component of V and we are given an orientation of the normal bundle to $F \cap V$ in V. We are going to describe a one to one correspondence between Pin^- structures on F and Pin^- structures on ∂E which do not extend across E. Furthermore, suppose that $U \subset F$ is a submanifold with trivialized normal bundle. Suppose that U intersects V transversally and let E_U denote the total space of the disk bundle for ξ restricted to U. Then over U we have our data. Notice that any Pin^- structure on F restricts to one on U, and any Pin^- structure on ∂E restricts to one on ∂E_U. Let $\mathcal{P}in^-(F, U)$ denote the set of Pin^- structures on F which restrict to a fixed one on U. Define $\mathcal{P}in^-(\partial E, \partial E_U)$ similarly except we require that the Pin^- structures do not extend across the disk bundles. Below we will define a 1–1 map $\Psi: \mathcal{P}in^-(\partial E, \emptyset) \to \mathcal{P}in^-(F, \emptyset)$. If we fix a Pin^- structure on U, which comes from one on F, and use Ψ for U to pick out a Pin^- structure on ∂E_U, then we also get a 1–1 map

$$\Psi: \mathcal{P}in^-(\partial E, \partial E_U) \to \mathcal{P}in^-(F, U) \quad .$$

There is an isomorphism, $p^*: H^1(F, U; \mathbf{Z}/2\mathbf{Z}) \to H^1(\partial E, \partial E_U \cup S^1; \mathbf{Z}/2\mathbf{Z})$, induced by the projection map, $p: \partial E \to F$, where S^1 denotes a fibre of the bundle (if $U \neq \emptyset$ then $\partial E_U \cup S^1 = \partial E_U$). The group $H^1(\partial E, \partial E_U \cup S^1; \mathbf{Z}/2\mathbf{Z})$ acts in a simply transitive fashion on $\mathcal{P}in^-(\partial E, \partial E_U)$ and the group $H^1(F, U; \mathbf{Z}/2\mathbf{Z})$ acts in a simply transitive fashion on $\mathcal{P}in^-(F, U)$. The map Ψ is equivariant with respect to these actions and p^*.

The relative version of the Pin^- Structure Correspondence gives the uniqueness result needed in Reduction 6.10 since any two choices are related by a picture with our data over $E \times I$ with structure fixed over $E \times 0$ and $E \times 1$.

Note first that F has a Pin^- structure by the calculations above.

Recall that there is a sub-bundle of $T_{\partial E}$, namely the bundle along the fibres, η. This is a line bundle which is tangent to the fibre circle at each point in ∂E. The quotient bundle, ρ, is naturally isomorphic to T_F, via the projection map, p. Our first task is to use our given data to describe an isomorphism between $\eta \oplus \det(T_{\partial E})$ and $\det(\rho) \oplus \epsilon^1$. To fix notation, let N be a tubular neighborhood of V in ∂E and fix an isomorphism between $\rho \oplus \eta$ and $T_{\partial E}$.

On $\partial E - V$ we have an orientation of $T_{\partial E}$. This describes an isomorphism between $\det(T_{\partial E})$ and ϵ^1. Furthermore, the orientation picks out an isomorphism between η and $\det(\rho)$ as follows. These two line bundles are isomorphic since they have the same w_1, and there are two distinct isomorphisms over each component of $\partial E - V$. Pick a point in each component of $\partial E - V$, and orient η at those points. The orientation of $T_{\partial E}$ picks out an orientation of ρ, and hence $\det(\rho)$, at each point. We choose the isomorphism between η and $\det(\rho)$ which preserves the orientations at each point. It is easy to check that if we reverse the orientation at a point for η, we reverse the orientation for $\det(\rho)$ and hence get the same isomorphism between these two bundles. The isomorphism between $\eta \oplus \det(T_{\partial E})$ and $\det(\rho) \oplus \epsilon^1$ is just the sum of the above two isomorphisms.

We turn our attention to the situation over N. Over $F \cap V$, ξ is the normal bundle to $F \cap V$ in V, and hence it is oriented. Hence so is $p^*(\xi)$ in ∂E, and $p^*(\xi)$ is isomorphic to $\eta \oplus \epsilon^1$. The outward normal to ∂E in E orients the ϵ^1, and hence η is oriented over $p^{-1}(F \cap V)$, and hence over N. This time $\det(\rho)$ and $\det(T_{\partial E})$ are abstractly isomorphic, and we can choose an isomorphism by choosing a local orientation. Since η is oriented and $0 \to \eta \to T_{\partial E} \to \rho \to 0$ is exact, there is a natural correspondence between orientations of $T_{\partial E}$ at a point and orientations of ρ at the same point. As before, if we switch the orientation on $T_{\partial E}$, we still get the same isomorphism between $\det(\rho)$ and $\det(T_{\partial E})$. As before, the orientation for η defines an isomorphism between η and ϵ^1, but this time we take the isomorphism which reverses the orientations. We take the sum of these two isomorphisms as our preferred isomorphism between $\eta \oplus \det(T_{\partial E})$ and $\det(\rho) \oplus \epsilon^1$.

Now over $N - V$, we have two isomorphisms between $\eta \oplus \det(T_{\partial E})$ and $\det(\rho) \oplus \epsilon^1$. If we restrict attention to a neighborhood of ∂N both bundles are the sum of two trivial bundles, and our two isomorphisms differ by composition with the matrix $\begin{pmatrix} 0 & -1 \\ 1 & 0 \end{pmatrix}$.

Parameterize a neighborhood of ∂N in N by $\partial N \times [0, \pi/2]$ and twist one bundle isomorphism by the matrix $\begin{pmatrix} \cos(t) & -\sin(t) \\ \sin(t) & \cos(t) \end{pmatrix}$. We can now glue our two isomorphisms together to get an isomorphism between $\eta \oplus \det(T_{\partial E})$ and $\det(\rho) \oplus \epsilon^1$ over all of ∂E.

Finally, we can describe our correspondence between Pin^- structures. Suppose that we have a Pin^- structure on F. This is a *Spin* structure on $T_F \oplus \det(T_F)$. Since ρ is isomorphic via p to T_F, we get a *Spin* structure on $\rho \oplus \det(\rho)$, and hence on $\rho \oplus \det(\rho) \oplus \epsilon^1$. Using our constructed isomorphism, this gives a *Spin* structure on $\rho \oplus \eta \oplus \det(T_{\partial E})$. Choose a splitting of the short exact sequence $0 \to \eta \to T_{\partial E} \to \rho \to 0$, and we get a *Spin* structure on $T_{\partial E} \oplus \det(T_{\partial E})$.

If we choose a different splitting, we get an automorphism of $T_{\partial E}$ and hence an automorphism of $T_{\partial E} \oplus \det(T_{\partial E})$ which takes one *Spin* structure to the other. But this automorphism is homotopic through bundle automorphisms to the identity, and so the *Spin* structure does not change.

Finally, let us consider the Pin^- structure induced on a fibre S^1. We will look at this situation for a fibre over a point in F where we have an orientation of $T_{\partial E}$. Restricted to S^1, the bundle $T_{\partial E}$ splits as η plus the normal bundle of S^1 in ∂E, so η is naturally identified as the tangent bundle of S^1 and the normal bundle of S^1 in ∂E is trivialized using the bundle map p. The trivialization of the normal bundle of S^1 in ∂E plus the $Spin$ structure on $T_{\partial E} \oplus \det(T_{\partial E})$ yields a trivialization of $\eta|_{S^1}$, which then yields a trivialization of the tangent bundle of S^1. Since $SO(1)$ is a point, any oriented 1–plane bundle has a unique framing, which in the case of the tangent bundle to the circle is the Lie group framing. The Pin^- structure that results from a framing of the tangent bundle of S^1 is therefore the one that does not extend across the disk, so our Pin^- structure on ∂E does not extend across E.

Recall that Pin^- structures on ∂E that do not extend across E are acted on by $H^1(F;\mathbf{Z}/2\mathbf{Z})$ in a simply–transitive manner by letting $p^*(x) \in H^1(\partial E;\mathbf{Z}/2\mathbf{Z})$ act as usual on Pin^- structures on ∂E. If we change Pin^- structures on F by $x \in H^1(F;\mathbf{Z}/2\mathbf{Z})$, we change the Pin^- structure that we get on ∂E by the $p^*(x)$ in $H^1(\partial E;\mathbf{Z}/2\mathbf{Z})$ so our procedure induces a one to one correspondence between Pin^- structures on F and Pin^- structures on ∂E which do not extend across E.

Next, we consider the effects of changing our orientations. We wish to study how the choices of orientations on $\partial E - V$ and on ξ effect the resulting map between Pin^- structures on F and Pin^- structures on ∂E which do not extend across E. Let us begin by considering the effect of changing the orientation on ξ. This switches the orientation on η and so our bundle map remains the same over $\partial E - N$ and over N it is multiplied by the matrix $\begin{pmatrix} -1 & 0 \\ 0 & -1 \end{pmatrix}$. This has the effect of putting s full twists into the framing around any circle that intersects $F \cap V$ geometrically t times where $s \equiv t \pmod 2$. Hence the class in $H^1(F;\mathbf{Z}/2\mathbf{Z})$ that measures the change in Pin^- structure is just the class dual to $F \cap V$. If $F \cap V$ has several components and we switch the orientation of ξ over only one of them then the class in $H^1(F;\mathbf{Z}/2\mathbf{Z})$ that measures the change in Pin^- structure is just the class dual to that component of $F \cap V$.

Now suppose that we switch the orientation on ξ and on $M - V$. This time the two bundle maps differ over all of ∂E by multiplication by the matrix $\begin{pmatrix} -1 & 0 \\ 0 & -1 \end{pmatrix}$. The effect of this is to change the Pin^- structure on F via $w_1(F)$. This follows from Lemma 1.6.

From the two results above the reader can work out the effect of the other possible changes of orientations. Finally, the diligent reader should work through the relative version.

This ends our description of the Pin^- Structure Correspondence. ∎

As an application of the Pin^- Structure Correspondence and Reduction 6.10 we present

Theorem 6.11. *There exists a homomorphism $R\colon \Omega^!_r \to \Omega^{Pin^-}_{r-2}(B_{O(2)})$. Given an object, $x \in \Omega^!_r$, let F denote the submanifold dual to $w_2 + w_1^2$. This manifold has a map $F \to B_{O(2)}$ classifying the normal bundle. Use the above construction to put a Pin^- structure on F: $R(x)$ is the bordism class of this Pin structure on F.*

Variants of this map enter into the discussions below.

Corollary 6.12. *If MFK_r denotes the r–th bordism group of Freedman–Kirby, then there exists a long exact sequence*

$$\cdots \to \Omega^{Spin}_r \xrightarrow{i} MFK_r \xrightarrow{R} \Omega^{Spin}_{r-2}(B_{SO(2)}) \xrightarrow{a} \Omega^{Spin}_{r-1} \to \cdots$$

where R takes the Spin bordism class of the classifying map for the normal bundle to F in M, and a takes the Spin structure we put on the total space of the associated circle bundle. The V we always take is the empty set.

Remark 6.13. There are definitely non–trivial extensions in this sequence.

Remark 6.14. The Freedman–Kirby bordism theory is equivalent to the bordism theory $Spin^c$, the theory of oriented manifolds with a specific reduction of w_2 to an integral cohomology class. This bordism theory has been computed, e.g. Stong [Stong], and is determined by Stiefel–Whitney numbers, Pontrjagin numbers, and rational numbers formed from products of Pontrjagin numbers and powers of the chosen integralization of w_2.

Remark 6.15. There are versions of this sequence for the bordism theory studied by Guillou–Marin and for our bordism theory. In both of these cases we replace Ω^{Spin} by the Pin^- bordism groups Ω^{Pin^-}. We also replace $\Omega^{Spin}_{r-2}(B_{SO(2)})$ by the bordism groups of $O(2)$–bundles over Pin^- manifolds with some extra structure. The bordism groups of $O(2)$–bundles over Pin^-manifolds can be identified with the homotopy groups of the Thom spectrum formed from $B_{Pin^-} \times B_{O(2)}$ using the universal bundle over B_{Pin^-} and the trivial bundle over $B_{O(2)}$. The associated bordism groups are denoted $\Omega^{Pin^-}_{r-2}(B_{O(2)})$. In the Guillou–Marin case we define BGM as the fibre of the map $B_{Pin^-} \times B_{O(2)} \to K(\mathbf{Z}/2\mathbf{Z}, 1)$ where the map is the sum of w_1 of the universal bundle over B_{Pin^-} and w_1 of the universal bundle over $B_{O(2)}$. In our case we let BE be the fibre of the map $B_{Pin^-} \times B_{O(2)} \to K(\mathbf{Z}/2\mathbf{Z}, 2)$ where the map is the product of two 1–dimensional cohomology classes: namely w_1 of the universal bundle over B_{Pin^-} and w_1 of the universal bundle over $B_{O(2)}$. Over either BGM or BE we can pull back the universal bundle over B_{Pin^-} plus the trivial bundle over $B_{O(2)}$ and form the associated Thom spectrum. The homotopy groups of these spectra fit into the analogous exact sequences for the bordism theory studied by Guillou–Marin and by us.

Remark 6.16. All the bordism groups defined in Theorem 6.11, Corollary 6.12 and its two other versions are naturally modules over the *Spin* bordism ring, and all the maps defined above are maps of Ω^{Spin}_*–modules.

§7. Geometric calculations of characteristic bordism.

In this section we will calculate the characteristic bordism introduced in the last section up through dimension 4.

The first remark is that any manifold M of dimension less than or equal to 4 has a characteristic structure. Hence !–bordism is onto unoriented bordism through dimension 4. We show next that

Theorem 7.1. *The forgetful map*

$$\Omega_r^! \to \Omega_r^O$$

is an isomorphism for $r = 0$, 1, *and* 2. *Hence* $\Omega_0^! \cong \Omega_2^! \cong \mathbf{Z}/2\mathbf{Z}$ *and* $\Omega_1^! \cong 0$.

Proof: Since the forgetful map is onto, it is merely necessary to show that the !–bordism groups are abstractly isomorphic to $\mathbf{Z}/2\mathbf{Z}$ or 0. We begin in dimension 0. The only connected manifold is the point and it has a unique characteristic structure: F and V are empty. Hence $\Omega_0^!$ is a quotient of $\mathbf{Z}$. It is easy to find a characteristic structure on $[0,1]$ which has 2 times the oriented point as its boundary: F is empty and $V = \{1/2\}$. Hence $\Omega_0^! \cong \mathbf{Z}/2\mathbf{Z}$ given by the number of points mod 2.

In dimensions at least 1, it is easy to add 1–handles to show any object is bordant to a connected one. Hence in dimension 1, the only objects we need to consider are characteristic structures on S^1. Here F is still empty, and V is an even number of points. The circle bounds B^2, the 2–disk, and it is easy to extend V to a collection of arcs in B^2 and to extend the orientation on $S^1 - V$. The Pin^- structure on the circle either bounds a 2–disk, in which case extend it over B^2, or it does not, in which case take F to be a point in B^2 which misses the arcs and extend the Pin^- structure over $B^2 - pt$. Hence $\Omega_1^! \cong 0$.

In dimension 2 we can assume that M is connected and that it bounds as an unoriented manifold. The goal is to prove that it bounds as a characteristic structure. Note V is a disjoint union of circles, and F is a finite set of points with $F \cap V$ being empty. Since every surface has a Pin^- structure, F is an even number of points. Let W be a collection of embedded arcs in $M \times [0,1]$ which miss $M \times 1$ and have boundary F. Since W is a dual to $w_2 + w_1^2$, there is a Pin^- structure on $M \times [0,1] - W$ which extends across no component of W. This induces such a structure on $M \times 0$. Since $H^1(M; \mathbf{Z}/2\mathbf{Z})$ acts on such structures, it is easy to adjust to get a Pin^- structure on $M \times [0,1] - W$ which extends across no component of W and which is our original Pin^- structure on $M \times 0$. Given $V \subset M \times 0$ we can extend to an embedding $V \times [0,1]$ in $M \times [0,1]$. The orientation on $M - V$ extends to one on $M \times [0,1] - V \times [0,1]$. Clearly this orientation extends across no component of $V \times [0,1]$, so this submanifold is dual to w_1. Hence we may assume our surface has empty F with no loss of generality: i.e. M has a fixed Pin^- structure.

Let E_K^3 denote the total space of the non–trivial 2–disk bundle over the circle. The boundary of E_K^3 is K^2, the Klein bottle and $H_1(K; \mathbf{Z}/2\mathbf{Z})$ is spanned by a fibre circle,C_f, and a choice of circle which maps non–trivially to the base, C_ℓ. Consider the Pin^- structure on K^2 whose quadratic enhancement satisfies $q(C_f) = 2$ and $q(C_\ell) = 1$. This structure does not extend across E_K so let F be the core circle in E_K. Let V be a fibre 2–disk. Orient the normal bundle to $V \cap F$ in F any way one likes. It is easy to check that this gives a characteristic structure on E^3_K extending the one on K^2 which does not bound as a Pin^- manifold. By adding copies of this structure on K^2 to M, we can assume that M is a Pin^- boundary, so let W^3 be a Pin^- boundary for M.

Inside W we find a dual to w_1, say X^2, which extends V in M. There is some orientation on $W - X$ which extends across no component of X and this structure restricts to such a structure on $M - V$. Since M is connected, there are only two such structures and both can be obtained from such a structure on $W - X$. Hence our original characteristic structure is a characteristic boundary assuming nothing more than that it was an unoriented boundary. ∎

The results in dimensions 3 and 4 are more complicated. We begin with the 3–dimensional result.

Theorem 7.2. *The homomorphism R of Theorem 6.11, followed by forgetting the map to $B_{O(2)}$ yields an isomorphism*

$$\hat{R}\colon \Omega^!_3 \to \Omega_1^{Pin^-} \cong \mathbf{Z}/2\mathbf{Z} \quad .$$

Proof: We first show that $\hat{R}$ is onto and then that it is injective.

Let E^3_K denote the disk bundle with boundary the Klein bottle as in the last proof. The Pin^- structure received by F in this structure is seen to be the Lie group Pin^- structure. There is a similar story for the torus, T^2. There is a 2–disk bundle over a circle, E^3_T, and a Pin^- structure on the torus which does not extend across the disk bundle so that the core circle receives the Lie group Pin^- structure. Indeed, E^3_T is just a double cover of E^3_K. If we take two copies of K^2 with its Pin^- structure and one copy of T^2 with its Pin^- structure, the resulting disjoint union bounds in $\Omega_2^{Pin^-}$. Let W^3 denote such a bordism. Let $M^3 = \overset{2}{\amalg} E^3_K \amalg E^3_T \amalg W^3$ with the boundaries identified. Let F be the disjoint union of the three core circles, and note F is a dual to $w_2 + w_1^2$ since the complement has a Pin^- structure which does not extend across any of the cores. Let V be a dual to w_1 and arrange it to meet F transversely. Indeed, with a little care one can arrange it so that $V \cap F$ consists of 2 points, one on each core circle in a E^3_K. This is our characteristic structure on M. Our homomorphism applied to M is onto the generator of $\Omega_1^{Pin^-}$.

It remains to show monicity. Let M be a characterized 3–manifold. By adding 1–handles, we may assume that M is connected. First we want to fix it so that $V \cap F$ is empty. In general, $V \cap F$ is dual to $w_2 w_1 + w_1^3$ and, for a 3–manifold, this vanishes. Hence $V \cap F$ consists of an even number of points. We explain how to remove a pair of such points.

Pick two points, p_0 and p_1, in $V \cap F$. Each point in F has an oriented normal bundle. The normal bundle to each point in V is also trivial and V is oriented, so the normal bundle to each point in V is oriented. Attach a 1–handle, $H = (B^1 \times [0,1]) \times B^2$ so as to preserve the orientations at p_0 and p_1. Let W^4 denote the resulting bordism. Inside W^4, we have embedded bordisms, V_1^3 and F_1^2 beginning at V and F in M. Notice that at the "top" of the bordism, the "top" of V_1 and the "top" of F_1 intersect in 2 fewer points. Moreover, the orientation of the normal bundle of $V \cap F$ in F clearly extends to an orientation of the normal bundle of $V_1 \cap F_1$ in F_1.

Since F_1 is a codimension 2 submanifold of W, it is dual to some 2–dimensional cohomology class. Since $H^*(W, M; \mathbf{Z}/2\mathbf{Z})$ is 0 except when $* = 1$ (in which case it is $\mathbf{Z}/2\mathbf{Z}$), this class is determined by its restriction to $H^2(M; \mathbf{Z}/2\mathbf{Z})$. Hence F_1 is dual to $w_2 + w_1^2$, so choose a Pin^- structure on $W - F_1$ which extends across no component of F_1. This restricts to a similar structure on M, and since $H^1(W; \mathbf{Z}/2\mathbf{Z}) \to H^1(M; \mathbf{Z}/2\mathbf{Z})$ is onto, we can adjust the Pin^- structure until it extends the given one on $M - F$.

The above argument does not quite work for V_1, but it is easy in this case to see that $W - V_1$ has an orientation extending the one on $M - V$. Any such orientation can not extend over any components of V_1. Hence we have a characteristic bordism as required.

We may now assume that $V \cap F$ is empty. Since F is a union of circles and $V \cap F = \emptyset$, F has a trivial normal bundle in M. If our homomorphism vanishes on our element, F is a Pin^- boundary, which, in this dimension, means that it is a *Spin* boundary: i.e. F bounds Q^2, an orientable Pin^- manifold. Glue $Q^2 \times B^2$ to $M \times [0,1]$ along $F \times B^2 \subset M \times 1$ to get a bordism X^4. Since Q is orientable, $V \times [0,1]$ is still dual to w_1, and it is not hard to extend the Pin^- structure on $M - F$ to one on $X - Q$ which extends across no component of Q. Since Q and $V \times [0,1]$ remain disjoint, the "top" of X is a new characteristic pair for which the dual to $w_2 + w_1^2$ is empty: i.e. the "top", say N^3, has a Pin^- structure. Since $\Omega_3^{Pin^-} = 0$, N^3 bounds a Pin^- manifold, Y^4. Since M was connected, so is N and there is no obstruction to extending the dual to w_1 in N, say V_1, to a dual to w_1 in Y, say U, and extending the orientation on $N - V_1$ to an orientation on $Y - U$ which extends across no component of U. The union of X^4 and Y^4 along N^3 is a characteristic bordism from M^3 to 0. ∎

The last goal of the section is to compute $\Omega_4^!$. Since the group is non–zero, we begin by describing the invariants which detect it. Given an element in $\Omega_4^!$, we get an associated surface F^2 with a Pin^- structure, and hence a quadratic enhancement, q. We may also consider η, the normal bundle to F in our original 4–manifold. We describe three homomorphisms. The first is $\beta\colon \Omega_4^! \to \mathbf{Z}/8\mathbf{Z}$ which just takes the Brown invariant of the enhancement q. The second homomorphism is $\Psi\colon \Omega_4^! \to \mathbf{Z}/4\mathbf{Z}$ given by the element $q(w_1(\eta)) \in \mathbf{Z}/4\mathbf{Z}$. The third homomorphism is $w_2\colon \Omega_4^! \to \mathbf{Z}/2\mathbf{Z}$ given by $\langle w_2(\eta), [F]\rangle \in \mathbf{Z}/2\mathbf{Z}$. We leave it to the reader to check that these three maps really are homomorphisms out of the bordism group, $\Omega_4^!$.

Theorem 7.3. *The sum of the homomorphisms*

$$\beta \oplus \Psi \oplus w_2\colon \Omega_4^! \to \mathbf{Z}/8\mathbf{Z} \oplus \mathbf{Z}/4\mathbf{Z} \oplus \mathbf{Z}/2\mathbf{Z}$$

is an isomorphism.

Proof: First we prove the map is onto and then we prove it is 1–1. Recall from Lemma 6.7 that a surface, M, with a Pin^- structure and a 2–plane bundle, η, can be completed to a characteristic bordism element iff $(w_1(M) + w_1(\eta)) \cup w_1(\eta) = 0$. Notice that this equation is always satisfied since cupping with $w_1(M)$ and squaring are the same. Hence we will only describe the surface with its Pin^- structure and the 2–plane bundle.

First note that $\mathbf{RP}^2$ with the trivial 2–plane bundle generates the $\mathbf{Z}/8\mathbf{Z}$ and maps trivially to the $\mathbf{Z}/4\mathbf{Z}$ and the $\mathbf{Z}/2\mathbf{Z}$.

The Hopf bundle over the 2–sphere maps trivially into the $\mathbf{Z}/8\mathbf{Z}$ and the $\mathbf{Z}/4\mathbf{Z}$ since S^2 is a Pin^- boundary and Ψ vanishes whenever the 2–plane bundle has trivial w_1. However, S^2 and the Hopf bundle maps non–trivially to the $\mathbf{Z}/2\mathbf{Z}$.

Let K^2 denote the Klein bottle, and fix a Pin^- structure for which K^2 is a Pin^- boundary. Let η be the 2–plane bundle coming from the line bundle with w_1 being the class in $H^1(K^2; \mathbf{Z}/2\mathbf{Z})$ with non–zero square. Since K^2 is a Pin^- boundary, $\beta(K^2) = 0$. Since η comes from a line bundle, $w_2(\eta) = 0$. However, $q(w_1(\eta))$ is an element in $\mathbf{Z}/4\mathbf{Z}$ of odd order and is hence a generator.

This shows that our map is onto. Before showing that our map is 1–1, we need a lemma.

Lemma 7.4. *There exists a 2–disk bundle B_{2n} over the punctured $S^1 \times S^2$, $S^1 \times S^2 -$ int D^3, whose restriction to the boundary S^2 has Euler class $2n$, $n \in \mathbf{Z}$.*

Proof: Start with the 2–disk bundle $\tilde{B}_n$ over S^2 with Euler number n and pull it back over the product $S^2 \times I$. Now add a 1–handle to $S^2 \times I$, forming $S^1 \times S^2 -$int B^3, and extend the bundle $\tilde{B}_n$ over the 1–handle so as to create a non–orientable bundle B_{2n}. Then $\chi(B_{2n}|_{S^2}) = 2n$. ∎

Suppose M^4, V^3, F^2, η^2 is a representative of an element of $\Omega_4^!$ and that $\beta(F^2) = 0$, $\Psi(w_1(\eta)) = 0$, and $w_2(\eta) = 0$. We need to construct a !–bordism to $\emptyset$.

Since we may assume that F, M and V are connected, there is a connected 1–manifold, an S^1, which is Poincaré dual to $w_1(\eta)$; then the normal vector to S^1 in F makes an even number of full twists in the Pin^- structure on F as S^1 is traversed. It follows that we can form a !–bordism by adding to F a $B^2 \times B^1$

where $S^1 \times B^1$ is attached to the dual S^1 to $w_1(\eta)$ and its normal B^1 bundle. Clearly the Pin^- structure on F extends across the bordism. Since the dual to S^1 has self–intersection zero in F, η restricted to S^1 is orientable, so η extends over $B^2 \times B^1$.

Since $w_2(\eta) = 0$, it follows that $\chi(\eta)[F] = 2n$ for some $n \in \mathbf{Z}$. By Lemma 7.4 there is a bundle B_{-2n} over a punctured $S^1 \times S^2$ with $\chi(B_{-2n}|_{S^2}) = -2n$. We form a 5–dimensional bordism to the boundary connected sum, i.e. in $M^4 \times 1 \subset M^4 \times I$, choose a 4–ball of the form $B^2 \times B^2$ where $B^2 \times 0 \subset F^2 - (V \cap F)$ and $p \times B^2$ is a normal plane of η over p, and identify $B^2 \times B^2$ with $B_{-2n}|_{S^2_-}$ where S^2_- is a hemisphere of S^2.

The new boundary to our !–bordism, which we shall denote $(M,\ V,\ F,\ \eta)$ now has a trivial normal bundle η.

Since $\beta(F^2) = 0$, F Pin^- bounds a 3–manifold N^3, so we add $N^3 \times B^2$ to $M \times 1$ along the normal bundle η to F, $F \times B^2$, where it does not matter how we trivialize η. The Pin^- structure on $M - F$ extends over the complement of N (using the Pin^- Correspondence Theorem, 6.9, and the Pin^- structure on N), so the new boundary to our !–bordism consists of a Pin^- manifold M with empty F^2. Since 4–dimensional Pin^- bordism, $\Omega_4^{Pin^-}$, is zero, we can complete our !–bordism by gluing on to $M \times 1$ a 5–dimensional Pin^- manifold. ▪

Remark. *It is worth comparing this argument with the argument in [F–K] showing that if $(M^4,\ F^2)$ is a characteristic pair with M^4 and F^2 orientable and with* sign$(M^4) = 0$ *and* $F \bullet F = 0$, *then $(M,\ F)$ is characteristically bordant to zero. The arguments would have been formally identical if we had also assumed that the Spin structure on F, obtained from the Pin^- Correspondence Theorem, bounded in 2–dimensional Spin bordism, $\Omega_2^{Spin} = \mathbf{Z}/2\mathbf{Z}$ (corresponding to $\beta(F) = 0$ above). However, it is possible to show that $\Omega_4^{\text{char}} = \mathbf{Z} \oplus \mathbf{Z}$ without the extra assumption on F, and this $\mathbf{Z}/2\mathbf{Z}$ improvement leads to Rochlin's Theorem (see [F–K], [Ki], ...).*

Further Remark. The image of the Guillou–Marin bordism in this theory can be determined as follows. The group is $Z \oplus Z$ generated by (S^4, RP^2) and (CP^2, S^2). Both β and Ψ vanish on (CP^2, S^2), but w_2 is non–zero. On (S^4, RP^2), w_2 evaluates 0 (the normal bundle comes from a line bundle): β is either 1 or -1 depending on which embedding one chooses. Moreover, Ψ is either 1 or -1 (the same sign as β).

§8. New knot invariants.

The goal here is to describe some generalizations of the usual Arf invariant of a knot (or some links) due to Robertello, [R].

We fix the following data. We have a 3–manifold M^3 with a fixed *Spin* structure and a link $L\colon \coprod_i S^1 \to M^3$. Since M is *Spin* , $w_2(M) = 0$ and we require that $[L] \in H_1(M; \mathbf{Z}/2\mathbf{Z})$ is also 0, hence dual to $w_2(M)$. We next require a characterization of the pair, (M, L): i.e. a *Spin* structure on $M - L$ which extends across no component of L. We call such a characterization *even* iff the Pin^- structure induced on each component of L by Lemma 6.2 is the structure which bounds. We say the link is *even* iff it has an even characterization.

One way to check if a link is even is the following. Each component of L has a normal bundle, and the even framing of this normal bundle picks out a mod 2 longitude on the peripheral torus. The link is even iff the sum of these even longitudes is 0 in $H_1(M - L; \mathbf{Z}/2\mathbf{Z})$

Remark. Not all links which represent 0 are even: the Hopf link in S^3 is an example where any structure which extends across no component of L induces the Lie group *Spin* structure on the two circles. We shall see later that a necessary and sufficient condition for a link in S^3 to be even is that each component of the link should link the other components evenly. This is Robertello's condition, [R].

Definition. A link, L, in M^3 with a fixed *Spin* structure on M and a fixed *Spin* structure on $M - L$ which extends across no component of L and induces the bounding Pin^- structure on each component of L is called a *characterized* link.

Given a characterized link, (M, L), we define a class $\gamma \in H^1(M - L; \mathbf{Z}/2\mathbf{Z})$: γ is the class which acts on the fixed *Spin* structure on $M - L$ to get the one which is the restriction of the one on M. The class γ is defined by the characterization and conversely a characterization is defined by a choice of class $\gamma \in H^1(M - L; \mathbf{Z}/2\mathbf{Z})$ so that, under the coboundary map, the image of γ in $H^2(M, M - L; \mathbf{Z}/2\mathbf{Z})$ hits each generator. (Recall that by the Thom isomorphism theorem, $H^2(M, M - L; \mathbf{Z}/2\mathbf{Z})$ is a sum of $\mathbf{Z}/2\mathbf{Z}$'s, one for each component of L.)

Let E be the total space of an open disk bundle for the normal bundle of L, and let S be the total space of the corresponding sphere bundle. Note S is a disjoint union of a peripheral torus for each component of L. The class γ is dual to an embedded surface $F \subset M - E$ and $\partial F \cap S$ is a longitude in the peripheral torus of each component of L. Let ℓ denote this set of longitudes. We will call ℓ a set of *even longitudes*. We will call F a *spanning surface* for the characterized link.

The set of even longitudes is not well–defined from just the characterized link. It is clear that two surfaces dual to the same γ must induce the same mod 2 longitudes. But if we act on one component of L by an even integer, we can find a new surface dual to γ which has the same longitudes on the other components and the new longitude on our given component differs from the old one via action by this even integer. Hence the characteristic structure only picks out the mod 2 longitudes and any set of integral classes which are longitudes and which reduce correctly mod 2 can be a set of even longitudes. Moreover, any set of even longitudes is induced by an embedded surface.

Since M is oriented, the normal bundle to any embedded surface, F, is isomorphic to the determinant bundle associated to the tangent bundle of F. The total space of the determinant bundle to the tangent bundle is naturally oriented. The total space to the normal bundle to F is M is oriented by the orientation on M. Choose the isomorphism between the normal bundle to F in M and the determinant bundle to the tangent bundle of F so that, under the induced diffeomorphism between the total spaces, the two orientations agree. Under these identifications, Corollary 1.15 picks out a Pin^- structure on F from the $Spin$ structure on M. We apply this to an F which is a spanning surface for our link. Of course we could apply the same result but use the $Spin$ structure on $M - L$. It is not hard to check that the two structures on F differ under the action of $w_1(F)$ since this is the restriction of γ to F. Hence it is not too crucial which structure we use but to fix things we use the structure on M.

We can restrict this structure on F to a component of L. If we put the $Spin$ structure on F coming from that on $M - L$ it is easy to see that we get the bounding Pin^- structure on each component of L. Hence this also holds for the Pin^- structure on F coming from the one on M. Hence, a spanning surface for a characterized link has an induced Pin^- structure which extends to the corresponding closed surface uniquely.

Our link invariant is a mod 8 integer which depends on the characterized link and the set of even longitudes.

Definition 8.1. Given a characterized link, (M, L), and a set of even longitudes, ℓ, pick a spanning surface F for L which induces the given set of longitudes. Then define

$$\beta(L, \ell, M) = \beta(\overline{F})$$

where $\overline{F}$ is F with a disk added to each component of L; the Pin^- structure is extended over each disk; and β is the usual Brown invariant applied to a closed surface with a Pin^- structure.

Remarks.

i) Notice that unlike Robertello's invariant, our invariant does not require that the link be oriented.

ii) It follows from the proof of Theorem 4.3 that a knot is even iff it is mod 2 trivial.

iii) If each component of L represents 0 in $H_1(M; \mathbf{Z}/2\mathbf{Z})$ then the mod 2 linking number of a component of L with the rest of the link is defined. If F is an embedded surface in M with boundary L, the longitude picked out for a component of L is even iff the mod 2 linking number of that component of L with the rest of the link is 0.

iv) If M is an oriented $\mathbf{Z}/2\mathbf{Z}$ homology 3 sphere, then it has a unique $Spin$ structure and there is a unique way to characterize an even link L.

v) Let M be an integral homology 3 sphere containing a link L. Orient each component of the link. Let ℓ_i be the linking number of the ith component of L

with the rest of the link. Each component of L has a preferred longitude, the one with self–linking 0, so ℓ_i also denotes a longitude. The link L is even iff each ℓ_i is even. Robertello's Arf invariant is equal to $\beta(L, -\ell, M)$, where the *Spin* structure and characterization are unique and ℓ is the set of longitudes obtained by using $-\ell_i$ on each component. Notice that ℓ_i depends on how the link is oriented.

It is not yet clear that our invariant really only depends on the characterizations and the even longitudes.

Theorem 8.2. *Let L be a link in a 3–manifold M. Suppose M has a Spin structure and that L is characterized. Let ℓ be a collection of even longitudes. Then $\beta(L, \ell, M)$ is well–defined. Let W^4 be an oriented bordism between M_1 and M_2. Let $L_i \subset M_i$, $i = 1, 2$ be characterized links. Let $F \subset W$ be a properly embedded surface with $F \cap M_i = L_i$. Suppose $W - F$ has a Spin structure which extends across no component of F and which gives a Spin bordism between the two structures on $M_i - L_i$, $i = 1, 2$, given by the characterizations.*

The normal bundle to F in W has a section over every non–closed component of F so pick one. This choice selects a longitude for each component of each link. Suppose the longitudes picked out for each L_i, say ℓ_i, are even. The surface F receives a Pin^- structure by Lemma 6.2. With this structure, each component of ∂F bounds and hence F has a β invariant. If we orient W so that M_1 receives the reverse Spin structure then the following formula holds.

$$\beta(L_2, \ell_2, M_2) - \beta(L_1, \ell_1, M_1) = -\beta(F) - \mathrm{sign}(W) - \mu(M_2) + \mu(M_1) \ .$$

Proof: We begin by discussing some constructions and results involving a *Spin* 3–manifold N and a spanning surface, V^2 for a characterized link, L. To begin, given $e: V^2 \subset N^3$, define $\hat{V} \subset N \times [0, 1]$ as the image of $e \times f$, where $f: V \to [0, 1/2]$ is any map with $f^{-1}(0) = \partial V$. If N has a *Spin* structure, $N \times [0, 1]$ receives one. The class represented by $[\hat{V}, L]$ in $H_2\left(N \times [0, 1], N \times 0 \amalg N \times 1; \mathbf{Z}/2\mathbf{Z}\right) \cong H_1\left(N \times 0; \mathbf{Z}/2\mathbf{Z}\right)$ is the same as that represented by $[L]$ in $H_1\left(N \times 0; \mathbf{Z}/2\mathbf{Z}\right)$. Hence it represents 0. Since $w_2(N \times [0, 1])$ is also trivial, there is a *Spin* structure on $N \times [0, 1] - \hat{V}$ which does not extend across any component of $\hat{V}$. Such structures are acted on simply transitively by $H^1\left(N; \mathbf{Z}/2\mathbf{Z}\right)$, so it is easy to construct a unique such *Spin* structure which restricts to the initial one on $N \times 1$.

We proceed to identify the *Spin* structure induced on $N \times 0 - L$. Let $X = V \times [0, 1]$ and embed two copies of V in the boundary so that $\partial X = V \cup V$ where the union is along ∂V thought of as $\partial V \times 1/2$. First observe that we can embed X in $N \times [0, 1]$ so that ∂X is $V \subset N \times 0$ union $V \times 1 = \hat{V}$. Since X has codimension 1, the Poincaré dual to W is a 1–dimensional cohomology class

$x \in H^1(N \times [0,1] - V; \mathbf{Z}/2\mathbf{Z})$. Suppose we take the *Spin* structure on $N \times [0,1]$ and restrict it to $N \times [0,1] - V$ and then act on it by x. This is a *Spin* structure on $N \times [0,1] - V$ which extends across no component of V and which is the original one on $N \times 1$. On $N \times 0 - L$ it can be described as the one obtained by taking the given *Spin* structure on $N \times 0$, restricting it, and then acting on it by the restriction of x. But the restriction of x is just the Poincaré dual of $F \subset N \times 0$ and so it is the *Spin* structure which characterizes the link. By Lemma 6.2, there is a preferred Pin^- structure on V, which is easily checked to be the same as the one we put on it in §4. The above *Spin* structure on $N \times [0,1] - \hat{V}$ will be called the *standard characterization* of the pair $(N \times [0,1], \hat{V})$.

With this general discussion behind us, let us turn to the situation described in the second part of the theorem. Recall W^4 is an oriented bordism between M_1 and M_2; $L_1 \subset M_1$ and $L_2 \subset M_2$ are characterized links; $F^2 \subset W$ be a properly embedded surface with $F \cap M_i = L_i$; and $W - F$ has a *Spin* structure which extends across no component of F and which gives a *Spin* bordism between the structures on $M_i - L_i$. Define sets of even longitudes ℓ_i as in the statement of the theorem.

Let $F_i \subset M_i$ be a spanning surface for L_i. Inside $\overline{W} = M_1 \times [-1,0] \cup W \cup M_2 \times [0,1]$ embed $\overline{F} = \hat{F_1} \cup F \cup \hat{F_2}$, where $\hat{F_1}$ is defined with function $f\colon F_1 \to [-1/2, 0]$ and still $f^{-1}(0) = \partial F_1$. There is a *Spin* structure on $\overline{W} - \overline{F}$ which extends across no component of $\overline{F}$. It is just the union of the standard characterization of $M_1 \times [-1,0], \hat{F_1}$, the given *Spin* structure on $W - F$ and the standard characterization of $M_2 \times [0,1], \hat{F_2}$.

By Lemma 6.2 again, there is a preferred Pin^- structure on $\overline{F}$, which agrees with the usual ones on F_1 and F_2. In particular, F also receives a Pin^- structure which only depends on W, not on the choice of F_1 or F_2. However, from F_1 and F_2, we see that the Pin^- structure induced on each component of each link is the bounding one. Moreover, $\beta(\overline{F}) = \beta(F) + \beta(F_2) - \beta(F_1)$.

By construction, $\overline{F} \bullet \overline{F}$ is 0, so 6.4 says that

$$\beta(F_2) - \beta(F_1) = -\big(\beta(F) + \operatorname{sign}(W) + \mu(M_2) - \mu(M_1)\big)$$

where the μ invariants arise because 6.4 only applies to closed manifolds.

Apply this to the case $W = M \times [0,1]$, $F = L \times [0,1]$ embedded as a product. Since we may use different spanning surfaces at the top and bottom, this shows β is well–defined. The formula in the theorem now follows from the formula immediately above. ∎

The next thing we wish to discuss is how our invariant depends on the longitudes. Given two different sets of even longitudes, ℓ and ℓ', for a characterized link $L \subset M^3$, there is a set of integers, one for each component of L defined as follows. The integer for the ith component acts on the longitude for ℓ to give the longitude for ℓ'. Since both these longitudes are even, so is this integer.

Theorem 8.3. *Let $L \subset M^3$ be a characterized link with two sets of even longitudes ℓ and ℓ'. Let $2r$ be the sum of the integers which act on the longitudes ℓ to give the longitudes ℓ'. Then*

$$\beta(L, \ell', M) = \beta(L, \ell, M) + r \qquad (\mathrm{mod}\ 8)\ .$$

Proof: Given F_1, a spanning surface for the longitude ℓ, we can construct a spanning surface for ℓ' as follows. Take a neighborhood of the peripheral torus, which will have the form $W = T^2 \times [0,1]$. Inside W embed a surface V which intersects $T^2 \times 0$ in the longitude ℓ, which intersects $T^2 \times 1$ in the longitude ℓ', which has no boundary in the interior of W; and which induces the zero map $H_2(V, \partial V; \mathbf{Z}/2\mathbf{Z}) \to H_2(W, \partial W; \mathbf{Z}/2\mathbf{Z})$. The *Spin* structure on M restricts to one on W which is easily described: it is the stabilization of one on T^2 and this can be described as the one which has enhancement 0 on the longitude and 0 on the meridian. Since the Pin^- structure induced from Corollary 1.15 is local, we see that $F_2 = V \cup F_1$ has invariant the invariant for F_1 plus the invariant for V. We further see that the invariant for V only depends on the surface and the *Spin* structure in W. But these are independent of the link and so we can calculate the difference of the β's using the unknot.

Furthermore, we see that the effect of successive changes is additive, so we only need to see how to go from the 0 longitude to the 2 longitude, and the 2 longitude is given by the Möbius band, which inherits a Pin^- structure. This Pin^- structure extends uniquely to one on $\mathbf{RP}^2$ and this $\mathbf{RP}^2$ has β invariant $+1$.

Remark. Even in the case of links in S^3, the longitudes used enter into the answer. It is just in this case that there is a unique set of longitudes given by using an orientable spanning surface.

Unfortunately, in general there is no natural choice of longitudes so it seems simplest to incorporate them into the definition. The drawback comes in discussing notions like link concordance. In order to assert that our invariant is a link concordance invariant, we need to describe to what extent a link concordance allows us to transport our structure for one link to another. Recall that a link concordance between $L_0 \subset M$ and $L_1 \subset M$ is an embedding of $(\amalg\, S^1) \times [0,1] \subset M \times [0,1]$ with is $(\amalg\, S^1) \times i$ being L_i for $i = 0, 1$. Suppose L_0 is an even link with ℓ_0 a set of even longitudes. There is a unique way to extend this framing of the normal bundle to L_0 in M to a framing of the normal bundle of $(\amalg\, S^1) \times [0,1]$ in $M \times [0,1]$. Hence the concordance picks out a set of longitudes for L_1 which we will denote by ℓ_1. There is a unique way to extend a characterization of L_0 to a *Spin* structure on $M \times [0,1] - (\amalg\, S^1) \times [0,1]$ and hence to $M - L_1$.

Corollary 8.4. *Let L_0 and L_1 be concordant links in M. Suppose L_0 is characterized and that ℓ_0 is a set of even framings. Then the transport of framings and*

Spin structures described above gives a characterization of L_1 and ℓ_1 is a set of even framings. Furthermore $\beta(L_0, \ell_0, M) = \beta(L_1, \ell_1, M)$.

Proof: The proof follows immediately from Theorem 8.2 and the fact that $(\amalg S^1) \times [0,1]$, when capped off with disks, is a union of S^2's and so has β invariant 0. ■

We do know one scheme to remove the longitudes which works in many cases. Suppose that each component of the link represents a torsion class in $H_1(M; \mathbf{Z})$. Each component has a self–linking and by Lemma 4.1 the framings, hence longitudes are in one to one correspondence with rational numbers whose equivalence class in $\mathbf{Q}/\mathbf{Z}$ is the self–linking number. There is a unique such number, say q_i for the ith component, so that q_i represents an even framing and $0 \le q_i < 2$. We say that this is the *minimal* even longitude. To calculate linking numbers it is necessary to orient the two elements one wants to link, but the answer for self–linking is independent of orientation.

Definition 8.5. Let L be a link in M so that each component of L represents a torsion class in $H_1(M; \mathbf{Z})$. Suppose L is characterized. Define

$$\hat{\beta}(L, M) = \beta(L, \ell, M)$$

where ℓ is the set of even longitudes such that each one is minimal.

Remark. It is not hard to check that $\hat{\beta}$ is a concordance invariant.

As we remarked above, β and $\hat{\beta}$ (if it is defined) do not depend on the orientation of the link. If we reverse the orientation of M, and also reverse the *Spin* structure on M and on $M - L$, it is not hard to check that the new Pin^- structure on F is the old one acted on by $w_1(F)$ so the new invariant is minus the old one.

The remaining point to ponder is the dependence on the two *Spin* structures. To do this properly would require a relative version of the β function 4.8. It does not seem worth the trouble.

Remark. We leave it to the reader to work out the details of starting with a characteristic structure on M^3 with the link as a dual to $w_2 + w_1^2$ (i.e. represents 0 in $H_1(M; \mathbf{Z}/2\mathbf{Z})$).

§9. Topological versions.

There is a topological version of this entire theory. Just as $Spin(n)$ is the double cover of $SO(n)$ and $Pin^{\pm}(n)$ are the double covers of $O(n)$, we can consider the double covers of $STop(n)$ and $Top(n)$. We get a group $TopSpin(n)$ and two groups $TopPin^{\pm}(n)$. A $Top(n)$ bundle with a $TopPin^{\pm}(n)$ structure and an $O(n)$ structure is equivalent to a $Pin^{\pm}(n)$ bundle.

Any manifold of dimension ≤ 3 has a unique smooth structure, so there is no difference between the smooth and the toplogical theory in dimensions 3 and less. The 3–dimensional bordism groups might be different because the bounding objects are 4–dimensional, but we shall see that even in bordism there is no difference.

We turn to dimension 4. First recall that the triangulation obstruction (strictly speaking, the stable triangulation obstruction) is a 4–dimensional cohomology class so evaluation gives a homomorphism, which we will denote κ, from any topological bordism group to $\mathbf{Z}/2\mathbf{Z}$. Since every 3–manifold has a unique smooth structure, the triangulation obstruction is also defined for 4–manifolds with boundary. Every connected 4–manifold M^4 has a smooth structure on $M - pt$, and any two such structures extend to a smoothing of $M \times [0,1] - pt \times [0,1]$.

Some of our constructions require us to study submanifolds of M. In particular, the definition of characteristic requires a submanifold dual to w_1 and a submanifold dual to $w_2 + w_1^2$. We require that these submanifolds be locally–flat and hence, by [Q], these submanifolds have normal vector bundles. Of course we continue to require that they intersect transversely. Hence we can smooth a neighborhood of these submanifolds. The complement of these smooth neighborhoods, say U, is a manifold with boundary, which may not be smooth. If we remove a point from the interior of each component of U, we can smooth the result. With this trick, it is not difficult to construct topological versions of all our "descent of structure" theorems. In particular, the $[\cap w_1^2]$, $[\cap w_1]$ and R maps we defined into low–dimensional $Pin^\pm$ bordism all factor through the corresponding topological bordism theories.

Theorem 9.1. *Let Smooth–bordism$_*$ denote Ω_*^{Spin}, $\Omega_*^{Pin^\pm}$, $\Omega_*^!$, or the Freedman–Kirby or Guillou–Marin bordism theories. Let Top – bordism$_*$ denote the topological version. The natural map*

$$Smooth - bordism_3 \to Top - bordism_3$$

is an isomorphism.

$$Smooth - bordism_4 \to Top - bordism_4 \xrightarrow{\kappa} \mathbf{Z}/2\mathbf{Z} \to 0$$

is exact.

Proof: The E_8 manifold, [F], is a *Spin* manifold with non–trivial triangulation obstruction. Suppose M^3 is a 3–manifold with one of our structures which is a topological boundary. Let W^4 be a boundary with the necessary structure. Smooth neighborhoods of any submanifolds that are part of the structure. This gives a new 4–manifold with boundary U^4. If the triangulation obstruction for a component of U is non–zero, we may form the connected sum with the E_8 manifold. Hence we may assume that U has vanishing triangulation obstruction. By [L–S] we can add some $S^2 \times S^2$'s to U and actually smooth it. The manifold W can now be smoothed so that all submanifolds that are part of the structure are smooth. Hence M^3 is already a smooth boundary.

The E_8 manifold has any of our structures, so the map $Top\text{-}bordism_4 \to \mathbf{Z}/2\mathbf{Z}$ given by the triangulation obstruction is onto.

Suppose that it vanishes. We can smooth neighborhoods of any submanifolds, so let U be the complement. Each component of U has a triangulation obstruction and the sum of all of them is 0. We can add E_8's and $-E_8$'s so that each component has vanishing triangulation obstruction and the new manifold is bordant to the old. Now we can add some $S^2 \times S^2$'s to each component of U to get a smooth manifold with smooth submanifolds bordant to our original one. ∎

Theorem 9.2. *The topological bordism groups have the following values.* $\Omega_4^{TopSpin} \cong \mathbf{Z}$; $\Omega_4^{TopPin^-} \cong \mathbf{Z}/2\mathbf{Z}$; $\Omega_4^{TopPin^+} \cong \mathbf{Z}/8\mathbf{Z} \oplus \mathbf{Z}/2\mathbf{Z}$; *and* $\Omega_4^{Top-!} \cong \mathbf{Z}/8\mathbf{Z} \oplus \mathbf{Z}/4\mathbf{Z} \oplus \mathbf{Z}/2\mathbf{Z} \oplus \mathbf{Z}/2\mathbf{Z}$. *The triangulation obstruction map is split in all cases except the Spin case: the smooth to topological forgetful map is monic in all cases except the TopPin$^+$ case where it has kernel* $\mathbf{Z}/2\mathbf{Z}$. *The triangulation obstruction map is split onto for the topological versions of the Freedman–Kirby and Guillou–Marin theories and the smooth versions inject.*

Proof: The TopPin^- case is easy from the exact sequence above. The Top$Spin$ case is well–known but also easy. The E_8 manifold has non–trivial triangulation obstruction and twice it has index 16 and hence generates Ω_4^{Spin}.

There is a $[\cap w_1^2]$ homomorphism from $\Omega_4^{TopPin^+}$ to $\Omega_2^{Pin^-} \cong \mathbf{Z}/8\mathbf{Z}$ which is onto. Consider the manifold $M = E_8 \# S^2 \times \mathbf{RP}^2$. The oriented double cover of M is *Spin* and has index 16, hence is bordant to a generator of the smooth *Spin* bordism group. It is not hard to see that the total space of the non–trivial line bundle over M has a Pin^+ structure, so the Kummer surface is a $TopPin^+$ boundary. Hence there is a $\mathbf{Z}/2\mathbf{Z}$ in the kernel of the forgetful map and the $[\cap w_1^2]$ map shows that this is all of the kernel. Furthermore, E_8 represents an element of order 2 with non–trivial triangulation obstruction.

The homomorphisms used to compute $\Omega_4^!$ factor through $\Omega_4^{Top-!}$, so $\Omega_4^{Top-!} \cong \Omega_4^! \oplus \mathbf{Z}/2\mathbf{Z}$.

Likewise, the homomorphisms we use to compute smooth Freedman–Kirby or Guillou–Marin bordism factor through the topological versions. ∎

Corollary 9.3. *Let M^4 be an oriented topological 4–manifold, and suppose we have a characteristic structure on the pair (M, F). The following formula holds:*

$$2 \cdot \beta(F) = F{\bullet}F - \mathrm{sign}(M) + 8 \cdot \kappa(M) \qquad (\mathrm{mod}\ 16)$$

where the Pin^- structure on F is the one induced by the characteristic structure on (M, F) via the topological version of the Pin^- Structure Correspondence, 6.2.

Proof: Generators for the topological Guillou–Marin group consist of the smooth generators, for which the formula holds, and the E_8 manifold, for which the formula is easily checked. ∎

Remark. The above formula shows that the generator of $H_2(\ ;\mathbf{Z})$ of Freedman's Chern manifold, [F, p. 378], is not the image of a locally–flat embedded S^2.

References

[ABP1] D. W. Anderson, E. H. Brown, Jr. and F. P. Peterson, *The structure of the Spin cobordism ring*, Ann. of Math., **86** (1967), 271–298.

[ABP2] ——————————, *Pin cobordism and related topics*, Comment. Math. Helv., **44** (1969), 462–468.

[ABS] M. F. Atiyah, R. Bott and A. Shapiro, *Clifford modules*, Topology, **3** (Suppl. 1) (1964), 3–38.

[Br] E. H. Brown, *The Kervaire invariant of a manifold*, in "Proc. of Symposia in Pure Math.", Amer. Math. Soc., Providence, Rhode Island, **XXII** 1971, 65–71.

[F] M. H. Freedman, *The topology of four–dimensional manifolds*, J. Differential Geom., **17** (1982), 357–453.

[F–K] ————— and R. C. Kirby, *A geometric proof of Rochlin's theorem*, in "Proc. of Symposia in Pure Math.", Amer. Math. Soc., Providence, Rhode Island, **XXXII,** Part 2 1978, 85–97.

[G–M] L. Guillou and A. Marin, *Une extension d'un theoreme de Rohlin sur la signature*, in "A la Recherche de la Topologie Perdue", edited by Guillou and Marin, Birkhauser, Boston - Basel - Stuttgart, 1986, 97–118.

[Ha] N. Habegger, *Une variété de dimension 4 avec forme d'intersection paire et signature* −8, Comment. Math. Helv., **57** (1982), 22–24.

[Ka] S. J. Kaplan, *Constructing framed 4–manifolds with given almost framed boundaries*, Trans. Amer. Math. Soc., **254** (1979), 237–263.

[Ki] R. C. Kirby, "The Topology of 4-Manifolds", Lecture Notes in Math. No. 1374, Springer–Verlag, New York, 1989.

[K–T] ————— and L. R. Taylor, *A calculation of* Pin^+ *bordism groups*, Comment. Math. Helv., to appear.

[L–S] R. Lashof and J. Shaneson, *Smoothing 4-manifolds*, Invent. Math., **14** (1971), 197–210.

[Mat] Y. Matsumoto, *An elementary proof of Rochlin's signature theorem and its extension by Guillou and Marin*, in "A la Recherche de la Topologie Perdue", edited by Guillou and Marin, Birkhäuser, Boston - Basel - Stuttgart, 1986, 119–139.

[M–S] J. W. Milnor and J. D. Stasheff, "Characteristic Classes", Annals of Math. Studies # 49, Princeton University Press, Princeton, NJ, 1974.

[Q] F. Quinn, *Ends of maps, III: dimensions 4 and 5*, J. Diff. Geom., **17** (1982), 502–521.

[R] R. A. Robertello, *An invariant of knot cobordism*, Comm. Pure and App. Math., **XVIII** (1965), 543–555.

[Ro] V. A. Rochlin, *Proof of a conjecture of Gudkov*, Funkt. Analiaz. ego Pril., **6.2** (1972), 62–24: translation; Funct. Anal. Appl., **6** (1972), 136–138.

[Stolz] S. Stolz, *Exotic structures on 4–manifolds detected by spectral invariants*, Invent. Math., **94** (1988), 147–162.

[Stong] R. E. Stong, "Notes on Cobordism Theory", Princeton Math. Notes, Princeton Univ. Press, Princeton, New Jersey, 1958.

[Ta] L. R. Taylor, *Relative Rochlin invariants*, Gen. Top. Appl., **18** (1984), 259–280.

[Tu] V. G. Turaev, *Spin structures on three–dimensional manifolds*, Math. USSR Sbornik, **48** (1984), 65–79.

Department of Mathematics
University of California, Berkeley
Berkeley, California 94720

Department of Mathematics
University of Notre Dame
Notre Dame, Indiana 46556